Coherence in Atomic Collision Physics

AF875896

PHYSICS OF ATOMS AND MOLECULES

Series Editors

P. G. Burke, *The Queen's University of Belfast, Northern Ireland*
H. Kleinpoppen, *Atomic Physics Laboratory, University of Stirling, Scotland*

Editorial Advisory Board

R. B. Bernstein *(New York, U.S.A.)*
J. C. Cohen-Tannoudji *(Paris, France)*
R. W. Crompton *(Canberra, Australia)*
J. N. Dodd *(Dunedin, New Zealand)*
W. Hanle *(Giessen, Germany)*
C. J. Joachain *(Brussels, Belgium)*
W. E. Lamb, Jr. *(Tucson, U.S.A.)*
P.-O. Löwdin *(Gainesville, U.S.A.)*
H. O. Lutz *(Bielefeld, Germany)*
M. R. C. McDowell *(London, U.K.)*
K. Takayanagi *(Tokyo, Japan)*

ATOMIC INNER-SHELL PHYSICS
Edited by Bernd Crasemann

ATOMS IN ASTROPHYSICS
Edited by P. G. Burke, W. B. Eissner, D. G. Hummer, and I. C. Percival

AUTOIONIZATION: Recent Developments and Applications
Edited by Aaron Temkin

COHERENCE IN ATOMIC COLLISION PHYSICS
Edited by H. J. Beyer, K. Blum, and R. Hippler

COLLISIONS OF ELECTRONS WITH ATOMS AND MOLECULES
G. F. Drukarev

DENSITY MATRIX THEORY AND APPLICATIONS
Karl Blum

ELECTRON-ATOM AND ELECTRON-MOLECULE COLLISIONS
Edited by Juergen Hinze

ELECTRON-MOLECULE COLLISIONS
Edited by Isao Shimamura and Kazuo Takayanagi

INNER-SHELL AND X-RAY PHYSICS OF ATOMS AND SOLIDS
Edited by Derek J. Fabian, Hans Kleinpoppen, and Lewis M. Watson

INTRODUCTION TO THE THEORY OF LASER-ATOM INTERACTIONS
Marvin H. Mittleman

ISOTOPE SHIFTS IN ATOMIC SPECTRA
W. H. King

PROGRESS IN ATOMIC SPECTROSCOPY, Parts A, B, C, and D
Edited by W. Hanle, H. Kleinpoppen, and H. J. Beyer

QUANTUM MECHANICS VERSUS LOCAL REALISM: The Einstein, Podolsky, and Rosen Paradox
Edited by Franco Selleri

RECENT STUDIES IN ATOMIC AND MOLECULAR PROCESSES
Edited by Arthur E. Kingston

THEORY OF MULTIPHOTON PROCESSES
Farhad H. M. Faisal

ZERO-RANGE POTENTIALS AND THEIR APPLICATIONS IN ATOMIC PHYSICS
R. K. Nesbet

A Continuation Order Plan is available for this series. A continuation order will bring delivery of each new volume immediately upon publication. Volumes are billed only upon actual shipment. For further information please contact the publisher.

Coherence in Atomic Collision Physics

For Hans Kleinpoppen on His Sixtieth Birthday

Edited by

H. J. Beyer
University of Stirling
Stirling, Scotland

K. Blum
University of Münster
Münster, Federal Republic of Germany

and

R. Hippler
University of Bielefeld
Bielefeld, Federal Republic of Germany

Springer Science+Business Media, LLC

Library of Congress Cataloging in Publication Data

Coherence in atomic collision physics: for Hans Kleinpoppen on his sixtieth birthday / edited by H. J. Beyer, K. Blum, and R. Hippler.
p. cm. – (Physics of atoms and molecules)
Includes bibliographies and index.
ISBN 978-1-4757-9747-3
1. Coherence (Nuclear physics) 2. Collisions (Nuclear physics) I. Beyer, H. J., 1940- . II. Blum, Karl, 1937- . III. Hippler, R. IV. Kleinpoppen, H. (Hans) V. Series.
[DNLM: 1. Nuclear Physics. QC 794.6.C6 C678]
QC794.6.C58C67 1988 88-9796
539.7′54 – dc 19 CIP

ISBN 978-1-4757-9747-3 ISBN 978-1-4757-9745-9 (eBook)
DOI 10.1007/978-1-4757-9745-9

© Springer Science+Business Media New York 1988
Originally published by Plenum Press, New York in 1988
Softcover reprint of the hardcover 1st edition 1988

All rights reserved

No part of this book may be reproduced, stored in a retrieval system, or transmitted in any form or by any means, electronic, mechanical, photocopying, microfilming, recording, or otherwise, without written permission from the Publisher

CONTRIBUTORS

K. BARTSCHAT • *Institute for Theoretical Physics I, University of Münster, D-4400 Münster, West Germany*

K. BLUM • *Institute for Theoretical Physics I, University of Münster, D-4400 Münster, West Germany*

P. G. BURKE • *Department of Applied Mathematics and Theoretical Physics, The Queen's University of Belfast, Belfast BT7 1NN, Northern Ireland*

A. J. DUNCAN • *Atomic Physics Laboratory, University of Stirling, Stirling FK9 4LA, Scotland*

G. F. HANNE • *Department of Physics and Astronomy, The University of Oklahoma, Norman, Oklahoma 73019. Present address: Physics Institute, University of Münster, D-4400 Münster, West Germany*

RAINER HIPPLER • *Fakultät für Physik, Universität Bielefeld, D-4800 Bielefeld, West Germany*

WOLFGANG JITSCHIN • *Physikalisch-Technische Bundesanstalt, Institut Berlin, D-1000 Berlin 10, West Germany*

W. R. MACGILLIVRAY • *School of Science, Griffith University, Nathan, Queensland, Australia 4111*

M. C. STANDAGE • *School of Science, Griffith University, Nathan, Queensland, Australia 4111*

UWE WILLE • *Bereich Kern- und Strahlenphysik, Hahn-Meitner Institut, D-1000 Berlin 39, West Germany*

H. WINTER • *Institute for Nuclear Physics, University of Münster, D-4400 Münster, West Germany*

R. ZIMNY • *Institute for Nuclear Physics, University of Münster, D-4400 Münster, West Germany*

Hans Kleinpoppen

PREFACE

During the last two decades the experimental investigation of atomic coherence phenomena has made rapid progress. Detailed studies have been performed of angular correlations, spin polarization effects, angular momentum transfer, and the alignment parameters which characterize the charge cloud of excited atoms. The enormous growth in the number of these investigations was made possible through substantial development and application of new experimental technology, the development of sophisticated theoretical models and numerical methods, and a fine interplay between theory and experiment. This interplay has resulted in a deeper understanding of the physical mechanisms of atomic collision processes. It is the purpose of the chapters in this book to provide introductions for nonspecialists to the various fields of this area as well as to present new experimental and theoretical results and ideas.

The interest in spin-dependent interactions in electron–atom scattering has a long history; it dates back to the early investigations of Mott in 1929. While the more traditional measurements in this field were concerned with the determination of spin polarization and asymmetries, the range of investigations has been expanded enormously during the last few years and now includes many observables sensitive to one or more of the various spin-dependent interactions.

The understanding of these effects requires a theoretical description of the orientation and alignment parameters of the target atoms, of the formation of resonances, of the influence of electron-exchange processes, and of the relativistic interactions inside the atom and between projectile and target. Sophisticated theoretical models and computational devices have been developed to describe electron–atom scattering. Only recently, however, has it become possible to include all relevant dynamical aspects in the calculations. In the case of scattering from molecules, additional structural effects have to be taken into account which give rise to interesting new phenomena.

The introduction of laser techniques has provided a powerful new tool for the study of collisions. Superelastic scattering enables detailed investigations to be carried out of the conditions under which atomic orientation and alignment are created during a collision. By combining electron and laser beams, stepwise excitation of atoms becomes possible. The narrow bandwidth of laser radiation allows fine and hyperfine structure of many transitions to be resolved so that the influence of the corresponding interactions can be studied. Further examples of the application of laser techniques are collisions involving Rydberg atoms, free–free transitions, and the study of spin polarization effects.

The interaction of energetic atomic particles with atoms has been the subject of investigations since the beginning of the century and is still one of the most active fields of research in atomic physics. Studies of elastic, inelastic, superelastic, and radiative collisions have provided increasing insight into the physical mechanisms governing these processes. Investigations of inner-shell vacancy production have been advanced over the past years by determining, e.g., the impact parameter dependence from the measurement of coincidences between projectiles and the decay products of the inner-shell vacancies (X rays, Auger electrons). The impact parameter dependence essentially reflects the effect on the excitation process of the minimum internuclear distance reached in a collision. Consequently, by scanning the impact parameter one may isolate excitation mechanisms which are effective in different ranges of the internuclear distance.

A relatively new field of research is the study of atomic collisions in the presence of external electric and magnetic fields. These effects are of particular interest for highly excited states like Rydberg atoms, where they are accessible to experimental investigations. Scattering processes influenced by large magnetic field strengths are of importance in astrophysical problems.

Coherence in ion–surface interactions was first observed after beam–foil excitation. It was soon realized that scattering processes at grazing incidence can give rise to exceptionally high orientation effects, which, in turn, result in a high degree of polarization of the light subsequently emitted by the excited atoms or ions. This large polarization can be used to perform high-resolution atomic spectroscopy and to produce polarized beams. Investigations into the origin of these large orientations led to the development of physical models which provide the basis for a deeper understanding of coherence phenomena in general.

Modern laser and photon correlation techniques have also been applied successfully to probe traditional quantum mechanical theories and to test the interpretations of quantum mechanics itself in novel situations. An interesting recent example of such an application of modern coherence techniques is the measurement of the polarization correlation of the two photons emitted in the simultaneous decay of metastable hydrogen. The results of this experiment are considered to provide a crucial test with regard to the rivalry between the traditional quantum mechanical interpretation and the so-called local realistic theories.

Hans Kleinpoppen has made substantial contributions to many of the areas discussed in this volume and to some which are not covered. It may thus be appropriate to attempt a short overview of his research career to date.

Having taken his first degree with Professor Wilhelm Hanle in Giessen, he joined Professor Hans Kopfermann in Heidelberg for his Ph.D. work, which he completed in 1961 with Professor Hubert Krüger in Tübingen. Thus his first work was in the field of spectroscopy, especially Lamb shift and fine structure measurements on hydrogenic systems. Over the years this interest expanded into a long series of level- and anticrossing measurements of H, He^+, and He, the study of the nonlinear low-field Stark effect in hydrogenic systems, and the two-photon decay of the metastable $2S$ state of H and D.

While working in Tübingen his interests widened, not into the transitional area between atomic and nuclear physics, as one might have expected under the influence of the Kopfermann school, but into electron–atom collision work. This led to a series of cross section and excitation function measurements and to the direct observation (in the scattered electron channel) of a narrow resonance in the elastic scattering cross section of hydrogen below the $2S$ threshold, for which high energy resolution electron monochromators and analyzers had to be used. Measurements of the threshold polarization of the atomic fluorescence following electron impact excitation resulted in the first experimental test of the Percival–Seaton theory and of the influence of fine and hyperfine structure on the line polarization.

Following the habilitation in Tübingen a spell in the United States in 1967–1968 resulted in a scattering experiment of electrons on polarized Na atoms carried out at JILA in Boulder and in the idea of an electron–photon coincidence experiment. Both types of experiment represented important steps toward quantum mechanically "complete" experiments since they allowed more detailed information to be extracted from the electron scattering process than had previously been possible by just measuring the electrons scattered into a given direction or the intensity or linear polarization of the line radiation. As it turned out, the coincidence technique of relating the electrons and photons of a single inelastic scattering event can in many cases provide complete information about the scattering amplitudes and their relative phases.

In 1968 the new university in Stirling, encouraged by the late Sir Harrie Massey, aimed to become a centre for theoretical and experimental atomic physics. Professor Kleinpoppen and his group took the chance of starting from empty laboratories not only to continue the spectroscopic and the classical electron scattering work but also to embark on new long-term projects.

The coincidence experiment, which had been started at Columbia University in New York, was transferred to Stirling, where it proceeded to provide the first angular and polarization correlation results in 1972 and 1975, both on ^1P states of helium. Even today the "Columbia experiment" is still going strong, modernized but otherwise unimpressed by attempts at metrification or by neighboring experiments. As is obvious from several chapters in this volume, coincidence experiments have greatly stimulated the progress in our understanding of scattering processes. Such experiments are now carried out in many places and on a multitude of scattering systems.

A very complex and laboratory-filling experiment was set up to scatter polarized electrons from polarized atomic beam targets. Much hard work was eventually rewarded by a series of important results, just before the experiment was nearly immobilized by the 1981 university spending cuts. It is now being refurbished to the current state of the art in collaboration with a group in Edinburgh.

A new system was constructed to detect the two-photon decay of the metastable $2S$ state of hydrogen. Despite dire predictions of failure, the experiment was finally made to work so well that it has now developed into an important test experiment for basic aspects of quantum mechanics, as is discussed in detail in Chapter 9.

Another field started in Stirling was the investigation of electron impact excitation of inner shells using free atoms rather than the more common solids. This experiment has developed far beyond the original Bremsstrahlung and characteristic X-ray work into a study of various types of coincidence between electrons, X rays, and ions following inner shell excitation and ionization processes. These investigations also form a link in the longstanding collaboration with researchers at Bielefeld University. The work in Bielefeld is mainly concerned with excitation and ionization processes in ion–atom collisions, as discussed in Chapters 5–7.

More recently Professor Kleinpoppen has also become interested in the study of photoionization processes. First measurements have been carried out at (and in collaboration with) the National Synchrotron Radiation Laboratory in Daresbury. By investigating a simultaneous photoionization and excitation process it is hoped that a "complete" photoionization experiment can eventually be carried out by polarization correlation measurements between photoelectrons and fluorescence photons. Further photoionization experiments are in the planning stage.

Apart from the experimental work, Professor Kleinpoppen has shown considerable interest in the theoretical background of the measured processes, as is borne out by his own and collaborative work with Belfast,

Freiburg, and Münster. He has been a driving force in the organization of a number of successful and well remembered conferences and summer schools, and he is coeditor of a highly regarded series of books on the physics of atoms and molecules.

The authors and editors are happy to dedicate this book to Professor Hans Kleinpoppen on the occasion of his sixtieth birthday on September 30, 1988 and wish him continued success in his many activities.

H. J. Beyer
Stirling
K. Blum
Münster
R. Hippler
Bielefeld

CONTENTS

CHAPTER 7

EXCITATION IN ION–ATOM COLLISIONS

Uwe Wille

CHAPTER 8

COHERENCE IN GRAZING ION–SURFACE COLLISIONS

H. Winter and R. Zimny

CHAPTER 9

THE TWO-PHOTON DECAY OF ATOMIC HYDROGEN AND TESTS OF BELL'S INEQUALITY

A. J. Duncan

CHAPTER 1

Electron Scattering from Heavy Atoms

K. Bartschat and P. G. Burke

1. INTRODUCTION

Electron-atom scattering has been studied experimentally and theoretically for many years. Experimentally, not only can elastic and inelastic cross sections now be measured, but the use of high-resolution polarized electron and atom beams enables observables such as the spin polarization, asymmetry functions, and Stokes' parameters to be determined. Reviews of these developments have been given by Blum and Kleinpoppen,[1] Hanne,[2] Slevin,[3] and Kessler.[4] In some cases, it is even possible to perform so-called "complete" experiments (Bederson[5,6]), where all the information about the scattering process—namely, the magnitudes and the relative phases of the scattering amplitudes—are determined. A fundamental experiment of this type was the first electron-photon coincidence experiment performed in Hans Kleinpoppen's group at Stirling (Eminyan *et al.*[7]). In this experiment the light emitted in the decay of the $(1s2p)\,^1P$ state of helium excited by electron impact was observed in coincidence with the scattered electron. This experiment enabled the relative phase and relative magnitude of the two relevant scattering amplitudes to be determined, although the absolute magnitudes of these amplitudes still have to await an absolute cross-section measurement.

The main topic of this review will be electron scattering from heavy atoms, where important work has also been performed in the Stirling group, particularly by observing the light emitted in the decay $(6s6p)\,^3P_1^o \rightarrow (6s^2)\,^1S_0^e$ of mercury after electron impact excitation. In these experiments

K. Bartschat • Institute for Theoretical Physics I, University of Münster, D-4400 Münster, West Germany. P. G. Burke • Department of Applied Mathematics and Theoretical Physics, The Queen's University of Belfast, Belfast BT7 1NN, Northern Ireland.

information about the classification of various resonances and the influence of spin dependent effects can be obtained.

The very detailed information obtainable by modern experimental techniques requires the development of sophisticated theoretical models and computational methods in order to obtain agreement between experimental and theoretical data. However, it is only recently that it has been possible to include all the important dynamical effects in low-energy electron scattering by heavy atoms, namely, electron exchange, channel coupling and resonances, electron-electron correlations, and finally relativistic effects. Furthermore, a close collaboration between experimental and theoretical groups in this field has been proved to be extremely important for detailed tests and further improvements of the theoretical model.

Besides the numerical calculations another important part of the theoretical analysis is to relate the experimental observables to the S, T, or K matrices generally calculated in the standard collision programs. In general, a given apparatus cannot perform a pure state selection of both the initial and final states of the combined [electron + atom] system, because this would require the initial and final momenta and the spin polarization of the electrons together with the charge cloud distribution of the target to be determined. Very often the initial state, for example, is an unpolarized mixture of various spin states, and also some quantities are not observed after the scattering, for example again the spin polarization of the scattered electrons. In the theoretical description of these restrictions the concept of the "reduced" density matrix has been introduced (see, for example, Blum[(8)]). This allows a systematic treatment of experiments where, for example, only the electrons or the atoms are observed after the scattering process or, in the case of photon coincidence experiments, where polarized or unpolarized incident electrons are used and where the spin of the final electron may or may not be analyzed. Furthermore, a detailed analysis of the reduced density matrix sometimes reveals a close connection between individual matrix elements, experimental observables, and specific dynamical aspects of the collision process such as the importance of relativistic or electron exchange effects. In recent years much work of this kind has been performed by Blum and collaborators (see, for example, Blum and Kleinpoppen[(1)]).

In this chapter we first review in Section 2 the reduced density matrix formalism for electron-atom scattering and introduce some observables that are particularly important in treating relativistic effects. These relativistic effects and their theoretical description in terms of the Dirac-Fock or the Breit-Pauli formalism will then be discussed in some detail in Section 3. The rest of this section deals with computational methods, and we will mainly concentrate on the R-matrix approach, which has been used very successfully for many multichannel scattering problems. In Section 4 we will then compare experimental and theoretical data for various observables in order to demonstrate the range of applicability of the method. Finally we draw some conclusions on the status of the theory and on possible future directions of research.

2. THE REDUCED DENSITY MATRIX FORMALISM FOR ELECTRON-ATOM SCATTERING

In electron-atom scattering one is interested in transitions from an initial state $|J_0M_0; \mathbf{k}_0m_0\rangle$ to a final state $|J_1M_1; \mathbf{k}_1m_1\rangle$, where J_0 and J_1 are the total electronic angular momenta of the target atom before and after the scattering, M_0 and M_1 are their corresponding z components, m_0 and m_1 denote the spin components of the incident and outgoing electrons, and finally $\mathbf{k}_0$ and $\mathbf{k}_1$ are their corresponding linear momenta. Theoretically such a transition is described by scattering amplitudes

$$f(J_1M_1, \mathbf{k}_1m_1; J_0M_0, \mathbf{k}_0m_0) \equiv \langle J_1M_1; \mathbf{k}_1m_1|T|J_0M_0; \mathbf{k}_0m_0\rangle \tag{1}$$

where $T = S - 1$ is the T matrix, and we normalize in such a way that the differential cross section is given by

$$\sigma(J_1M_1, \mathbf{k}_1m_1; J_0M_0, \mathbf{k}_0m_0) = \frac{k_1}{k_0}|f(J_1M_1, \mathbf{k}_1m_1; J_0M_0, \mathbf{k}_0m_0)|^2 \tag{2}$$

We choose the quantization axis parallel to the incident beam direction defined by $\mathbf{k}_0$, and for simplicity we will also assume that $\mathbf{k}_0$ and $\mathbf{k}_1$ are sufficiently well defined, which means that any angle- or energy-dependent features are resolved by the apparatus used in the measurement. If necessary the theory can be generalized in a straightforward way by convoluting the results with the momentum distribution in the initial state as well as with the resolution of the apparatus.

An important problem that now arises is the fact that equations (1) and (2) cannot be used immediately for the description of a real experiment, because in general there is neither a pure initial state ($J_0, M_0, \mathbf{k}_0$, and m_0 well defined) nor can the observables $J_1, M_1, \mathbf{k}_1$, and m_1 all be measured simultaneously after the scattering process. In order to allow for a possible mixture of states before the scatttering we theorefore define a density matrix ρ_{in} as follows:

$$\rho_{\text{in}} = \sum_{\substack{J_0'M_0'm_0' \\ J_0M_0m_0}} (\rho_{\text{in}})^{J_0'M_0'J_0M_0}_{m_0'm_0} |J_0'M_0'; \mathbf{k}_0m_0'\rangle\langle J_0M_0; \mathbf{k}_0m_0| \tag{3}$$

The density matrix elements $(\rho_{\text{in}})^{J_0'M_0'J_0M_0}_{m_0'm_0}$ contain all the information about the preparation of the initial state, and we note that pure states can also be described in this formalism. With the common normalization $\text{tr}(\rho_{\text{in}}) = 1$ such pure states are characterized by (see also Blum[8,9])

$$\text{tr}(\rho_{\text{in}}) = \text{tr}(\rho_{\text{in}}^2) = 1 \tag{4}$$

As an example we note that the initial density matrix for the scattering of polarized electrons [polarization vector $\mathbf{P} = (P_x, P_y, P_z)$] from unpolarized target atoms with total angular momentum J_0 is given by

$$\rho_{\text{in}} = \frac{1}{2J_0+1} \sum_{M_0 m_0' m_0} (\rho_{\text{in}}^{\text{el}})_{m_0'm_0} |J_0M_0; \mathbf{k}_0 m_0'\rangle\langle J_0M_0; \mathbf{k}_0 m_0| \tag{5}$$

The factor $1/(2J_0 + 1)$ originates from averaging over the $(2J_0 + 1)$ magnetic sublevels, and the spin density matrix of the incident electrons is given by (see, for example, Kessler[4])

$$\rho_{\text{in}}^{\text{el}} = \frac{1}{2}\begin{bmatrix} 1+P_z & P_x - iP_y \\ P_x + iP_y & 1-P_z \end{bmatrix} \tag{6}$$

After the scattering the density matrix ρ_{out} is given by (Blum,[8] Appendix E)

$$\rho_{\text{out}} = T\rho_{\text{in}}T^{\dagger} \tag{7}$$

where the dagger denotes the adjoint operator. The complete information about the scattering process is now contained in the matrix elements of ρ_{out}, which are given by

$$\begin{aligned}(\rho_{\text{out}})^{J_1'M_1'J_1M_1}_{\mathbf{k}_1 m_1' m_1} \equiv \langle J_1'M_1'; \mathbf{k}_1 m_1' | \rho_{\text{out}} | J_1M_1; \mathbf{k}_1 m_1\rangle = &\sum_{\substack{J_0'M_0'm_0' \\ J_0M_0m_0}} \\ &\times f(J_1'M_1', \mathbf{k}_1 m_1'; J_0'M_0', \mathbf{k}_0 m_0') f^*(J_1M_1, \mathbf{k}_1 m_1; J_0M_0, \mathbf{k}_0 m_0) \\ &\times (\rho_{\text{in}})^{J_0'M_0'J_0M_0}_{m_0'm_0}\end{aligned} \tag{8}$$

with the asterisk denoting the complex conjugate quantity.

Equation (8) shows the relationship between the density matrices in the initial and final states and the scattering amplitudes. A complete experiment requires the measurement of all independent density matrix elements, which is equivalent to the determination of all independent scattering amplitudes. Such highly sophisticated experiments are, except for some special cases, impossible to perform. In general only the sum of contributions to various outgoing channels are experimentally detected, and in a quantum mechanical description this corresponds to performing an incoherent sum over discrete variables or an integral over continuous variables, respectively. This procedure yields the so-called "reduced" density matrix, whose elements contain all the information about the scattering process that can be obtained under given experimental restrictions.

In the following we will discuss in some detail the reduced density matrices of three kinds of experiments, and it will be shown how various observables can be defined that are particularly sensitive to certain dynamical aspects of the collision process, for example, the presence of spin-dependent forces. The measurement of these parameters often allows immediate conclusions to be drawn about the collision dynamics, and the comparison of experimental and theoretical data provides a detailed test of the theoretical model.

2.1. *The Reduced Density Matrix of the Scattered Electrons*

We now assume that the target atoms are not observed after the scattering. In this case we have to perform an incoherent sum over $J_1 = J_1'$ and $M_1 = M_1'$ in equation (8). For a given value of $\mathbf{k}_1$ we then obtain the spin density matrix of the scattering electrons with the matrix elements

$$(\rho_{\text{out}}^{\text{el}})_{m_1' m_1, \Theta_e} = \sum_{J_1 M_1} \langle J_1 M_1; \mathbf{k}_1 m_1' | \rho_{\text{out}} | J_1 M_1; \mathbf{k}_1 m_1 \rangle \tag{9}$$

where the scattering angle Θ_e is the angle between $\mathbf{k}_1$ and the z axis defined by $\mathbf{k}_0$. All the observables which can be measured by looking only at the scattered electrons can be expressed in terms of the above reduced density matrix elements. A general discussion has been given by Kessler,[4] and we will only concentrate here on two observables, which are illustrated in Figure 1.

If unpolarized electrons are scattered from unpolarized target atoms, the scattered electrons can be spin polarized with a polarization vector perpendicular to the scattering plane (xz plane in Figure 1a). The value of the polarization is determined by the so-called Sherman function S_P, which is given by

$$S_P = \text{tr}(\rho_{\text{out}}^{\text{el}} \sigma_y)/\text{tr}\, \rho_{\text{out}}^{\text{el}} \tag{10}$$

where

$$\sigma_y = \begin{pmatrix} 0 & -i \\ i & 0 \end{pmatrix} \tag{11}$$

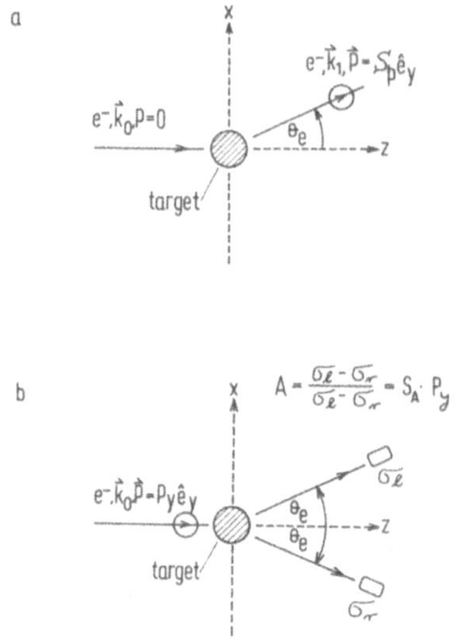

FIGURE 1. (a) Scattering of unpolarized electrons from unpolarized targets; the polarization vector after the scattering process is perpendicular to the scattering plane and is given by $\mathbf{P} = S_p \hat{e}_y$. (b) Scattering of polarized electrons from unpolarized targets; the left-right asymmetry in the differential cross section is determined by the asymmetry function S_A.

is the corresponding Pauli matrix. From equations (5) and (9)-(11) we obtain the explicit expression

$$S_P(\Theta_e) = \frac{k_1}{k_0} \frac{1}{2J_0 + 1} \frac{1}{\sigma_u(\Theta_e)} \operatorname{Im}\left\{ \sum_{M_1 M_0 m_0} f(J_1 M_1, \mathbf{k}_1 m_1' = -\tfrac{1}{2}; J_0 M_0, \mathbf{k}_0 m_0) \right.$$
$$\left. \times f^*(J_1 M_1, \mathbf{k}_1 m_1 = \tfrac{1}{2}; J_0 M_0, \mathbf{k}_0 m_0) \right\} \tag{12}$$

In equation (12) Im$\{x\}$ denotes the imaginary part of the complex quantity x, and the differential cross section for the scattering of unpolarized electrons is given by

$$\sigma_u(\Theta_e) = \frac{k_1}{k_0} \frac{1}{(2J_0 + 1)} \sum_{M_1 M_0 m_1 m_0} |f(J_1 M_1, \mathbf{k}_1 m_1; J_0 M_0, \mathbf{k}_0 m_0)|^2 \tag{13}$$

On the other hand, if spin polarized electrons with polarization vector $\mathbf{P} = P_y \hat{e}_y$ are scattered from an unpolarized target (see Figure 1b), in general a left-right asymmetry is found in the differential cross section, which is obtained by summing over $m_1' = m_1$, since no spin analysis is performed for the scattered electrons [see also equation (13)]. The asymmetry is due to the spin density matrix of the initial state, which is given by

$$\rho_{\text{in}} = \frac{1}{2J_0 + 1} I_A \times \frac{1}{2}\left[\begin{pmatrix} 1 & 0 \\ 0 & 1 \end{pmatrix} + P_y \begin{pmatrix} 0 & -i \\ i & 0 \end{pmatrix}\right] \tag{14}$$

where I_A is the $(2J_0+1)$-dimensional unit matrix, and $\times$ denotes the outer matrix product. The differential cross section can now be expressed as

$$\sigma(\Theta_e) = \sigma_u(\Theta_e)[1 + S_A P_y] \tag{15}$$

with the asymmetry function

$$S_A(\Theta_e) = \frac{k_1}{k_0}\frac{1}{2J_0+1}\frac{1}{\sigma_u(\Theta_e)}\,\mathrm{Im}\left\{\sum_{M_1 M_0 m_1} f(J_1M_1, \mathbf{k}_1 m_1; J_0 M_0, \mathbf{k}_0 m_0' = \tfrac{1}{2}) \right.$$
$$\left. \times f^*(J_1M_1, \mathbf{k}_1 m_1; J_0 M_0, \mathbf{k}_0 m_0 = -\tfrac{1}{2})\right\} \tag{16}$$

We note here that although for elastic scattering S_P and S_A are identical for the same target because of the time reversal properties, they are different in the general case and contain independent information about the scattering process. There are, however, some important exceptions, such as the following:

1. For scattering from targets with $J_0 = J_1 = 0$, one obtains from the reflection invariance of the scattering amplitudes $S_P = \Pi_0\Pi_1 S_A$ [see equations (9) and (18) of Bartschat *et al.*[10]], where Π_0 and Π_1 are the parities of the initial and final atomic states.
2. If additional assumptions are made about the scattering dynamics—e.g., the two-step-model of the DWBA discussed by Bartschat and Blum[11] or the "fine-structure effect" (Hanne[2,12])—these will often lead to further relations between S_P and S_A as well as their values within a fine-structure multiplet.

The importance of S_P and S_A as spin-dependent observables is twofold: in some cases direct information about the validity of the above assumptions can be obtained from experimental data; and secondly the comparison between experimental and theoretical results for these parameters allows a detailed test of the theoretical model to be made, in particular with regard to the treatment of spin-dependent effects.

2.2. *The Reduced Density Matrix for Electron–Photon Coincidence Experiments*

As mentioned in the introduction the first electron–photon coincidence experiment was performed in Hans Kleinpoppen's group (Eminyan *et al.*[7]) for electron impact excitation of helium. Since this classic experiment, data for other targets like hydrogen, the noble gases, and mercury have been measured (for a review of recent experimental work we refer to Slevin[3]). These experiments are of course restricted to optically allowed transitions, and apart from atomic hydrogen as the target most of the work has been done for ${}^1S_0^e \to {}^{1,3}P_1^o \to {}^1S_0^e$ transitions, i.e., $J_1 = 1$ and $J_0 = 0$. We will therefore concentrate on these transitions, and for clarification we will omit the arguments J_1 and J_0 as well as $\mathbf{k}_1$ and $\mathbf{k}_0$ in the following.

In all electron–photon coincidence experiments performed so far no spin analysis of the scattered electrons has been performed. One therefore has to sum over $m_1' = m_1$ in equation (8) in order to get the reduced density matrix elements

$$(\rho_{\text{out}}^{\text{coin}})_{M_1'M_1,\Theta_e} = \sum_{m_1 m_0' m_0} f(M_1' m_1 m_0'; \Theta_e) f^*(M_1 m_1 m_0; \Theta_e)(\rho_{\text{in}})_{m_0' m_0} \quad (17)$$

For our special case of $J_0 = 0$ the elements $(\rho_{\text{in}})_{m_0' m_0}$ just describe the spin state of the incident electrons.

It is important to note that equations (9) and (17) show very clearly how the matrix elements $(\rho_{\text{out}}^{\text{el}})_{m_1' m_1, \Theta_e}$ and $(\rho_{\text{out}}^{\text{coin}})_{M_1' M_1, \Theta_e}$ contain different and in general independent information about the scattering process. The elements $(\rho_{\text{out}}^{\text{el}})_{m_1' m_1, \Theta_e}$ describe the scattered electrons and only contain bilinear products of scattering amplitudes with $M_1' = M_1$, whereas $m_1' \neq m_1$ is possible; the elements $(\rho_{\text{out}}^{\text{coin}})_{M_1' M_1, \Theta_e}$, however, describe the atoms that scattered the electrons in the direction defined by Θ_e, and they contain only products with $m_1' = m_1$ but off-diagonal elements with $M_1' \neq M_1$.

2.2.1. Parametrization of the Density Matrix for Unpolarized Incident Electrons. If the incident electron beam is unpolarized we have $(\rho_{\text{in}})_{m_0' m_0} = \frac{1}{2}\delta_{m_0' m_0}$. Following Blum *et al.*[13] and da Paixao *et al.*[14] we define the spin-averaged quantities

$$\langle f(M_1') f^*(M_1)\rangle \equiv \frac{1}{2} \sum_{m_1 m_0} f(M_1' m_1 m_0) f^*(M_1 m_1 m_0) \quad (18)$$

and we will omit the parameter Θ_e for the rest of this section. In the above papers it has been shown that in the case of $J_1 = 1$ the atomic density matrix is fully described by five independent parameters, which are defined as follows: the absolute differential cross section

$$\sigma = \sum_{M_1} \sigma(M_1) = 2\sigma(1) + \sigma(0) = 2\langle f(1) f^*(1)\rangle + \langle f(0) f^*(0)\rangle \quad (19a)$$

together with the relative parameters

$$\lambda = \frac{\sigma(0)}{\sigma} \quad (19b)$$

$$\bar{\chi} = \arg\{\langle f(1) f^*(0)\}$$ (19c)

$$\cos \varepsilon = -\frac{\langle f(1) f^*(-1)\rangle}{\sigma(1)}$$ (19d)

$$\cos \Delta = \frac{|\langle f(1) f^*(0)|}{[\sigma(1)\sigma(0)]^{1/2}}$$ (19e)

It has further been shown by Blum and collaborators that in the limit of LS coupling (no explicit spin-dependent forces) and a ${}^1S_0^e \to {}^1P_1^o$ transition the parameter $\bar{\chi}$ reduces to the phase χ between two scattering amplitudes (see Eminyan *et al.*[7]), and $\cos \varepsilon = \cos \Delta = 1$. For the important intercombination line $(6s6p)\ {}^3P_1^o - (6s^2)\ {}^1S_0^e$ in mercury the situation is more complicated. In this case the excited ${}^3P_1^o$ state is often described in an intermediate coupling scheme as

$$(6s6p)\ {}^3P_1 = \alpha(6s6p)\ {}^3\bar{P}_1 + \beta(6s6p)\ {}^1\bar{P}_1$$ (20)

where the bars denote pure Russell-Saunders states and $\alpha = 0.985$, $\beta = 0.171$ (Lurio[15]). If the spin-orbit interaction for the continuum electron is neglected (which corresponds to the approximation used to describe the fine-structure effect (Hanne[2,12]), it follows (Bartschat[16]) that

$$-\frac{\lambda}{1-\lambda} \leqslant \cos \varepsilon \leqslant 1$$ (21)

which gives a nontrivial lower limit for $\cos \varepsilon$ if $\lambda \leqslant 0.5$. For $J = 0$, even $\to$ $J = 1$, odd transitions we can therefore expect $\cos \varepsilon$ and $\cos \Delta$ to play a similar role in determining the importance of spin-dependent effects as the parameters S_P and S_A introduced in the preceding section.

2.2.2. Electron-Photon Coincidence Experiments at Zero Degree Scattering Angle. In this section we discuss an electron-photon coincidence experiment with polarized electrons that has recently been performed in the Münster group (Wolcke *et al.*[17]). Transversely polarized electrons ($\mathbf{P} = P_y \hat{e}_y$) are used to excite the $(6s6p)\ {}^3P_1^o$ state of mercury, and the electrons scattered in the forward direction ($\Theta_e = 0^0$) are observed in coincidence with the light emitted in the subsequent decay to the ground state $(6s^2)\ {}^1S_0^e$. The light detector is placed parallel to the initial electron spin polarization vector (y direction), and we note that in this special arrangement there is no symmetry plane defined by the momenta $\mathbf{k}_0$ and $\mathbf{k}_1$, which is usually the case when the vectors are not parallel. However, the axial polarization vector of the incident electrons defines the xz plane as a symmetry plane, and in general the following three light polarizations (Stokes' parameters) can be expected to be nonzero (see Figure 2):

1. The degree of linear polarization with respect to the z and x axes,

$$\eta_3 = \frac{I(0) - I(90)}{I(0) + I(90)} \tag{22a}$$

   where $I(\beta)$ denotes the light intensity transmitted by a Nicol prism orientated at an angle β with respect to the z axis;
2. The degree of linear polarization with respect to axes at 45° and 135° to the z axis,

$$\eta_1 = \frac{I(45) - I(135)}{I(45) + I(135)} \tag{22b}$$

3. The degree of circular polarization,

$$\eta_2 = \frac{I_+ - I_-}{I_+ + I_-} \tag{22c}$$

   where $I_+(I_-)$ is the light transmitted by polarization filters which only admit light with positive (negative) helicity.

It can be shown by general arguments (Bartschat and Blum[18]) that the Stokes' parameters η_1 and η_2 vanish, if the incoming electron beam is unpolarized (the arguments of the above paper have been derived for the so-called "integrated" Stokes' parameters discussed in the following section, but the symmetry properties of the two experiments are identical). Therefore η_1 and η_2 can be expected to contain information about spin-dependent effects (including electron exchange) on the excitation process and to be suitable observables to test theoretical models to describe these phenomena.

Moreover, as has been shown by Bartschat *et al.*,[19] the particular electron–photon coincidence experiment discussed in this section can also

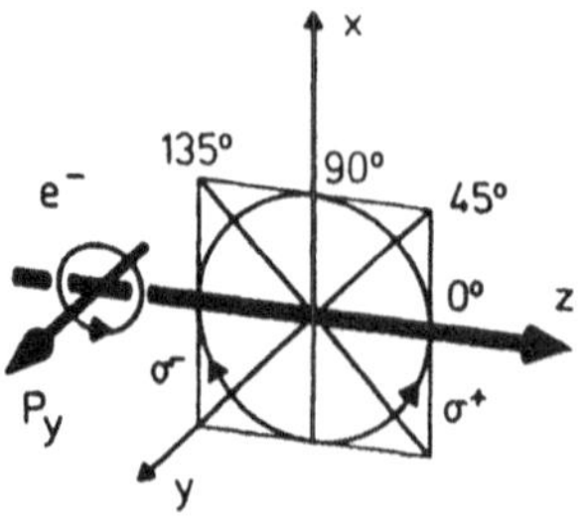

FIGURE 2. Geometrical arrangement of the Münster experiment (Wolcke *et al.*[17,21]) to measure the Stokes' parameters after excitation with transversally polarized electrons.

provide information about the role of the nuclear spin in the excitation and decay process. The Percival-Seaton hypothesis (Percival and Seaton[20]) assumes that the nuclear spin I is frozen during the excitation and decay process. However, during the lifetime of the excited state, angular momentum orientation can be transferred from the electronic to the nuclear system and vice versa, which in general leads to a depolarization of the emitted radiation. If the above approximation is valid, the excitation of the $(6s6p)\,^3P_1^o$ state of mercury by electrons scattered in the forward direction can be described in terms of the two amplitudes f_2 and f_5 introduced by Bartschat *et al.*,[10] since all other scattering amplitudes have to vanish for zero degree scattering angle because of angular momentum selection rules.

The important consequence in this case is the fact that the three Stokes' parameters are no longer independent of each other, but are related by the following equations, which are valid for the natural isotope mixture of mercury:

$$(\eta_2/P_y)^2 = \frac{1.573x}{0.286x^2 + 0.995x + 0.865} - 1.263(\eta_1/P_y)^2 \tag{23a}$$

with

$$x = \frac{|f_2|^2}{|f_5|^2} = \frac{0.789 - 0.930\eta_3}{0.395 + 0.535\eta_3} \tag{23b}$$

Therefore equations (23) can be used as a consistency check for theoretical calculations that make use of the Percival-Seaton hypothesis, but in principle they can also provide direct evidence from experimental data about the validity of this approximation.

2.3. *The Reduced Density Matrix of the Excited Atoms*

We will now discuss experiments where the scattered electrons are not observed at all. In this case we sum over $m_1' = m_1$ and integrate over all possible directions of $\mathbf{k}_1$ in equation (8) to obtain the reduced matrix elements

$$\begin{aligned}(\rho_{\rm out}^{\rm atom})_{M_1'M_1}^{J_1'J_1} &= \sum_{\substack{M_0'M_0\\ m_0'm_0m_1\\ J_0'J_0}} \int d\Omega_{\hat{k}_1} f(J_1'M_1', \mathbf{k}_1 m_1; J_0'M_0', \mathbf{k}_0 m_0')\\ &\quad \times f^*(J_1M_1, \mathbf{k}_1 m_1; J_0M_0, \mathbf{k}_0 m_0)(\rho_{\rm in})_{m_0'm_0}^{J_0'M_0'J_0M_0}\\ &\equiv \sum_{M_0'M_0m_0'm_0} \langle F(J_1'M_1', J_0'M_0'; m_0')F^*(J_1M_1, J_0M_0; m_0)\rangle\\ &\quad \times (\rho_{\rm in})_{m_0'm_0}^{J_0'M_0'J_0M_0}\end{aligned} \tag{24}$$

Again, all relevant parameters related to the observation of the atoms alone can be expressed in terms of the integrated bilinear products $\langle F(J_1'M_1', J_0'M_0'; m_0')F^*(J_1M_1, J_0M_0; m_0)\rangle$ defined by equation (24).

2.3.1. Integrated Stokes' Parameters. In this section we discuss again the explicit example where the $(6s6p)\,^3P_1^o$ state of mercury is excited by polarized electron impact. For transversely polarized electrons ($\mathbf{P} = P_y\hat{e}_y$) the explicit form of the reduced atomic density matrix is given by

$$\rho_{\text{out}}^{\text{atom}} \quad \begin{array}{c|ccc} M_1' \backslash M_1 & 1 & 0 & -1 \\ \hline 1 & \rho_{11} & \rho_{10}P_y & 0 \\ 0 & \rho_{10}^*P_y & \rho_{00} & -\rho_{10}^*P_y \\ -1 & 0 & -\rho_{10}P_y & \rho_{11} \end{array} \tag{25}$$

where we have used the reflection invariance of the system with regard to the xz plane. The density matrix elements ρ_{1-1} and ρ_{-11} must vanish because of a special selection rule derived in equations (29)–(34) of Bartschat *et al.*[10] Using $\rho_{10} = |\rho_{10}|e^{i\delta}$ and taking into account the natural isotope mixture of mercury by applying the Percival–Seaton hypothesis, we obtain the following expressions for the integrated Stokes' parameters:

$$\eta_1/P_y = \frac{-6.69|\rho_{10}|\cos\delta}{3.21\rho_{11} + 2.79\rho_{00}} \tag{26a}$$

$$\eta_2/P_y = \frac{-7.53|\rho_{10}|\sin\delta}{3.21\rho_{11} + 2.79\rho_{00}} \tag{26b}$$

$$\eta_3 = \frac{-2.37(\rho_{11} - \rho_{00})}{3.21\rho_{11} + 2.79\rho_{00}} \tag{26c}$$

Since the symmetry properties of this experiment are identical to those discussed in the previous section for electron–photon coincidence experiments at zero degree scattering angle, η_1 and η_2 are again proportional to the transverse polarization component P_y of the incident electron beam. The physical importance of these parameters has been discussed by Bartschat and Blum,[18] and we only quote the main results:

1. η_1 vanishes for $P_y \neq 0$ if no explicit spin-dependent forces like the spin–orbit interaction influence the excitation process.

2. η_2 can be different from zero if only electron exchange processes take place; however, in the absence of explicit spin-dependent forces, η_2/P_y will always be positive.
3. The integrated Stokes' parameters can be used for the analysis of resonances (Wolcke *et al.*[21]). In particular, measurements of η_2 will in some cases allow an optical determination of the electron polarization (Farago and Wykes,[22] Wykes,[23] Eminyan and Lampel,[24] Gay[25]).

In conclusion we can also expect these parameters to be very suitable observables to test theoretical models describing spin-dependent effects (including exchange) as well as resonances.

2.3.2. Total Magnetic Sublevel Cross Sections. It follows from equation (26c) that the integrated Stokes' parameter η_3 can be used to determine the ratio ρ_{11}/ρ_{00}, which corresponds to the ratio of the total excitation cross sections to the magnetic sublevels with $M_1 = 1$ and $M_1 = 0$, respectively. This method, however, is restricted to optically allowed transitions and cannot be directly applied, for example, to the excitation of the $(6s6p)\,^3P_2^o$ state of mercury. Nevertheless, the ratios of partial cross sections can be determined by a combined stepwise excitation by electron impact plus laser radiation. The method has been described in detail by McLucas *et al.*[26,27] and Webb *et al.*[28,29] and has been applied by the authors (Webb *et al.*[30]) and by Hanne *et al.*[31] to the above transition. We only note here that these ratios can also be used to obtain information about the spin dependence of the process, which is shown as follows.

If explicit spin-dependent forces are neglected in the excitation process we can express the total sublevel cross section $Q(M_1)$ as (Percival and Seaton[20])

$$Q(M_1) = \tfrac{1}{3} \sum_{M_L M_S} (1M_S, 1M_L|2M_1)^2 Q_{M_L} \tag{27}$$

where Q_{M_L} is the partial cross section for excitation of the sublevel with orbital angular momentum component M_L and $(1M_S, 1M_L|2M_1)$ is a standard Clebsch-Gordan coefficient. Reflection invariance yields $Q(M_1) = Q(-M_1)$ and $Q_{M_L} = Q_{-M_L}$, which means that in this approximation the three magnetic sublevel cross sections $Q(0)$, $Q(1)$, and $Q(2)$ can be expressed in terms of the two independent parameters Q_1 and Q_0. For our case of interest the explicit relations are

$$\frac{Q(2)}{Q(0)} = \frac{3x}{2+x} \tag{28a}$$

$$\frac{Q(1)}{Q(0)} = \frac{3}{2}\frac{1+x}{2+x} \tag{28b}$$

with $x = Q_1/Q_0$. It is clear that a comparison of x values obtained from equations (28a) and (28b) allows a direct test of the validity of the above approximation using only experimental data. A similar analysis can be performed to check the influence of the nuclear spin on the excitation process by comparing the results of $Q(1)/Q(0)$ obtained directly from equations (26c) or using the stepwise excitation with the $(6s6p)\,^3P_1^o$ state of mercury as an intermediate state (McLucas *et al.*[26,27]).

2.4. Conclusions

In this section we have shown how the concept of the reduced density matrix allows a systematic description of various experiments performed on elastic and inelastic electron–atom scattering. Many of the observables such as the spin polarization, the scattering asymmetry or the Stokes' parameters of the light emitted after electron impact excitation contain information about the scattering dynamics such as spin-dependent effects including electron exchange processes and resonances. Furthermore, the role of the nuclear spin can be investigated in order to test the validity of the Percival–Seaton hypothesis. In conclusion, a comparison of experimental and theoretical data for the various independent observables that can now be measured using highly sophisticated apparatus, where polarized electron (and atom) beams are combined with electron–photon coincidence and laser techniques, should enable a detailed test of theoretical models to be performed. Some of these theoretical models, particularly the R-matrix method using the Breit–Pauli Hamiltonian to include relativistic effects, will be discussed in the next section.

3. RELATIVISTIC EFFECTS AND COMPUTATIONAL METHODS

3.1. Inclusion of Relativistic Effects

As the nuclear charge Z increases, relativistic effects become important even for low-energy electron scattering. There are two important effects:

1. The direct effect, that is, the distortion of the wave function of the scattered electron by the strong nuclear potential;
2. The indirect effect, that is, the modification of the target wave function, which changes the potential seen by the scattered electron.

In the theoretical description these effects can be represented in three ways:

(*a*) *Intermediate Z Atoms and Ions.* For these systems relativistic effects are small and the calculation of S matrices using LS coupling, that is, neglecting these effects, can be expected to be a good approximation. In order to describe fine-structure transitions, the S matrix is then transformed using purely algebraic recoupling techniques. This approach has been widely used for electron-ion collisions using a program of Saraph.[32,33]

(*b*) *High Z Atoms and Ions—Breit-Pauli Hamiltonian.* The Dirac equation and the Breit interaction can be reduced to Pauli form giving the Breit-Pauli Hamiltonian (e.g., Bethe and Salpeter[34])

$$H^{\mathrm{BP}} = H^{\mathrm{NR}} + H^{\mathrm{REL}} \tag{29}$$

where H^{NR} is the nonrelativistic part and H^{REL} is the sum of one- and two-body relativistic terms. This approach has been considered by Jones,[35] Scott and Burke,[36] and Sin Fai Lam and Baylis.[37] The one-body terms retained in H^{REL} in the work discussed later in this article are as follows:

1. The spin-orbit term:

$$H^{\mathrm{SO}} = \tfrac{1}{2}\alpha^2 Z \sum_{i=1}^{N+1} r_i^{-3}(\mathbf{l}_i \cdot \mathbf{s}_i) \tag{30a}$$

2. The mass-correction term:

$$H^{\mathrm{MASS}} = -\tfrac{1}{8}\alpha^2 \sum_{i=1}^{N+1} \nabla_i^4 \tag{30b}$$

3. The Darwin term:

$$H^{\mathrm{D}} = -\tfrac{1}{8}\alpha^2 Z \sum_{i=1}^{N+1} \nabla_i^2\left(\frac{1}{r_i}\right) \tag{30c}$$

where l_i and s_i denote the orbital and spin angular momenta of the ith electron and where N is the number of electrons in the target atom. The inclusion of the terms (30) also implies the use of a J-dependent coupling scheme; in general this leads to many more coupled channels than in an LS-coupling calculation and therefore increases the computational effort significantly.

(*c*) *High Z Atoms and Ions-Dirac Hamiltonian.* The Dirac Hamiltonian for a $(N + 1)$-electron system can be written (e.g., Grant[38])

$$H^{\mathrm{Dirac}} - \sum_{i=1}^{N+1}\left(c\boldsymbol{\alpha} \cdot \mathbf{p}_i + \beta' c^2 - \frac{Z}{r_i}\right) + \sum_{i>j=1}^{N+1} \frac{1}{r_{ij}} \tag{31}$$

where α and $\beta' = \beta - 1$ are the usual Dirac matrices. The bound and continuum orbitals are written as four component spinors

$$\Phi(\mathbf{r}, \sigma) = \frac{1}{r}\begin{bmatrix} P(r) & \chi_{\kappa m}(\hat{r}, \sigma) \\ iQ(r) & \chi_{-\kappa m}(\hat{r}, \sigma) \end{bmatrix} \tag{32}$$

where $\chi_{\kappa m}$ are functions formed by coupling the orbital and spin angular momentum eigenfunctions and P and Q are the large and small components of the radial functions.

An important question that can be asked is whether the approach based on the Breit–Pauli Hamiltonian can give reliable results for low-energy scattering of electrons by heavy atoms such as Xe, Cs, or Hg or whether a fully relativistic approach based on the Dirac equation is required. Some evidence on this question is presented in the next section.

3.2. *Elastic Scattering*

Elastic scattering of electrons by heavy atoms has been studied experimentally and theoretically for many years. Early reviews have been written by Kessler[39] and by Walker,[40] and a more recent discussion has been given by Kessler.[4]

For a closed shell $^1S_0$ target atom such as mercury, elastic scattering is described by two amplitudes: the direct scattering amplitude f and the spin-flip scattering amplitude g. Hence three independent parameters $|f|$, $|g|$, and the relative phase between f and g completely describe the collision. Experimentally, as well as the differential scattering cross section σ, it is convenient to introduce additional observables S, T, and U which describe the change of the electron spin polarization vector by the scattering process (Kessler[4]). We show in Figure 3 measurements of T and U at a scattering angle of 90° made by Berger *et al.*[41] compared with calculations by Walker,[40] who solved the continuum Dirac–Fock equations. It is seen that the agreement is excellent over a very wide range of energies.

In general, the agreement between elastic scattering experiments and the fully relativistic theory based on the Dirac–Fock equation is very good over a wide range of energies. Detailed comparisons between different ways of including relativistic effects can be most easily made for elastic scattering, and such a comparison has been presented by Sin Fai Lam and Baylis.[37] They compared phase shifts for e^-–Hg scattering obtained by reducing the Dirac–Fock equation to nonrelativistic form with a fully relativistic treatment. We show their results for the $p_{1/2}$ and $p_{3/2}$ phase shifts below 5 eV in Figure 4. The nonrelativistic calculations included a second-order Dirac potential V_R (Mott and Massey,[42] Browne and Bauer,[43] Sin Fai Lam[44]), and were carried out using both nonrelativistic Hartree–Fock and relativistic

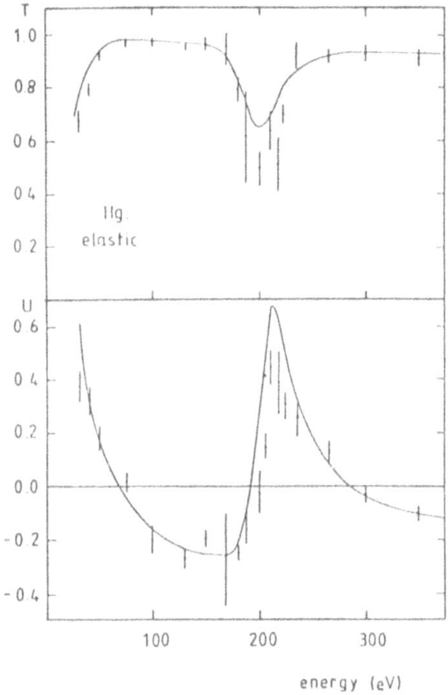

FIGURE 3. Results for the parameters T and U at a scattering angle $\Theta_e = 90°$ as a function of the incident electron energy; |, experiment (Berger *et al.*[41]); ——, theory (Berger *et al.*[41] based on the calculations of Walker[40]).

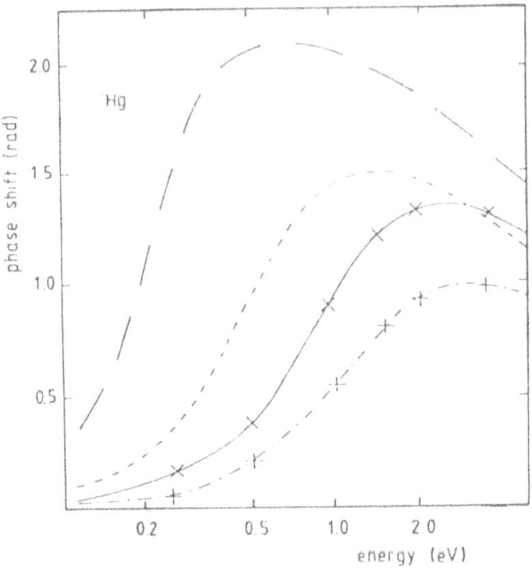

FIGURE 4. *p*-wave phase shifts for e^--Hg scattering: —— and -·-·- are $p_{1/2}$ *and* $p_{3/2}$ phase shifts using a nonrelativistic form of the theory with relativistic target wave functions; — — and - - - are $p_{1/2}$ and $p_{3/2}$ phase shifts using a nonrelativistic form of the theory with nonrelativistic target wave functions; + and × are $p_{1/2}$ and $p_{3/2}$ phase shifts using a fully relativistic theory (Sin Fai Lam and Baylis[37]).

Dirac–Fock target wave functions. The relativistic calculations were carried out using relativistic target wave functions. It is seen that the agreement between the two treatments is excellent provided that relativistic target wave functions are used in the nonrelativistic formulation. This shows that both the direct effect, which gives the $p_{3/2}$–$p_{1/2}$ splitting, and the indirect effect, which gives the dependence on the target wave functions, are important. However, we see that both can be accurately included in a formulation where the Dirac–Fock equation is reduced to nonrelativistic form.

The main uncertainty in the above treatment of elastic scattering is one common to scattering by light atoms, namely, an appropriate representation of the distortion or charge cloud polarization of the target as well as the loss of flux into inelastic channels. The charge cloud polarization effect has usually been represented by a local adiabatic potential with a parametrized cutoff at short distances. However, as pointed out by McEachran and Stauffer,[45] the use of such a polarization potential requires special care in order to correctly represent exchange. Also, in most of the calculations performed so far, the effect due to loss of flux has been completely neglected, which in general results in absolute cross sections that can be too large by up to a factor of 2 at higher energies. As an example we present in Figure 5 the differential cross section for elastic scattering of electrons from xenon atoms at a collision energy of 40 eV. We compare the experimental data of Holtkamp and Jost[46] with three calculations (Hasenburg *et al.*[47]) where the same static and charge cloud polarization potentials have been used. As in the work of Sin Fai Lam and Baylis, the Dirac–Fock equation was reduced to nonrelativistic form, which can be done exactly provided all the potentials used are spherically symmetric. The differences between the three calculations consist (1) in the different treatment of electron exchange either by calculating the one-electron exchange integrals explicitly (McEachran and Stauffer[45]) or by simulating this effect by the local energy-dependent exchange potential of Furness and McCarthy,[48] and (2) in the inclusion

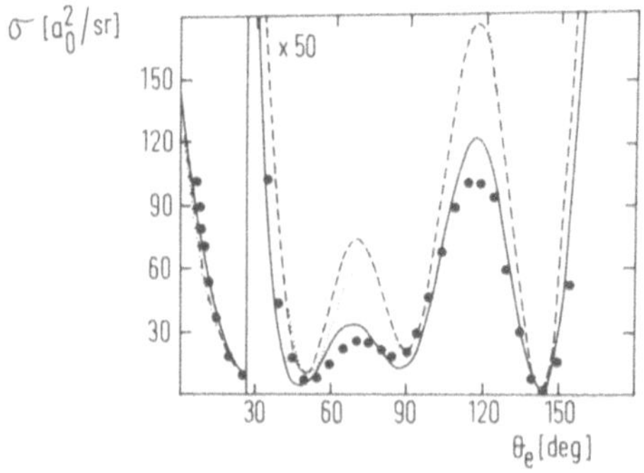

FIGURE 5. Differential cross section for elastic e^-–Xe scattering for a collision energy of 40 eV. – – –, full relativistic calculation including the one-electron exchange integrals (McEachran and Stauffer[45]); · · ·, full relativistic calculation treating exchange by a local model potential (Hasenburg *et al.*[47]); ——, full relativistic calculation including local exchange and absorption potentials; • • •, experimental data of Holtkamp and Jost.[46]

of the quasifree scattering absorption potential of Staszewska *et al.*[49] in one of the calculations (Hasenburg *et al.*[50]). It can be seen from Figure 5 that the local exchange approximation gives results in very good agreement with the exact treatment at this rather high energy. However, the inclusion of the absorption potential, representing the loss of flux into inelastic channels, is essential in order to obtain good agreement with absolute experimental cross-section data. We note here that the calculations by McEachran and Stauffer[45] gave very satisfactory agreement with the relative S, T, and U data of Berger and Kessler[51] for elastic e–Xe scattering over a wide range of energies. However, the local exchange approximation is expected to break down at lower collision energies, where these effects and their correct treatment become very important.

3.3. Multichannel Scattering

Low-energy elastic and inelastic scattering of electrons by atoms and ions can be most accurately described in terms of the close coupling plus correlation expansion of the total wave function. For N target electrons and one continuum electron this expansion is written as

$$\Psi = \mathscr{A} \sum_{i=1}^{n} \bar{\Phi}_i(\mathbf{x}_1, \ldots, \mathbf{x}_N, \hat{r}_{N+1}\sigma_{N+1})F_i(r_{N+1}) + \sum_{i=1}^{m} \chi_i(\mathbf{x}_1, \ldots, \mathbf{x}_{N+1})c_i \tag{33}$$

where $\mathscr{A}$ is the antisymmetrization operator, the $\bar{\Phi}_i$ are channel functions formed by coupling a few low-lying target states with the spin-angle function for the scattered electron, the F_i describe the radial motion of the scattered electron, and the χ_i are L^2 integrable correlation functions. The latter functions allow for short-range electron–electron correlation effects that are only partly represented by the few terms in the first expansion in equation (33). Substituting equation (33) into the Schrödinger equation for the $(N+1)$-electron system and projecting onto the channel functions $\bar{\Phi}_i$ and onto the correlation functions χ_i leads immediately to a set of coupled integrodifferential equations for the functions F_i. The solution of these equations requires advanced numerical techniques, which have been discussed by Burke and Seaton[52] and Burke and Eissner.[53]

A computationally very efficient method for solving these equations is the so-called R-matrix approach, which was introduced in nuclear physics by Wigner[54] and Wigner and Eisenbud[55] and recently applied widely to problems in atomic physics (see, for example, the review by Burke and Robb[56]). The main idea of this method is to partition configuration space into two regions by a sphere of radius $r = a$ (where a is the relative coordinate of the colliding electron and the center of gravity of the target). This partitioning is illustrated in Figure 6. In the internal region $(r < a)$ electron exchange and correlations between the scattered electron and the N-electron target are important and the $(N+1)$-electron complex behaves very much like a bound state. Consequently a configuration interaction expansion of the $(N+1)$-electron wave function similar to that used in bound state calculations is appropriate. In the external region $(r > a)$ electron exchange between the scattered electron and the target can be neglected if the radius a of the sphere is chosen to just envelope the charge distribution of the target states of interest. The scattered electron then moves in the long-range multipole potential of the target and the solution in this region can be obtained using, for example, an asymptotic expansion or a perturbative approach.

We now expand the wave function describing the $(N+1)$-electron system in the internal region in analogy with equation (33) in terms of the basis

$$\Psi_k = \mathcal{A} \sum_{i,j} \bar{\Phi}_i(\mathbf{x}_1, \ldots, \mathbf{x}_N, \hat{r}_{N+1}\sigma_{N+1}) u_{ij}(r_{N+1}) a_{ijk} + \sum_i \chi_i(\mathbf{x}_1, \ldots, \mathbf{x}_{N+1}) b_{ik} \quad (34)$$

where as before the $\bar{\Phi}_i$ are channel functions and $\mathcal{A}$ is the antisymmetrization operator. The u_{ij} are radial basis functions describing the motion of the scattered electron which are nonzero on the boundary of the internal region. These functions, which provide the link between the wave functions in the internal and external regions, are taken to be solutions of the differential

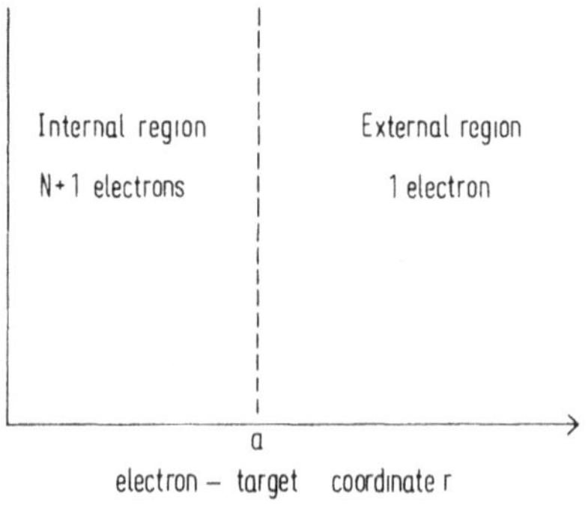

FIGURE 6. Partitioning of configuration space in the R-matrix method.

equation

$$\left[\frac{d^2}{dr^2} - \frac{l_i(l_i+1)}{r^2} + V(r) + k_{ij}^2\right] u_{ij}(r) = \sum_k \lambda_{ijk} P_k(r) \tag{35a}$$

subject to the boundary conditions

$$u_{ij}(0) = 0 \tag{35b}$$

and

$$\left.\frac{a}{u_{ij}} \frac{du_{ij}}{dr}\right|_{r=a} = b \tag{35c}$$

$V(r)$ is usually chosen to be the spherically symmetric part of the static potential of the target, b is an arbitrary constant normally set to zero, and the λ_{ijk} are Lagrange multipliers, which are chosen so that the orthogonality conditions

$$\int_0^a u_{ij}(r) P_k(r)\, dr = 0 \tag{36}$$

are satisfied, where the P_k are the reduced radial components of the occupied orbitals of the target with angular momentum l_i. We note here that the functions $\chi_i(\mathbf{x}_1, \ldots, \mathbf{x}_{N+1})$ in equation (34) not only allow for short-range correlation effects, but some also have to be included in the expansion for completeness of the basis because of the above orthogonality requirement.

The expansion coefficients a_{ijk} and b_{ik} in equation (34) are determined by diagonalizing the $(N+1)$ electron Hamiltonian in the internal region giving

$$\langle \Psi_k | H^{N+1} | \Psi_{k'} \rangle = E_k^{N+1}\, \delta_{kk'} \tag{37}$$

The R matrix is then defined by

$$R_{ij} = \frac{1}{2a} \sum_k \frac{w_{ik} w_{jk}}{E_k^{N+1} - E} \tag{38}$$

where E is the total energy and the surface amplitudes w_{ik} are given by

$$w_{ik} = \sum_j u_{ij}(a) a_{ijk} \tag{39}$$

The R matrix provides the starting condition for the determination of the functions $F_i(r_{N+1})$ in equation (33) in the external region, by relating their values and derivatives on the boundary a by

$$F_i(a) = \sum_j R_{ij}(E)\left(a\frac{dF_j}{dr} - bF_j\right)_{r=a} \tag{40}$$

The final step is then to solve the collision problem in the external region. In the case of atoms and atomic ions the total wave function is expanded in this region in the form

$$\Psi = \sum_i \bar{\Phi}_i F_i(r) \tag{41}$$

where the $\bar{\Phi}_i$ are the channel functions used in equations (33) and (34). Substituting equation (41) into the Schrödinger equation and projecting onto the channel functions $\bar{\Phi}_i$ yields the following n coupled differential equations for the reduced radial functions F_i:

$$\left[\frac{d^2}{dr^2} - \frac{l_i(l_i+1)}{r^2} + k_i^2\right]F_i(r) = 2\sum_{j=1}^{n} V_{ij}(r)F_j(r),$$
$$i = 1, \ldots, n, \qquad r > a \tag{42}$$

where n is the number of coupled channels retained in equations (34) and (41), l_i are the channel angular momenta, k_i^2 are the channel energies, and the potential matrix can be expanded in inverse powers of the radius as

$$V_{ij}(r) = \sum_{\lambda=0}^{\lambda_{\max}} a_{ij}^{\lambda} r^{-\lambda-1}, \qquad r \geqslant a \tag{43}$$

Equations (42) can be solved by one of a number of standard methods yielding n_a linearly independent solutions, where n_a is the number of open channels ($k_i^2 \geqslant 0$) at the energy of interest. We can choose these solutions to satisfy the asymptotic boundary conditions

$$\begin{aligned} F_{ij} &\underset{r\to\infty}{\sim} k_i^{-1/2}(\sin\Theta_i\delta_{ij} + \cos\Theta_i K_{ij}) \text{ open channels} \\ F_{ij} &\underset{r\to\infty}{\sim} 0 \text{ closed channels} \end{aligned} \tag{44}$$

The second index j on F_{ij} distinguishes the n_a linearly independent solutions and

$$\Theta_i = k_i r - \tfrac{1}{2}l_i\pi - \eta_i \ln 2k_i r + \sigma_i$$

$$\eta_i = -\frac{Z-N}{k_i} \tag{45}$$

$$\sigma_i = \arg \Gamma(l_i + 1 + i\eta_i)$$

where Z is the nuclear charge. Substituting these solutions evaluated at $r = a$ into equation (40) enables the $n_a \times n_a$ K-matrix elements K_{ij} to be determined in terms of the $n \times n$ R-matrix elements R_{ij}. The $n_a \times n_a$ S matrix and T matrix are then given by the matrix equations

$$S = \frac{1+iK}{1-iK}$$
$$T = S - 1 \tag{46}$$

The various observables introduced in Section 2 can be expressed in terms of the S- or T-matrix elements and are then calculated in a straightforward way.

In summary we note that the computational advantage of the R-matrix method is the fact that the solution of a system of coupled integrodifferential equations at each collision energy is replaced by a single matrix-diagonalization given by equation (37). This feature is particularly important for the analysis of resonance structures where the solution is required for many collision energies. We also point out that this approach includes the main effects that are expected to be important in low-energy scattering by heavy atoms. Electron exchange is clearly represented since the total wave function is properly antisymmetrized as required by the Pauli exclusion principle. Electron–electron correlation effects, giving rise in many cases to resonances, are represented by both terms in equation (34). Finally in the Breit–Pauli R-matrix approach relativistic effects are represented both in the choice of the target states, which are obtained using the Breit–Pauli Hamiltonian in the structure calculation, and in the use of this Hamiltonian in equation (37) in the scattering calculation. Furthermore the theory can be extended by using the Dirac–Fock Hamiltonian in equation (37), and test calculations performed with the fully relativistic R-matrix method have been reported by Norrington and Grant.[57]

4. RESULTS

In this section we present some selected results for the various observables introduced in Section 2 where we will mainly concentrate on electron scattering from cesium and mercury atoms. Most of the theoretical data have been obtained using the Breit–Pauli R-matrix method introduced by Scott and Burke[36] and programmed by Scott and Taylor.[58] The various electron scattering and photoionization calculations that have recently been performed using this approach are summarized in Table I.[59,60,61,62,63,64]

4.1. Cross Sections, Spin Polarization, and Scattering Asymmetry

4.1.1. Scattering by Cesium Atoms. Several calculations of low-energy electron cesium atom cross sections based on equation (33) have been carried out (e.g., by Karule,[65,66] Burke and Mitchell,[67,68] and Scott *et al.*[69,70]). In addition, measurements have been made by Brode,[71] Visconti *et al.*,[72] and Gehenn and Reichert[73] and other experiments are planned by the Bielefeld, Mainz, Münster, Stirling, and N.B.S. Washington groups.

The most recent R-matrix calculation by Scott *et al.*[69,70] assumed that the scattered electron and the valence target electron were interacting in the average field of the 54 closed shell core electrons. The corresponding Hamiltonian is

$$H^{\mathrm{BP}} = \sum_{i=1}^{2}\left[-\tfrac{1}{2}\nabla_i^2 - V(r) + \tfrac{1}{2}\alpha^2 Z\frac{\mathbf{l}_i\cdot\mathbf{s}_i}{r_i^3}\right] + \frac{1}{r_{12}} \tag{47}$$

where $V(r)$ is taken to be a Thomas–Fermi model potential. The five lowest target states given in Table I were then included in equation (34). We give

TABLE I
R-matrix Calculations for Electron Scattering from and Photoionization of Heavy Atoms Including Relativistic Effects

Target	Terms retained in equation (30)	Target states retained in equation (34)
		(*a*) *Electron scattering*
Cs	$\mathbf{l}\cdot\mathbf{s}$ only	$6s\,{}^2S^e_{1/2}, 6p\,{}^2P^o_{1/2,3/2}, 5d\,{}^2D^e_{3/2,5/2}$
Hg	(i) $\mathbf{l}\cdot\mathbf{s}$ only (ii) All one-body terms	$(6s^2)\,{}^1S^e_0, (6s6p)\,{}^3P^o_{0,1,2}, (6s6p)\,{}^1P^o_1$
Tl	$\mathbf{l}\cdot\mathbf{s}$ only	$6p\,{}^2P^o_{1/2,3/2}, 7s\,{}^2S^e_{1/2}, 6d\,{}^2D^e_{3/2,5/2}$
Pb	$\mathbf{l}\cdot\mathbf{s}$ only	$(6p^2)\,{}^3P^e_{0,1,2}, (6p^2)\,{}^1D^e_2, (6p^2)\,{}^1S^e_0$
		(*b*) *Photoionization*
Hg	All one-body terms	$(5d^{10}6s)\,{}^2S^e_{1/2}, (5d^96s^2)\,{}^2D^e_{5/2,3/2}, (5d^{10}6p)\,{}^2P^o_{1/2,3/2}$
Ba	$\mathbf{l}\cdot\mathbf{s}$ only	$(5p^66s)\,{}^2S^e_{1/2}, (5p^66p)\,{}^2P^o_{1/2,3/2}, (5p^65d)\,{}^2D^e_{3/2,5/2}$

in Figure 7 the total cross section for elastic scattering and for the $6s\ ^2S^e_{1/2} \rightarrow 6p\ ^2P^o_{1/2,3/2}$ transitions. Over the energy range presented, the cross sections are dominated by a very complicated resonance structure, the most important features being due to a $6p5d\ ^3F$ resonance close to the $^2P^o_{1/2}$ threshold at 1.39 eV that affects both the elastic scattering and the $^2P^o_{1/2}$ excitation cross sections.

In Figure 8 we present the differential cross section σ and the Sherman function $S(=S_P = S_A)$ for elastic electron scattering from unpolarized cesium atoms at collision energies of 0.82, 1.63, and 2.04 eV. The theoretical data show a lot of structure, but up until now experimental data in this energy region are still required in order to test the theoretical predictions. We note that since Cs is an open-shell "one-electron" target, electron polarization measurements alone are not sufficient for a "complete" experiment. Instead, the determination of the six independent scattering amplitudes, introduced by Burke and Mitchell,[68] is necessary to completely describe the scattering process. In this case, measurements with spin polarized atomic beams, which are currently in progress in the Bielefeld group,[74] are also required.

In Figure 9 we show results for the differential cross section $d\sigma/d\Omega$ and the asymmetry function S_A for the $6s\ ^2S^e_{1/2} \rightarrow 6p\ ^2P^o_{1/2,3/2}$ transitions (Scott *et al.*[69,70]). S_A describes the left-right asymmetry in the differential cross section that occurs if a polarized electron beam is scattered by unpolarized target atoms (for details see Kessler[4]). Of special interest in

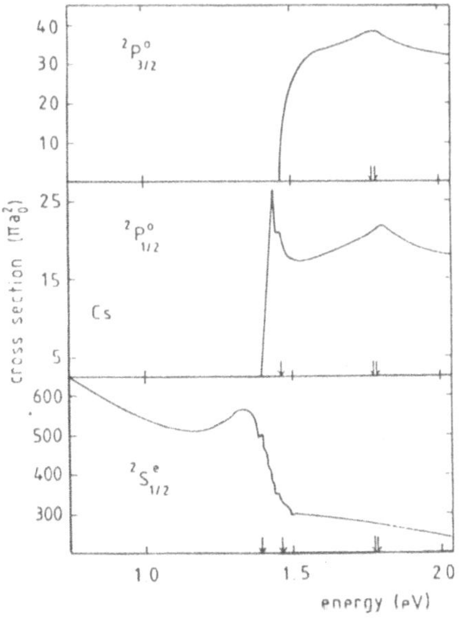

FIGURE 7. Total cross sections for the $^2S^e_{1/2} \rightarrow {}^2S^e_{1/2}, {}^2P^o_{1/2,3/2}$ transitions in Cs. The $^2P^o_{1/2}, {}^2P^o_{3/2}, {}^2D^e_{3/2}$, and $^2D^e_{5/2}$ thresholds at 1.39, 1.46, 1.80, and 1.81 eV, respectively, are indicated by arrows.

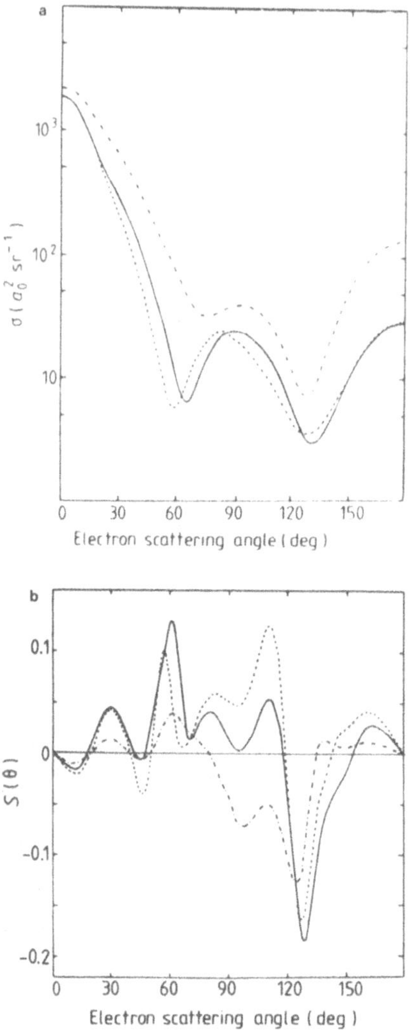

FIGURE 8. Differential cross section σ and Sherman function S for elastic electron-cesium scattering at collision energies of 0.82 eV (- . . . . -), 1.63 eV (——), and 2.04 eV (- - -).

Figure 9 are the following approximate symmetry relations between the results for the excitation of the two fine-structure levels $6p\ ^2P^o_{1/2}$ and $6p\ ^2P^o_{3/2}$. We see that

$$\sigma(^2P^o_{3/2}) \approx 2\sigma(^2P^o_{1/2}) \tag{48a}$$

and

$$S_A(^2P^o_{1/2}) \approx -2S_A(^2P^o_{3/2}) \tag{48b}$$

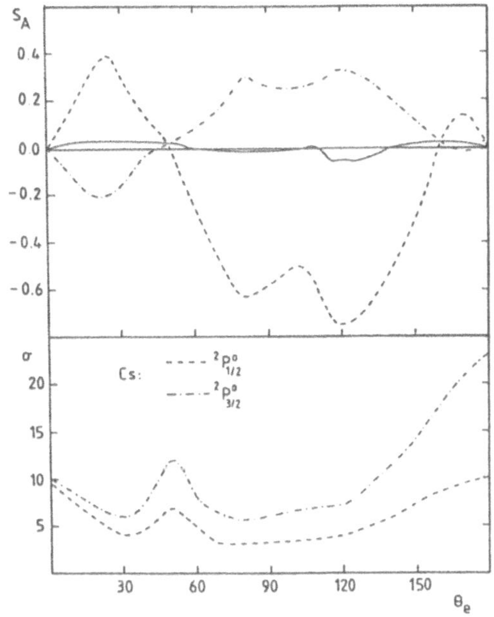

FIGURE 9. Differential cross section $\sigma(a_0^2/sr)$ and asymmetry function S_A for the transition $6s\ ^2S^e_{1/2} \to 6p\ ^2P^o_{1/2}$ (- - -) and $6s\ ^2S^e_{1/2} \to 6p\ ^2P^o_{3/2}$ (- · - · -) in Cs for an incident electron energy of 2.04 eV. ——, mean value of S_A averaged over the fine-structure levels.

It has been shown by Hanne[2] that equations (48) hold exactly if relativistic effects are taken into account by recoupling techniques to obtain fine-structure transitions as described above. The fact that relations (48) are approximately satisfied means that the direct effect of the spin-orbit interaction is small in this case. This is an example of the so-called "fine structure effect" (Hanne[2,12]).

4.1.2. Scattering by Mercury Atoms. In the case of e^--Hg scattering the calculations based on equation (34) assumed that the scattered electron and the two valence electrons were interacting in the average field of the remaining 78 closed shell core electrons. We show in Figure 10 the total cross sections for elastic scattering and for the $(6s^2)\ ^1S^e_0 \to (6s6p)\ ^3P^o_{0,1,2}$ transitions calculated by Scott *et al.*[75] They included the states given in Table I in equation (34) and retained only the spin-orbit term in H^{REL}.

The cross sections are dominated by resonances arising from the $6s6p^2$ configuration (e.g., Albert *et al.*,[76] Burrow and Michejda,[77] Fano and Cooper,[78] Heddle,[79,80] Kuyatt *et al.*,[81] Ottley and Kleinpoppen,[82] and Shpenik *et al.*[83]). The theoretical positions and assignments of these resonances are given in Table II. The calculated positions of the lowest four resonances are in good accord with experiment. However, the location of the $(6s6p^2)\ ^2D_{3/2,5/2}$ resonances requires further discussion. These are almost degenerate at 5.5 eV in the present calculation because the $(nsnp^2)\ ^2D$ terms

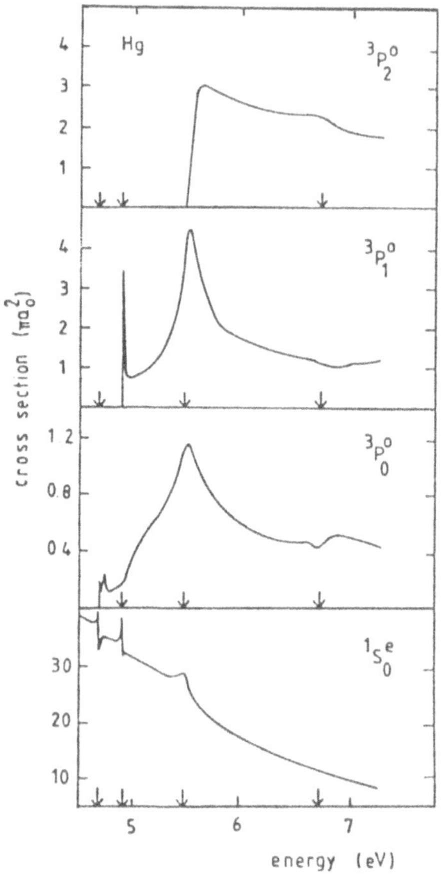

FIGURE 10. Total cross sections for the ${}^1S_0^e \rightarrow {}^1S_0^e, {}^3P_0^o, {}^3P_1^o, {}^3P_2^o$ transitions in Hg. The ${}^3P_0^o, {}^3P_1^o, {}^3P_2^o$, and ${}^1P_1^o$ thresholds at 4.67, 4.89, 5.46, and 6.70 eV, respectively, are indicated by arrows.

TABLE II
Calculated Positions and Assignments of the $6s6p^2$ Resonances in Hg[a]

$J\pi = \frac{1}{2}+$	$J\pi = \frac{3}{2}+$	$J\pi = \frac{5}{2}+$
4.7 eV(${}^4P$)	4.7 eV(${}^4P$)	
		4.9 eV(${}^4P$)
	5.0 eV(${}^4P$)	
	5.5 eV(${}^2D$)	5.5 eV(${}^2D$)
5.8 eV(?)		
6.7 eV(?)	6.7 eV(?)	

[a] From Scott *et al.*[75]

are not split by the spin-orbit interaction (Condon and Shortley[84]). Ottley and Kleinpoppen[82] had already suggested a possible resonance with total angular momentum $J = 3/2$ near 5.2 eV from their analysis of the optical excitation function and the integrated Stokes' parameter η_3 for the excitation of the $(6s6p)\,{}^3P_1^o$ state of mercury (see Section 4.3). However, since the observed structure was quite small, no definite conclusions could be drawn at that time. However, Hanne *et al.*[31] have recently measured the cross section for the $(6s^2)\,{}^1S_0^e \rightarrow (6s6p)\,{}^3P_0^0$ transition and found a broad maximum around 5.2 eV. In order to understand this discrepancy with the theoretical data of Figure 10 Bartschat and Burke[85] included additional correlation functions in equation (34) belonging to the configuration $6p^26d$. It was found that interaction between the $6s6p^2$ and $6p^26d$ configurations split the ${}^2D_{3/2,5/2}$ resonances, moving the ${}^2D_{3/2}$ resonances down from 5.5 eV to 4.9 eV. Thus this simple calculation showed that configuration interaction effects are important in determining the positions and widths of resonances, although in order to obtain accurate results more sophisticated calculations will have to be carried out.

Further calculations were carried out by Bartschat and Burke[85] including all the one-body relativistic terms defined by equation (30). We show in Figure 11 the asymmetry function S_A and the function S_P, which describes the polarization of an initially unpolarized beam after scattering from an unpolarized target, for the $(6s^2)\,{}^1S_0^e \rightarrow (6s6p)\,{}^3P_1^o$ and $(6s^2)\,{}^2S_0^e \rightarrow (6s6p)\,{}^3P_2^o$ transitions. Also shown in this figure are calculations including only the spin-orbit term (Bartschat *et al.*[60]). Inclusion of all the one-body relativistic terms modifies the results appreciably but only marginally improves the agreement with the experimental measurements of S_A by Bartschat *et al.*[86] Nevertheless the overall agreement between theory and experiment is quite good, bearing in mind the limited number of target states included in equation (34) and the neglect of core polarization.

At higher energies the results using equation (34) can be compared with calculations using the "Distorted Wave Born Approximation" (DWBA) by Bartschat *et al.*[87] and Bartschat and Madison.[88] The DWBA method (Madison and Shelton[89,90]) basically corresponds to a two-channel calculation where distorted waves are calculated in the incident and outgoing channels and the interaction between the channels is included to first order. Relativistic effects are included via correction potentials obtained from the transformation of the Dirac equation to Schrödinger form as discussed in Section 3.2. The transition matrix elements are then obtained by evaluating the integral of the Coulomb interaction between the continuum electron and the target for the initial and final $(N + 1)$-electron states. This approximation does not take into account strong coupling effects between the channels, but it can be expected to be a reasonably good approximation for sufficiently high collision energies.

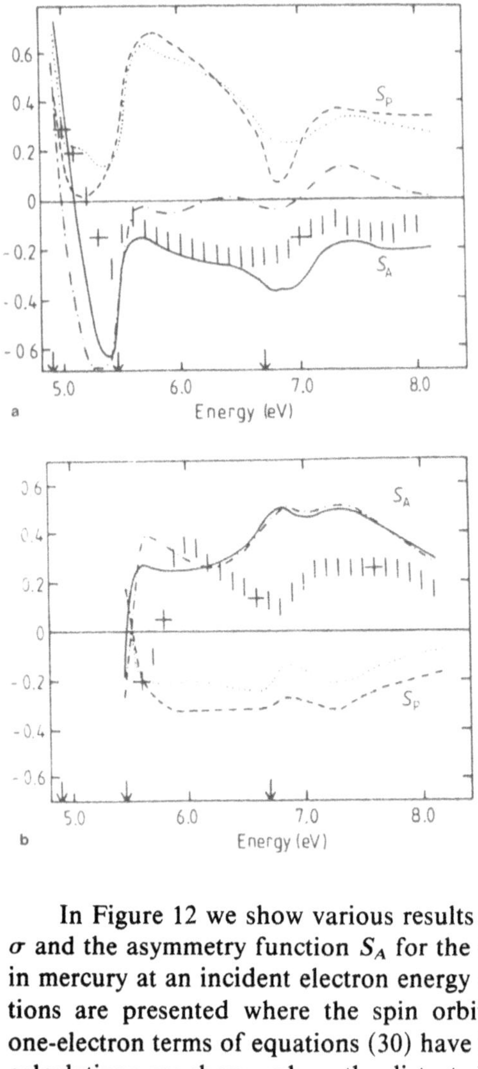

FIGURE 11. Spin polarization S_P and asymmetry function S_A for the scattering angle $\Theta_e = 90°$ as a function of the incident electron energy for the (a) $(6s^2)\,^1S_0^e \rightarrow (6s6p)\,^3P_1^o$, (b) $(6s^2)\,^1S_0^e \rightarrow (6s6p)\,^3P_2^o$ transitions in Hg. ——, S_A (only $\mathbf{l} \cdot \mathbf{s}$ explicitly included, Bartschat *et al.*[60]); - · - · -, S_A [all one-body terms of equations (30) included]; - - -, S_P (only $\mathbf{l} \cdot \mathbf{s}$ explicitly included); · · ·, S_P [all one-body terms of equations (30) included]: +experimental data for S_A (Bartschat *et al.*[86]).

In Figure 12 we show various results for the differential cross section σ and the asymmetry function S_A for the $(6s^2)\,^1S_0^e \rightarrow (6s6p)\,^3P_2^0$ transition in mercury at an incident electron energy of 15 eV. Two R-matrix calculations are presented where the spin orbit $\mathbf{l} \cdot \mathbf{s}$ term only and where all one-electron terms of equations (30) have been retained. Also, two DWBA calculations are shown where the distorted waves have been calculated (1) by just using the static potential of the target plus the relativistic correction terms (Bartschat *et al.*[87]) and (2) by adding in addition the local energy-dependent adiabatic exchange potential of Furness and McCarthy[48] (Bartschat and Madison[88]). In all four calculations the same atomic wave functions and static potentials obtained from the program SUPERSTRUCTURE of Eissner *et al.*[91] were used. It can be seen that the two R-matrix calculations approximately reflect the features seen in the experimental data of Borgmann *et al.*,[92] but it is clear from Figure 12 that more work will have to be done in this intermediate energy region. Furthermore

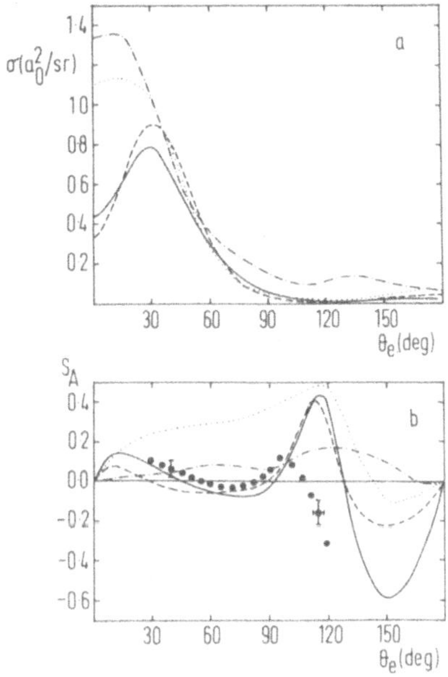

FIGURE 12. (*a*) Differential cross section σ and (b) asymmetry function S_A for electron impact excitation of the $(6s6p)\ ^3P_2^o$ state of Hg at 15 eV; ——, *R* matrix [with all one-body terms of equations (30) included]; - - -, *R* matrix (only $\mathbf{l}\cdot\mathbf{s}$ included); - · · · -, DWBA with relativistic corrections to the scattering potential (Bartschat *et al.*[87]): · · ·, DWBA with energy dependent exchange potential included (Bartschat and Madison[88]). $\overline{\underline{\Phi}}$, experimental data of Borgmann *et al.*[92]

the importance of electron exchange effects and channel coupling at this intermediate energy is clearly demonstrated by the difference between the *R* matrix and the DWBA calculations and the fact that the approximate inclusion of electron exchange significantly improves the DWBA results.

As expected, the DWBA approximation yields better results at still higher collision energies. This is demonstrated in Figure 13, where we again show the results for the asymmetry function S_A for the $(6s^2)\ ^1S_0^e \to (6s6p)\ ^3P_2^o$ excitation in mercury, but this time for an incident electron energy of 30 eV. Once more, the exchange potential clearly improves the calculation, and excellent agreement with the experimental data of Borgmann *et al.*[92] is obtained.

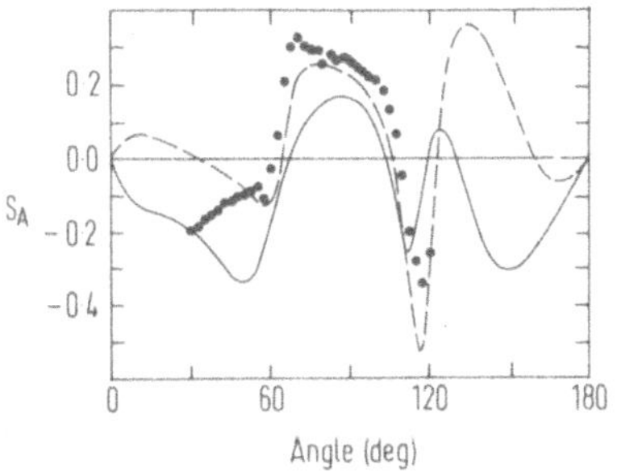

FIGURE 13. Asymmetry function S_A for the electron impact excitation of the $(6s6p)\ ^3P_2^o$ state of Hg at 30 eV; ——, DWBA with relativistic corrections to the scattering potential; - - -, DWBA with energy-dependent exchange potential included (Bartschat and Madison[88]); •, experimental data of Borgmann *et al.*[92]

4.2. Electron-Photon Coincidences

We now turn to electron-photon coincidence experiments involving heavy atoms. As a first example we present in Figure 14 theoretical data for the parameter $\cos \varepsilon$ introduced in equation (19d). It can be seen that the calculations predict a violation of the inequality (21), which indicates that in this case the direct effect, that is, the spin-orbit interaction for the scattered electron, is important. The first experiments to determine the new parameters $\cos \varepsilon$ and $\cos \Delta$ in equations (19) for electron-mercury excitation have been performed in Hans Kleinpoppen's group (Zaidi *et al.*[93]), but more experimental data are desirable in order to test the theoretical predictions.

In Figure 15 we show the R-matrix results for the electron-photon coincidence experiment with polarized electrons discussed in Section 2.2.2 and compare them with the experimental data of Wolcke *et al.*[17] for η_1/P_y and η_2/P_y and Hanne *et al.*[94] for η_3. The agreement between experiment and theory is very satisfactory for the polarization-dependent parameters η_1/P_y and η_2/P_y, whereas the experimental data for η_3 clearly lie below the theoretical results. This indicates, however, that the experimental data violate the Percival-Seaton hypothesis, which for this particular arrangement relates the three parameters to each other by equations (23). Further tests are currently being undertaken by the experimentalists, and data closer to threshold should also allow an interesting check of the resonance structures predicted by the calculations.

4.3. Integrated Stokes Parameters and Magnetic Sublevel Cross Sections

In Figure 16 we present the results for the integrated Stokes parameters discussed in Section 2.3.1. The experimental data of Wolcke *et al.*[21] for the $(6s6p)\,^3P_1^o$ excitation of Hg are compared with two R-matrix calculations, both including only the spin-orbit interaction. The first calculation

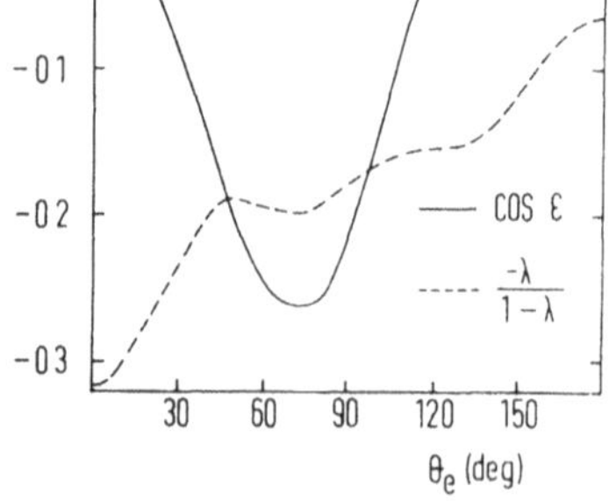

FIGURE 14. $\cos \varepsilon$ and $-\lambda/(1-\lambda)$ for electron impact excitation of the $(6s6p)\,^3P_1^0$ state of Hg for a collision energy of 6 eV (Bartschat[16]).

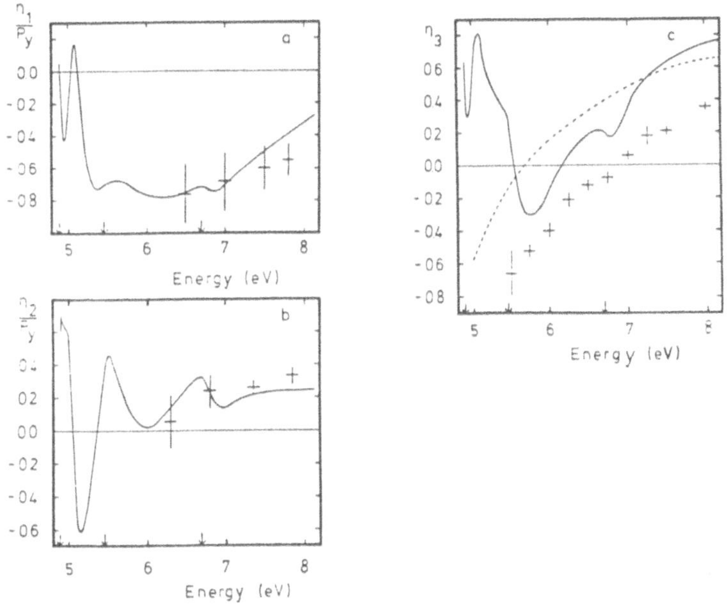

FIGURE 15. Stokes parameters for electron-photon coincidence experiments with transversally polarized electrons. Excitation of the $(6s6p)\,^3P_1^o$ state of Hg with electron scattering angle $\Theta_e = 0°$ and observation of the emitted light in the y direction. (a) linear light polarization η_1 normalized to $P_y = 100\%$; (b) circular light polarization η_2 normalized to $P_y = 100\%$; (c) linear light polarisation η_3. Calculations: ——, R matrix (Bartschat *et al.*[19]); - - - (Bonham[99]). +, experimental data of Wolcke *et al.*[17] and Hanne *et al.*[94] The arrows mark the excitation thresholds for the $^3P_{1,2}^o$ and $^1P_1^o$ states.

corresponds to the resonance positions given in Table II, while in the second calculation the R-matrix pole corresponding to the $(6s6p^2)\,^2D_{3/2}$ resonance was shifted so that the resonance occurred at 5.2 eV, which is suggested by various experimental data (see above). The overall agreement between experiment and theory is further improved by this shift; it is excellent for η_2/P_y and further confirms the position and assignment of this resonance. We also note that the nonvanishing values of η_1/P_y and the negative values of η_2/P_y in this energy region underline the importance of the direct relativistic effect, particularly near resonances, and the values of $\eta_2/P_y \approx -0.3$ and $\eta_3 \approx -0.35$ eV at 5.2 eV are in close agreement with the values of -0.47 and -0.40, respectively, which would be expected for a pure $(6s6p^2)\,^2D_{3/2}$ resonance if background scattering is entirely neglected (Bartschat,[96] Wolcke *et al.*[21]).

In Figure 17 we show the results for another integrated Stokes parameter, this time for excitation of the $6p\,^2P_{1/2}^o$ state of cesium by longitudinally polarized electrons followed by the decay back to the ground state $6s\,^2S_{1/2}^o$.

In this case only the circular light polarization η_2 can be different from zero. It is proportional to the longitudinal spin polarization component P_z, and in contrast to Figure 2 the photon detector is now placed along the z direction as well. The experimental data shown in Figure 17 have been obtained by Nass *et al.*[97] and we compare them with the results calculated by Nagy *et al.*[98] using the K matrices of Scott *et al.*[69,70] The theoretical data are only available near threshold, but in this energy region the agreement between experiment and theory is again quite satisfactory.

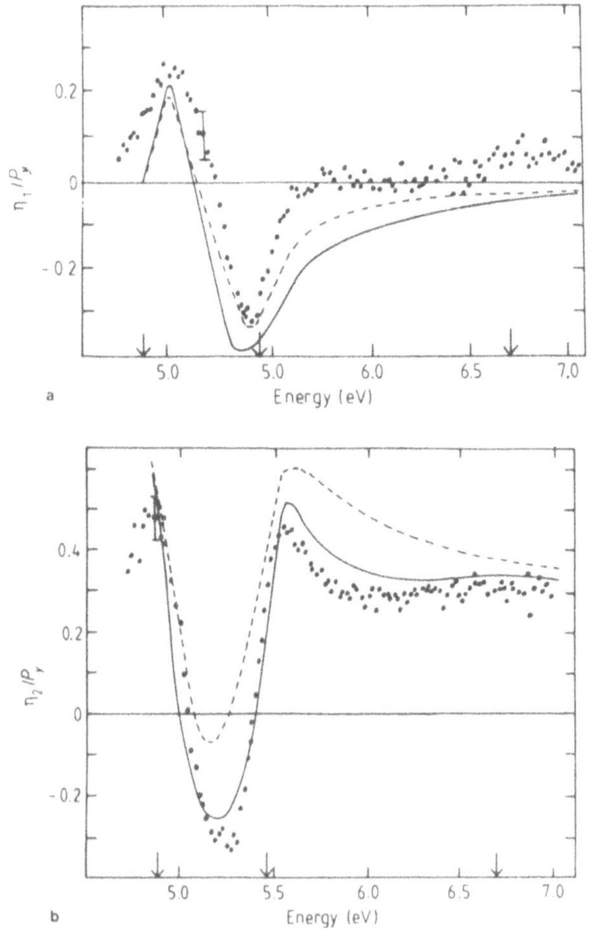

FIGURE 16. "Integrated" Stokes parameters η_1, η_2 (normalized to $P_y = 100\%$) and η_3 for the $(6s6p)\,^3P_1^o$ excitation of Hg. ⧫, experimental results (Wolcke *et al.*[21]); - - -, R-matrix calculation (Bartschat *et al.*[95], only $\mathbf{l}\cdot\mathbf{s}$ included); ——, R-matrix calculation with $(6s6p^2)\,^2D_{3/2}$ pole shifted from 5.5 to 5.2 eV.[85]

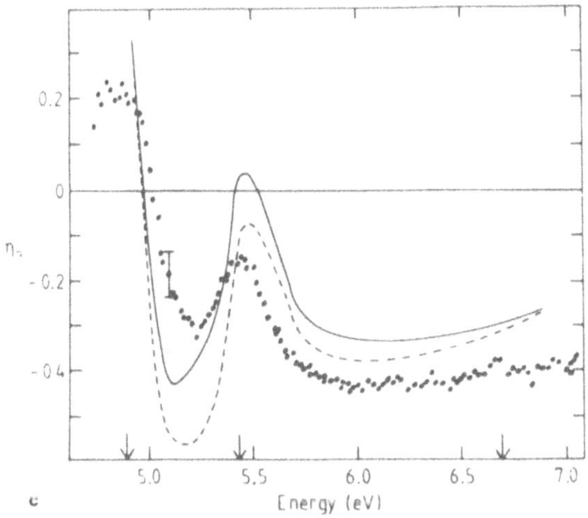

FIGURE 16. (*Continued*)

We conclude this section by comparing in Figure 18 our R-matrix results for the ratios of the partial cross sections $Q(M = 2)$, $Q(M = 1)$, and $Q(M = 0)$ for the excitation of the $(6s6p)\,^3P_2^o$ state of mercury with the experimental data of Webb *et al.*[30] and Hanne *et al.*[31] For both $Q(2)/Q(0)$ and $Q(1)/Q(0)$ the theoretical results lie between the data obtained in the

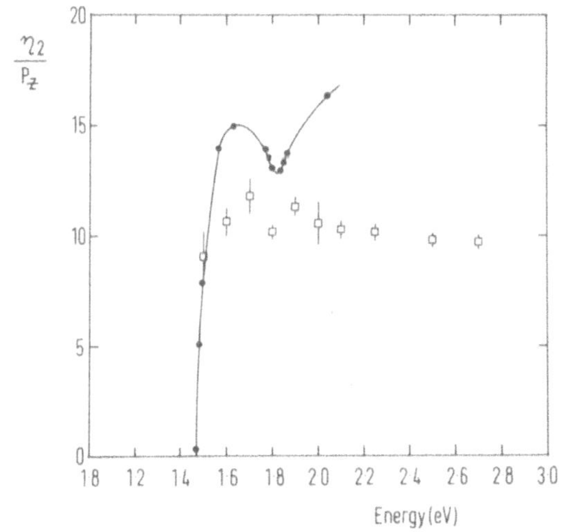

FIGURE 17. "Integrated" Stokes parameter η_2 (normalized to $P_z = 100\%$) for the excitation of the $6p\,^2P_1^o$ state of Cs by longitudinally polarized electrons; ⏀, experimental data of Nass *et al.*[97]; —●—, R-matrix calculation.[69,70,98]

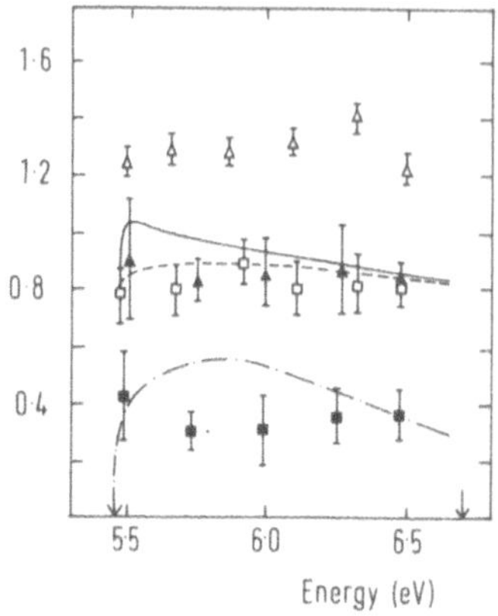

FIGURE 18. Ratios of partial cross sections for the excitation of the $(6s6p)\,^3P_2^o$ state of Hg; ⌽ and Δ, experimental data of Hanne *et al.*[31] for $Q(2)/Q(0)$ and $Q(1)/Q(0)$; ■ and ▲, experimental data of Webb *et al.*[30] for $Q(2)/Q(0)$ and $Q(1)/Q(0)$; -·-·-·, *R*-matrix results for $Q(2)/Q(0)$ (Bartschat[16]); ——, *R*-matrix results for $Q(1)/Q(0)$; - - -, theoretical values for $Q(1)/Q(0)$ obtained from $Q(2)/Q(0)$ by using equations (28).

two independent experiments, which, however, do not agree within the given error bars. Nevertheless the general structure is clearly reproduced by the calculation, which of course shows a sharp fall at threshold (5.46 eV), which is not seen in the experiments because of the energy resolution. It is interesting to note that the theoretical data fulfill the approximate relations (28) fairly well, indicating that by integrating over all electron scattering angles much of the spin dependence in the excitation process is averaged out.

5. SUMMARY

In this review we have discussed the present status of low-energy scattering of electrons by heavy atoms. We have shown how the reduced density matrix formalism can be used for a unified description of various different experiments, where elastic and inelastic electron–atom scattering is studied with the help of spin-polarized electron beams as well as electron–photon coincidence and laser techniques. The recent advances in experimental techniques enable the determination of various new observables, in particular those measured in electron–photon coincidence experiments, which contain very detailed information about coherence and correlation effects in the scattering process. The development of new theoretical methods including advanced computational techniques was necessary to enable detailed comparisons to be made between theory and experiment. In general the agreement for cross sections, spin polarizations, scattering asymmetries, and Stokes parameters is satisfactory, bearing in mind the difficulties of the experiments and the complexity of the atomic targets being studied. However, there is further work to be done, particularly theoretically in extending the fully relativistic theory based on the Dirac equation to allow inelastic scattering to be calculated, and computationally in developing techniques that will enable a more complete treatment of electron–electron correlation effects to be included for complex atoms. For many of the above parameters some fundamental experiments have now been performed, but in most cases more experimental data are still required for a more complete test of the theoretical models.

ACKNOWLEDGMENTS. The theoretical work reported here has been greatly stimulated by the experimental studies carried out by the Stirling group led by Hans Kleinpoppen, by the Münster group led by J. Kessler and G. F. Hanne, and by the Mainz group of E. Reichert. We also wish to thank K. Blum for many stimulating discussions on various aspects of the formal theory and N. S. Scott and W. Eissner for collaborating on some of the calculations reported in this work. Finally we wish to acknowledge with thanks support from the British Council, the Science and Engineering Research Council, and the Deutsche Forschungsgemeinschaft.

REFERENCES

1. K. BLUM AND H. KLEINPOPPEN, *Adv. At. Mol. Phys.* **19**, 187 (1983).
2. G. F. HANNE, *Phys. Rep.* **95**, 95 (1983).
3. J. SLEVIN, *Rep. Prog. Phys.* **47**, 461 (1984).
4. J. KESSLER, *Polarized Electrons*, 2nd ed. (Springer-Verlag, Berlin, 1985).
5. B. BEDERSON, *Comments At. Mol. Phys.* **1**, 71; **2**, 65 (1969).
6. B. BEDERSON, in *Atomic Physics*, edited by S. J. Smith and G. K. Walters (Plenum Press, New York, 1973).
7. M. EMINYAN, K. MCADAM, J. SLEVIN, AND H. KLEINPOPPEN, *J. Phys. B* **7**, 1519 (1974).
8. K. BLUM, *Density Matrix Theory and Applications* (Plenum Press, New York, 1981).
9. K. BLUM, in *Fundamental Processes in Atomic Collision Physics*, edited by H. Kleinpoppen, J. S. Briggs, and H. O. Lutz (Plenum Press, New York, 1985).
10. K. BARTSCHAT, K. BLUM, G. F. HANNE, AND J. KESSLER, *J. Phys. B.* **14**, 3761 (1981).
11. K. BARTSCHAT AND K. BLUM, *J. Phys. B.* **15**, 2747 (1982).
12. G. F. HANNE, *Comments At. Mol. Phys.* **14**, 163 (1984).
13. K. BLUM, F. DA PAIXAO, AND G. CSANAK, *J. Phys. B* **13**, L257 (1980).
14. F. DA PAIXAO, N. T. PADIAL, G. CSANAK, AND K. BLUM, *Phys. Rev. Lett.* **45**, 1164 (1980).
15. A. LURIO, *Phys. Rev. A* **140**, 1505 (1965).
16. K. BARTSCHAT, Ph.D. thesis, University of Münster (1984), unpublished.
17. A. WOLCKE, J. GOEKE, G. F. HANNE, J. KESSLER, W. VOLLMER, K. BARTSCHAT, AND K. BLUM, *Phys. Rev. Lett.* **52**, 1108 (1984).
18. K. BARTSCHAT AND K. BLUM, *Z. Phys. A* **304**, 85 (1982).
19. K. BARTSCHAT, K. BLUM, P. G. BURKE, AND N. S. SCOTT, in *Fundamental Processes in Atomic Collision Physics*, edited by H. Kleinpoppen, J. S. Briggs, and H. O. Lutz (Plenum Press, New York, 1985), p. 579.
20. I. C. PERCIVAL AND M. J. SEATON, *Phil. Trans. R. Soc. London Ser. A* **251**, 113 (1958).
21. A. WOLCKE, K. BARTSCHAT, K. BLUM, H. BORGMANN, G. F. HANNE, AND J. KESSLER, *J. Phys. B.* **16**, 639 (1983).
22. P. S. FARAGO AND J. WYKES, *J. Phys. B.* **2**, 747 (1969).

23. J. WYKES, *J. Phys. B.* **4**, L91 (1971).
24. M. EMINYAN AND G. LAMPEL, *Phys. Rev. Lett.* **45**, 1171 (1980).
25. T. J. GAY, *J. Phys. B* **16**, L553 (1983).
26. G. W. MCLUCAS, W. R. MACGILLIVRAY, AND M. C. STANDAGE, *Phys. Rev. Lett.* **48**, 88 (1982).
27. C. W. MCLUCAS, H. J. E. WEHR, W. R. MACGILLIVRAY, AND M. C. STANDAGE, *J. Phys. B* **15**, 1883 (1982).
28. C. J. WEBB, W. R. MACGILLIVRAY, AND M. C. STANDAGE, *J. Phys. B* **17**, 1675 (1984).
29. C. J. WEBB, W. R. MACGILLIVRAY, AND M. C. STANDAGE, *J. Phys. B* **17**, 2577 (1984).
30. C. J. WEBB, W. R. MACGILLIVRAY, AND M. C. STANDAGE, *J. Phys. B* **18**, L259 (1985).
31. G. F. HANNE, V. NICKICH, AND M. SOHN, *J. Phys. B* **18**, 2037 (1985).
32. H. E. SARAPH, *Comput. Phys. Commun.* **3**, 256 (1972).
33. H. E. SARAPH, *Comput. Phys. Commun.* **15**, 247 (1978).
34. H. A. BETHE AND E. E. SALPETER, *Quantum Mechanics of One and Two-Electron Systems* (Springer-Verlag, Berlin, 1972).
35. M. JONES, *Phil. Trans. R. Soc. London Ser. A* **277**, 582 (1975).
36. N. S. SCOTT AND P. G. BURKE, *J. Phys. B* **13**, 4299 (1980).
37. L. T. SIN FAI LAM AND W. E. BAYLIS, *J. Phys. B* **14**, 559 (1981).
38. I. P. GRANT, *Adv. Phys.* **19**, 747 (1970).
39. J. KESSLER, *Rev. Mod. Phys.* **41**, 3 (1969).
40. D. W. WALKER, *Adv. Phys.* **20**, 257 (1971).
41. O. BERGER, J. KESSLER, K. J. KOLLATH, R. MÖLLENKAMP, AND W. WÜBKER, *Phys. Rev. Lett.* **46**, 768 (1981).
42. N. F. MOTT AND H. S. W. MASSEY, *The Theory of Atomic Collisions*, 3rd ed (Oxford University Press, London, 1965).
43. H. N. BROWNE AND E. BAUER, *Phys. Rev. Lett.* **16**, 495 (1966).
44. L. T. SIN FAI LAM, *Aust. J. Phys.* **33**, 261 (1980).
45. R. P. MCEACHRAN AND A. D. STAUFFER, *J. Phys. B* **19**, 3523 (1986).
46. G. HOLTKAMP AND K. JOST, private communication (1986).
47. K. HASENBURG, K. BARTSCHAT, R. P. MCEACHRAN, AND A. D. STAUFFER, *J. Phys. B* **20**, 5165 (1987).
48. J. B. FURNESS AND I. E. MCCARTHY, *J. Phys. B* **6**, 2280 (1973).
49. G. STASZEWSKA, D. W. SCHWENKE, AND D. G. TRUHLAR, *Phys. Rev. A* **29**, 3078 (1984).
50. K. HASENBURG, D. H. MADISON, K. BARTSCHAT, AND K. BLUM, *J. Phys. B* **19**, 1803 (1986).
51. O. BERGER AND J. KESSLER, *J. Phys. B* **19**, 3539 (1986).
52. P. G. BURKE AND M. J. SEATON, *Meth. Comp. Phys.* **10**, 1 (1971).
53. P. G. BURKE AND W. B. EISSNER, in *Atoms in Astrophysics*, edited by P. G. Burke, W. B. Eissner, D. G. Hummer, and I. C. Percival (Plenum Press, New York, 1983).
54. E. P. WIGNER, *Phys. Rev.* **70**, 15 and 606 (1946).
55. E. P. WIGNER AND L. EISENBUD, *Phys. Rev.* **72**, 29 (1947).
56. P. G. BURKE AND W. D. ROBB, *Adv. At. Mol. Phys.* **11**, 143 (1975).
57. P. H. NORRINGTON AND I. P. GRANT, *J. Phys. B* **14**, L261 (1981).
58. N. S. SCOTT AND K. T. TAYLOR, *Comput. Phys. Commun.* **25**, 347 (1982).
59. K. BARTSCHAT AND N. S. SCOTT, *J. Phys. B* **17**, 3787 (1984).
60. K. BARTSCHAT, K. BLUM, P. G. BURKE, G. F. HANNE, AND N. S. SCOTT, *J. Phys. B* **17**, 3797 (1984).
61. K. BARTSCHAT, *J. Phys. B* **18**, 2519 (1985).
62. K. BARTSCHAT AND M. P. SCOTT, *J. Phys. B* **18**, L191 (1985).
63. K. BARTSCHAT AND M. P. SCOTT, *J. Phys. B* **18**, 3725 (1985).
64. K. BARTSCHAT, M. R. H. RUDGE, AND M. P. SCOTT, *J. Phys. B* **19**, 2469 (1986).

65. E. M. KARULE, in *Atomic Collisions III*, edited by V. Ia. Veldre (Latvian Academy of Sciences, Riga, 1965) (Translation TT-66-12939 available through SLA Translation Centre, Chicago).
66. E. M. KARULE, *J. Phys. B* **5**, 2051 (1972).
67. P. G. BURKE AND J. F. B. MITCHELL, *J. Phys. B* **6**, L161 (1973).
68. P. G. BURKE AND J. F. B. MITCHELL, *J. Phys. B* **7**, 214 (1974).
69. N. S. SCOTT, K. BARTSCHAT, P. G. BURKE, W. B. EISSNER, AND O. NAGY, *J. Phys. B* **17**, L191 (1984).
70. N. S. SCOTT, K. BARTSCHAT, P. G. BURKE, O. NAGY, AND W. B. EISSNER, *J. Phys. B* **17**, 3775 (1984).
71. R. B. BRODE, *Phys. Rev.* **34**, 673 (1929).
72. P. J. VISCONTI, J. A. SLEVIN, AND R. RUBIN, *Phys. Rev. A* **3**, 1310 (1971).
73. W. GEHENN AND E. REICHERT, *J. Phys. B* **10**, 3105 (1977).
74. G. BAUM, W. RAITH, AND H. STEIDL, private communication (1986).
75. N. S. SCOTT, P. G. BURKE, AND K. BARTSCHAT, *J. Phys. B* **16**, L361 (1983).
76. K. ALBERT, C. CHRISTIAN, T. HEINDORFF, E. REICHERT, AND S. SCHÖN, *J. Phys. B* **10**, 3733 (1977).
77. P. D. BURROW AND J. D. MICHEJDA, Int. Symp. on Electron and Photon Interactions with Atoms, Stirling (1974).
78. U. FANO AND J. W. COOPER, *Phys. Rev. A* **138**, 400 (1965).
79. D. W. O. HEDDLE, *J. Phys. B* **8**, L33 (1975).
80. D. W. O. HEDDLE, *J. Phys. B* **11**, L711 (1978).
81. C. E. KUYATT, J. A. SIMPSON, AND S. R. MIELCZAREK, *Phys. Rev. A* **138**, 385 (1965).
82. T. W. OTTLEY AND H. KLEINPOPPEN, *J. Phys. B* **8**, 621 (1975).
83. O. SHPENIK, V. V. SOUTER, A. N. ZAVILOPULO, I. P. ZAPESOCHNYI, AND E. E. KONTROSH, *Sov. Phys. JETP* **42**, 23 (1976).
84. E. U. CONDON AND G. H. SHORTLEY, *The Theory of Atomic Spectra* (Cambridge University Press, 1935), p. 268.
85. K. BARTSCHAT AND P. G. BURKE, *J. Phys. B* **19**, 1231 (1986).
86. K. BARTSCHAT, G. F. HANNE, A. WOLCKE, AND J. KESSLER, *Phys. Rev. Lett.* **47**, 997 (1981).
87. K. BARTSCHAT, D. H. MADISON, AND G. F. HANNE, *J. Phys. B* **18**, 1847 (1985).
88. K. BARTSCHAT AND D. H. MADISON, *J. Phys. B* **20**, 1609 (1987).
89. D. H. MADISON AND W. N. SHELTON, *Phys. Rev. A* **7**, 499 (1973).
90. D. H. MADISON AND W. N. SHELTON, *Phys. Rev. A* **7**, 514 (1973).
91. W. EISSNER, M. JONES, AND H. NUSSBAUMER, *Comput. Phys. Commun.* **8**, 270 (1974).
92. H. BORGMANN, J. GOEKE, G. F. HANNE, J. KESSLER, AND A. WOLCKE, *J. Phys. B* **20**, 1619 (1987).
93. A. A. ZAIDI, S. M. KHALID, I. MCGREGOR, AND H. KLEINPOPPEN, *J. Phys. B* **14**, L503 (1981).
94. G. F. HANNE, K. WEMHOFF, A. WOLCKE, AND J. KESSLER, *J. Phys. B* **14**, L503 (1981).
95. K. BARTSCHAT, N. S. SCOTT, K. BLUM, AND P. G. BURKE, *J. Phys. B* **17**, 269 (1984).
96. K. BARTSCHAT, diploma thesis, University of Münster (1981), unpublished.
97. P. NASS, N. LUDWIG, AND E. REICHERT, private communication (1986).
98. O. NAGY, K. BARTSCHAT, K. BLUM, P. G. BURKE, AND N. S. SCOTT, *J. Phys. B* **17**, L527 (1984).
99. R. A. BONHAM, *J. Phys. B* **15**, L361.

CHAPTER 2

Spin Polarization in Electron-Atom Scattering

G. F. Hanne

1. INTRODUCTION

The investigation of spin effects is an effort to probe the electron-atom collision process in greater detail than is possible through conventional cross-section measurements. This effort is part of a broad study of coherence in atomic collision physics that has been reviewed by a number of authors in a volume edited by Kleinpoppen and Williams.[1] The present chapter is part of a review of the progress made since then. Studies of coherence and polarization in electron-collision physics have been reviewed previously also by Hanne,[2] Blum and Kleinpoppen,[3] Slevin,[4] and Kessler.[5]

A study of atomic collisions provides a direct approach to the solution of the problem of many simultaneous interactions, and the study of electron collisions is basic for our understanding of atomic collisions: Although the electron is a pointlike, structureless particle (except for its spin) and thus easy to describe, the electron collision retains the richness afforded by the target structure and the various interactions and phenomena by which it is governed. Let us mention here the attractive Coulomb interaction with the nucleus, the repulsive electron-electron interaction, the target distortion, exchange effects, i.e., inclusion of the Pauli principle, and the spin-orbit interaction. The creation of oriented and aligned target states by the collision, the formation of temporary negative ion states (resonances), and the threshold behavior of observables are very interesting and basic phenomena of elastic and inelastic collision processes. Figures 1 and 2 illustrate the significance of electron exchange and target distortion or spin effects to the collision process.

G. F. Hanne • Department of Physics and Astronomy, The University of Oklahoma, Norman, Oklahoma 73019. *Present address*: Physics Institute, University of Münster, D-4400 Münster, West Germany.

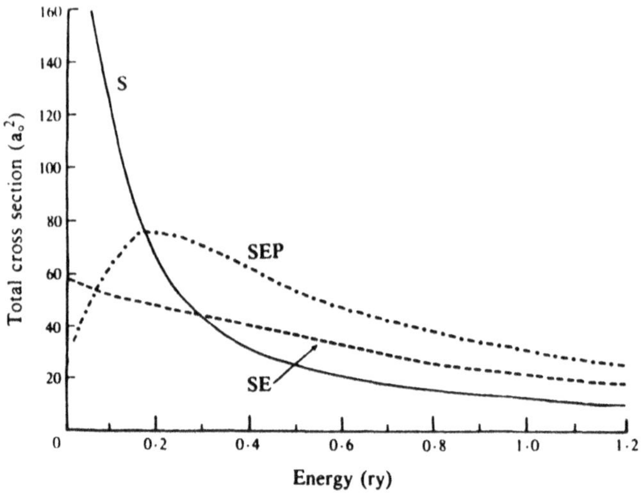

FIGURE 1. Total cross sections for e–H_2 collisions employing three model interaction potentials: S, static only; SE, static and exchange; SEP, static, exchange and charge-cloud polarization. Figure taken from Morrison.[6]

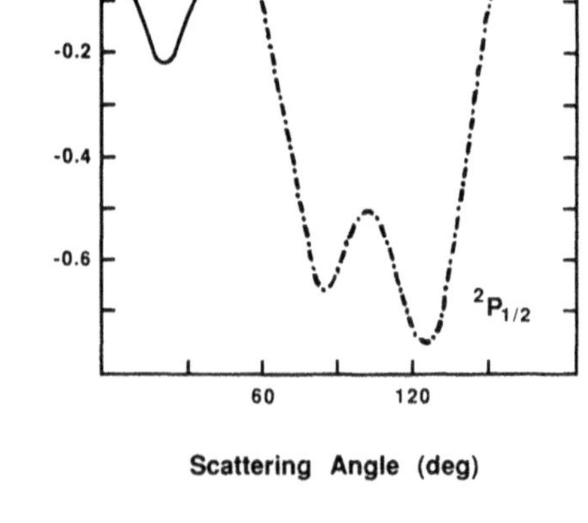

FIGURE 2. Calculated spin-asymmetry function S_A for excitation of $6\,^2P_{1/2}$ and $6\,^2P_{3/2}$ states in Cs versus scattering angle at 2.04 eV.[7] S_A describes, e.g., the spin up–down asymmetry (see Section 2). A value of $S_A = -0.67$ means that the cross section for spin-down electrons is five times larger than for spin-up electrons.

Details of such interaction and correlation phenomena in electron collisions can be studied by determining the spin polarization of scattered electrons and the orientation and alignment of the target after scattering, where initially polarized electron and state-selected target beams may be used (cf. Figure 3). Such experimental work is feasible now by using intense sources of polarized electrons,[8] lasers for target-state selection,[9,10] and the electron–photon coincidence method.[11] Whereas little progress has been made in improving spin-polarization detectors,[5,12] efficient methods have been developed to produce highly polarized ground-state[13,14] or state-selected excited Na beams.[9]

These experiments will provide the most stringent test of theoretical models of the collision process because they will measure many—or, in simpler cases, all—of the "quantum mechanical observables" in the particular experiment. Numerically, distorted-wave[15,16] and R-matrix close-coupling[17] calculations of these observables have been improved to include effects from spin–orbit interactions. These calculations have been performed for electron collisions with heavy atoms.

In the following we review recent experimental results and illustrate the models used to describe electron–atom collisions by simple pictures. Spin effects in electron–atom collisions are certainly seen in experiments where polarized electrons are not involved,[2] e.g., by using only polarized or state-selected laser-excited atoms, or by observing effects of incoherence in electron–photon coincidence studies with unpolarized electrons. Following the intention of the editors, however, this review focuses on recent progress in experiments where polarized free electrons are involved.

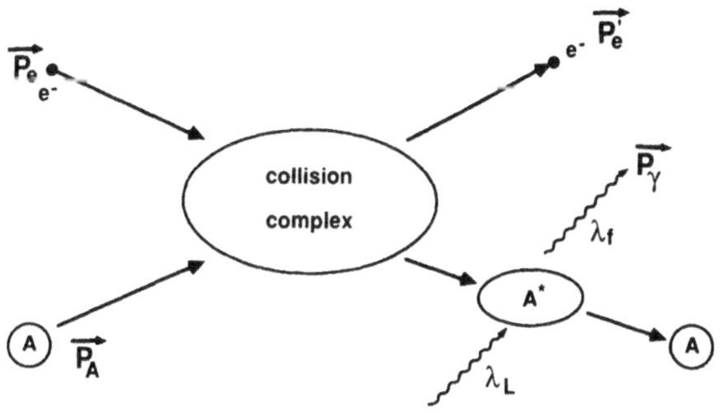

FIGURE 3. Schematic diagram of a detailed study of electron–atom collisions. Polarized electrons or atoms (more generally: state-selected atoms) may be prepared before or detected after the collision process. If an inelastic process is studied, the polarization of fluorescence emission (λ_f) from the short-lived excited target state A^* may be obtained as well, even in coincidence with scattered electrons. Metastable states may be analyzed by laser-excited (λ_L) fluorescence.

2. DISCUSSION OF SPIN EFFECTS

In most calculations the quantization axis is chosen along the direction of the incident electron beam ("collision" frame). But simple pictures of various collision processes are more conveniently described in a frame that uses the normal to the scattering plane as the "natural" quantization axis. The transformation from the "collision" frame to the "natural" frame is a purely mathematical one. Therefore in the following we will use the frame that will provide the simplest picture of the collision process.

Another general remark refers to the term "spin-flip." Throughout this paper we will call any process in which the spin component of the continuum electron has changed (by ± 1) after scattering a spin-flip process irrespective of whether this change is caused by a spin-dependent interaction or by exchange with a target electron of opposite spin.

2.1. Spin-Orbit Interaction in the Continuum

Any magnetic field **B** that is present during an electron collision process will interact with the magnetic moment $\boldsymbol{\mu}_e$ of an electron by the potential energy $V = -\boldsymbol{\mu}_e \cdot \mathbf{B}$. External (macroscopic) magnetic fields are not considered here, but there exist intrinsic fields that need to be considered. Of these, the most important is the magnetic field $\mathbf{B}_l = -(\mathbf{v} \times \mathbf{E})/c$ that is observed in the rest frame of an electron due to its motion (with velocity $\mathbf{v}$) in the electric field **E** of the atomic core (predominantly the field of the positively charged nucleus; see Figure 4). This field gives rise to the spin-orbit energy, which increases with increasing nuclear charge and is strong if the electron comes close to the nucleus. The influence of the spin-orbit interaction of the target electrons will be discussed in Section 2.3.

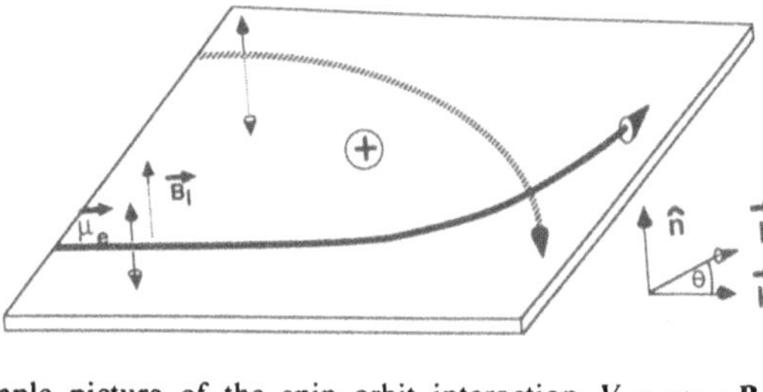

FIGURE 4. Simple picture of the spin-orbit interaction $V = -\boldsymbol{\mu}_e \cdot \mathbf{B}_l$. The sign of the spin-orbit interaction depends on the relative orientation of the magnetic moment $\boldsymbol{\mu}_e$ of an electron with respect to the magnetic field $\mathbf{B}_l$ arising from the relative motion of the electron in the field of the nucleus. Shown are the relative orientations of magnetic moments up and down for scattering to the left where $\mathbf{B}_l$ is up, and for magnetic moment up for scattering to the right where $\mathbf{B}_l$ is down. The scattering plane is defined by $\hat{n} = \mathbf{k} \times \mathbf{k}'/|\mathbf{k} \times \mathbf{k}'|$, where **k** and **k**$'$ are the wave vectors of the continuum electron before and after scattering, respectively.

The following spin effects for the continuum electron can be obtained if it comes close enough to the nucleus, i.e., for large-angle scattering.

1. Initially unpolarized electrons are polarized after scattering. This statement may be made plausible by the following simple picture: A beam of unpolarized electrons can be thought of as being composed of 50% spin up and 50% spin down with respect to the normal of the scattering plane. According to Figure 4 the spin–orbit energy for electrons with spin up and spin down is different, because their magnetic moments have opposite orientations with respect to the magnetic field $\mathbf{B}_l$. Therefore, in general, the cross sections $\sigma(\uparrow)$ and $\sigma(\downarrow)$ for electrons whose spins initially are up or down will be different, as will be the number $N_\uparrow$ and $N_\downarrow$ of electrons with spin up and spin down after scattering. As a result the polarization

$$P' = \frac{N_\uparrow - N_\downarrow}{N_\uparrow + N_\downarrow} = S_P \tag{1}$$

of the scattered electron beam is different from zero. S_P is a parameter which, for elastic scattering from spinless targets, is called the Sherman function S.[2,5] Its relation to the scattering amplitudes is shown in equations (22).

2. Transversely polarized electrons show a left–right or a spin up–down scattering asymmetry. Let us assume, for the sake of simplicity, that the initial electron beam is completely polarized, i.e., all spins or magnetic moments are oriented normal to the scattering plane. Since the magnetic field $\mathbf{B}_l \propto \mathbf{v} \times \mathbf{E}$ for electrons scattered to the left (cross section σ_l) is opposite (opposite $\mathbf{E}$ vector) to that for electrons scattered symmetrically to the right (σ_r), the spin–orbit energy for these cases is different, as are, in general, the cross sections σ_l and σ_r and the number of electrons scattered to the left (N_l) and to the right (N_r), respectively.

As a result we obtain a left–right scattering asymmetry

$$A = \frac{N_l - N_r}{N_l + N_r} = S_A P \tag{2}$$

where $S_A = A/P$ (P is the initial electron polarization) is the normalized parameter describing this asymmetry. S_A also describes the spin up–down asymmetry, which is obtained for electrons scattered to the left only in collisions where the initial polarization is changed from up to down ($P \to -P$)

$$A = \frac{\sigma(P) - \sigma(-P)}{\sigma(P) + \sigma(-P)} = S_A P \tag{3}$$

Note that only for elastic scattering from unpolarized atoms are $N_\uparrow \propto \sigma(\uparrow)$ and $N_\downarrow \propto \sigma(\downarrow)$ so that $S_P = S_A = S$. For inelastic scattering or scattering from polarized targets the parameters S_P and S_A are different functions in general.[2,5]

3. *The polarization vector of an electron beam rotates if its initial direction is not perpendicular to the scattering plane.* If the polarization vector of an electron beam is not parallel or antiparallel to $\mathbf{B}_l$, the magnetic moment of the beam experiences a torque that causes the magnetic-moment (or polarization) vector to change its direction and to precess. In formal terms the polarization $\mathbf{P}'$ after scattering from unpolarized targets is[117]

$$\mathbf{P}' = \frac{T_n(\theta)\mathbf{P}_n + S_p(\theta)\hat{n} + T_k(\theta)\mathbf{P}_k + T_t(\theta)\mathbf{P}_t + \hat{n} \times [U_k(\theta)\mathbf{P}_k + U_t(\theta)\mathbf{P}_t]}{1 + \mathbf{P} \cdot \hat{n} S_A(\theta)} \tag{4}$$

where we have resolved the initial polarization vector $\mathbf{P} = \mathbf{P}_n + \mathbf{P}_k + \mathbf{P}_t$ into components $\mathbf{P}_k$ parallel to the incident beam direction, $\mathbf{P}_t$ perpendicular to $\mathbf{P}_k$ in the scattering plane, and $\mathbf{P}_n$ perpendicular to it (i.e., parallel to $\hat{n}$). $T_n(\theta)$, $T_k(\theta)$, and $T_t(\theta)$ are parameters that reduce and change the magnitude ($|T| \leq 1$) of the initial polarization components, $S_p(\theta)\hat{n}$ is added to the component perpendicular to the scattering plane and $U_k(\theta)$ and $U_t(\theta)$ describe new components perpendicular to $\hat{n}$, and $\mathbf{P}_k$ and $\mathbf{P}_t$, respectively. All components are modified by the common denominator $1 + \mathbf{P} \cdot \hat{n} S_A(\theta)$. For elastic collisions from spinless targets, $T_n(\theta) = 1$, $T_k(\theta) = T_t(\theta) = T$, $U_k(\theta) = U_t(\theta) = U$, and $S_p = S_A = S$. The experimental determination of S, T, and U for this case is described in Section 3.2. A much more complicated formula is obtained if scattering of polarized electrons from polarized atoms is considered, as has been demonstrated by Burke and Mitchell for electron–Cs collisions.[18]

2.2. *Exchange Collisions*

According to the Pauli principle, the continuum electron and the target electrons need to be described by a completely antisymmetrized state. This gives automatically transition matrix elements describing exchange collisions. The "reaction"

$$e(\mathbf{r}) + \phi_i(\ldots, \mathbf{r}_\nu, \ldots) \to \phi_f(\ldots, \mathbf{r}, \ldots) + e(\mathbf{r}_\nu) \tag{5}$$

describes schematically how an incoming electron (with coordinate $\mathbf{r}$) changes place by a collision with a target electron (with coordinate $\mathbf{r}_\nu$) which is observed as the scattered electron.

The "exchange interaction" is short range in nature because it is confined to a volume roughly the size of the "charge-cloud" probability density of the target. Thus exchange effects are more significant for scattering at small impact parameters, i.e., scattering by larger angles.

Exchange processes leaving closed shells unchanged are attractive since the antisymmetric wave function has the effect of preventing two electrons of like spin—which can exchange only in this case—from being found near one another. The "Fermi hole" that surrounds the target electron weakens the repulsive Coulomb interaction between the two electrons of like spin.

Finally we mention that exchange scattering is significant only when the continuum electron is not much faster than the bound electron, as can be shown by a rough approximation (Bonham-Ochkur approximation[19,20] or similar treatments[21,22]). Exchange is significant at collision energies as large as ten times the binding energy of the target electron that is exchanged.

We can only observe an exchange process directly if spin is transferred to the atom, i.e., if the spin projection quantum number M_s of the target has been changed by the collision. This will result in a change of polarization of scattered electrons,

$$P' = (\omega_D - \omega_{SF})P = (1 - 2\omega_{SF})P \tag{6}$$

where P and P' are the polarizations before and after scattering and ω_D and ω_{SF} are the probabilities for direct (non-spin-flip) and "spin-flip" collisions, respectively ($\omega_D + \omega_{SF} = 1$). A straightforward interpretation of observed "spin-flips" (i.e., $\omega_{SF} > 0$) as pure exchange processes—which conserve spin projections according to

$$m_s' - m_s = M_s - M_s' \tag{7}$$

where m_s, M_s and m_s', M_s' are the spin quantum numbers of continuum electron and atom before and after scattering, respectively—is possible only if spin-flips by the spin-orbit interaction can be excluded. This may be assumed for light atoms or forward scattering.

A change of the atomic-spin quantum number may occur in the following situations:

1. *The atom has unsaturated spin ($S \neq 0$).* An example of this situation is scattering of electrons from one-electron atoms like H and alkalis. Here exchange with the unsaturated valence electron can be directly observed; in a schematic reaction such a process can be written as

$$e(\uparrow) + A(\downarrow) \rightarrow A(\uparrow) + e(\downarrow) \tag{8}$$

Obviously such a reaction will reverse ("flip") the preferential orientation of the spins, so that, according to equation (6), the polarization P' of scattered electrons will be different from the initial polarization P. A polarized target may not only change the magnitude of the initial polarization, it may also rotate the polarization vector of the electron beam if $\mathbf{P}_e$ and $\mathbf{P}_A$ are not parallel or antiparallel.[5]

2. The multiplicity of the atom is changed by the collision. This phenomenon occurs, for example, in singlet–triplet excitation in electron scattering from a two-electron atom like He or Hg. Figure 5 gives a simple picture of such a process.[23] It illustrates that a singlet–triplet transition in a two-electron atom is possible by electron exchange (with and without spin-flip) only, as long as the spin-conservation rule (7) holds.

It is worthwhile mentioning that in the situations presented here, where the spin–orbit coupling is neglected, exchange effects cannot lead to polarized scattered beams unless these beams are initially polarized. Moreover, left–right scattering asymmetry is not observed even if the electrons and atoms are initially polarized. This is because exchange is not represented by a spin-dependent operator in the Hamiltonian and thus the potential energy cannot depend on the relative spin state.

From these statements it seems as though exchange scattering of unpolarized beams cannot produce spin-polarized scattered beams. This is, however, not generally true, as will be discussed in the following.[24]

2.3. Spin–Orbit Interaction of Target Electrons

2.3.1. Inelastic Collisions: Electron-Impact Excitation of Atomic Fine-Structure Levels. The spin–orbit interaction causes atoms with unsaturated

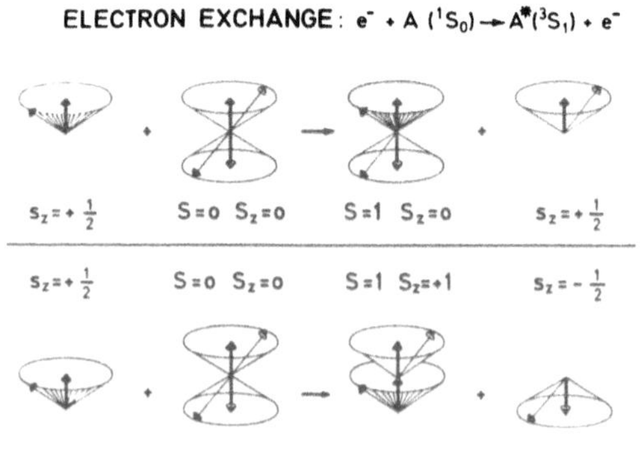

FIGURE 5. Simple picture of an exchange collision leading to a singlet–triplet excitation of a two-electron atom. The incident electron is depicted by hatching. The process shown in the lower part is a spin-flip process that changes the polarization of continuum electrons (from Hanne and Kessler[23]).

spin and orbital angular momentum to relax into fine-structure states. It is weak for atoms with small nuclear charge. The fine-structure splitting of the $2p$ state in H, for example, is only $\Delta E = 4.5 \times 10^{-5}$ eV. Correspondingly we may define a relaxation time $\Delta t = h/\Delta E$ that is roughly the period in which, in a semiclassical vector model, the spin and the orbital angular momentum precess around the total angular momentum. For a H atom this time is about 10^{-10} s. It should be compared with the collision time for an electron-scattering process, which is the time in which the "pointlike" continuum electron moves past an atom with a typical diameter of the order of 10^{-10} m. For an electron whose kinetic energy is not too small ($E \geqslant 0.1$ eV), this time is easily calculated to be less than 10^{-15} s.

These different time scales are the justification for the sudden approximation or Percival–Seaton hypothesis[25]: During the short collision time, atomic spin and orbital angular momentum are considered to be at rest because of their slow precession. As a consequence the spin–orbit interaction of the target electrons can be completely neglected during the collision. This approximation is even better for the hyperfine interaction, which, in general, is still weaker.

In an experiment, however, a splitting, even as small as $\Delta E = 4.5 \times 10^{-5}$ eV can be resolved in principle, though not in practice for H atoms at present. State selection by laser excitation is a powerful way to achieve such resolution. Thus we may ask what happens if a collision process is observed in which the target is left in a certain well-defined fine-structure state. At first glance one might think that because no spin-dependent force acts during scattering from very light targets, the arguments given in Section 2.2 hold, and polarization of scattered beams will not be obtained unless the beams are initially polarized.

Surprisingly, calculations and experiments show that in such a situation unpolarized beams may become significantly polarized. As an example we show in Figure 6 the result of a calculation by Shelton[26] for electron-impact excitation of the $2p$ states of H atoms at an energy of 50 eV. Figure 6 is a plot of the spin polarization versus scattering angle for inelastically scattered electrons that excite separately the $2p$ fine-structure levels of the H atom. A significant polarization transverse to the scattering plane is observed, though initially unpolarized beams are assumed, and spin–orbit interaction is not included in the calculation, except that the final states are fine-structure states obtained by standard coupling procedures.

What is the origin for the spin polarization shown in Figure 6? The H atom is left in the $2p$ state after the $1s$–$2p$ electron-impact excitation process. It is well established that the orbital angular momentum transferred from the electron to the atom has, in general, a preferential orientation in the scattering plane. There has been, however, a controversy in the literature concerning the relationship in semiclassical pictures of the sign of this orientation to attractive and repulsive parts of the scattering potential.[27–32]

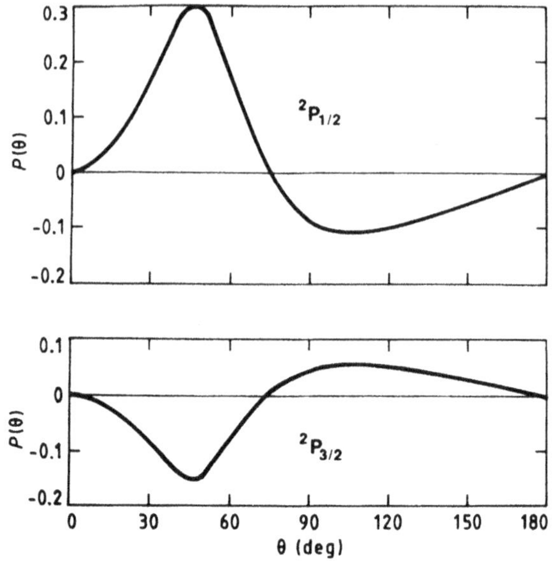

FIGURE 6. Calculated spin polarization versus scattering angle for inelastically scattered electrons exciting separately the $2p$ fine-structure levels of the H atom.[26]

No convincing model has yet been established. Nevertheless, significant orbital angular momentum orientation has been observed in various systems.

Quantum mechanically, an ensemble of oriented atoms has a nonvanishing expectation value $\langle \mathbf{L} \rangle$ of the orbital angular momentum (operator) **L**. Orientation in the scattering plane means a nonvanishing component of $\langle \mathbf{L} \rangle$ perpendicular to the scattering plane. A semiclassical picture of a collision-induced orientation is shown in Figure 7a, and Figure 7b shows a vector representing, for the sake of simplicity, the maximum possible

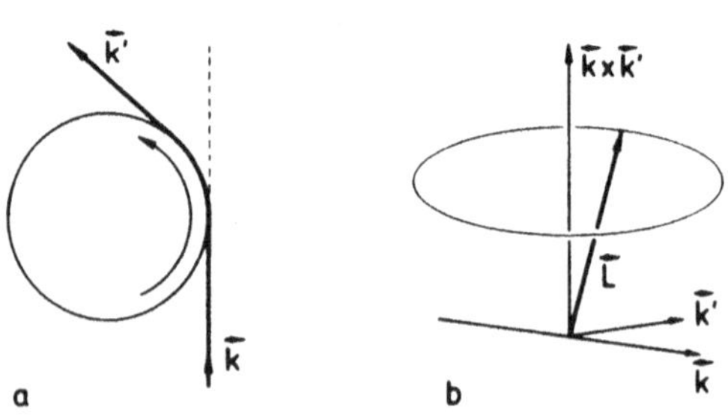

FIGURE 7. Semiclassical description of orientation by collision. (A) Orientation in the scattering plane caused by a grazing collision. The angular momentum transferred from the electron to the atom has a preferential orientation, i.e., the atom is oriented after the collision as indicated. (b) Maximum orientation of the atom described by the vector model.[24]

orientation with respect to an axis normal to the scattering plane. For an ensemble of H atoms in a pure oriented state the **L** of an individual atom may be found somewhere on a cone around this axis as indicated in Figure 7b. For an ensemble of unoriented atoms no such preferential direction of **L** exists.

The H atom has an unsaturated spin with $S = 1/2$. Let us assume that we definitely observe the excitation of one of the $J = 1/2$, $J = 3/2$ states in the experiment. As shown in Figure 8a, not all atomic spin vectors can couple with the given **L** to give the total angular momentum **J** with $J = 1/2$. The allowed direction of the spin vector **S** is "opposite" to that of **L**. Accordingly a $J = 3/2$ state can be formed only by atomic spin vectors **S** that are oriented parallel to **L** (Figure 8b). Because the precession of **S** and **L** about **J** is very slow, Figure 8 is a simple picture of the atom immediately after the collision: Collisionally oriented atoms that are selected for different fine-structure states are also spin polarized after the collision (transverse to the scattering plane).

Consider now the effect of exchange on the excitation of a $J = 1/2$ state. An unpolarized electron (or atom) beam can be regarded as consisting of electrons (atoms) half of whose spins are up and half down with respect to the normal to the scattering plane. Provided **L** is oriented by the collision, as in Figure 8a, electrons of either spin orientations can be scattered without spin-flip, if the atomic spin is initially oriented in the right way, namely, down. But only spin-down electrons can undergo an exchange collision, because spin-up electrons leave the atoms in a state with $J = 3/2$.

Note that the term "spin down" characterizes only the preferential orientation of the spin rather than a pure spin-down state. With this precaution we can summarize the cross sections for the basic reactions in the following scheme, where "up" and "down" for electron and atomic spin are symbolized by arrows and f and g are scattering amplitudes for direct and exchange collisions, respectively. For initially unpolarized electrons

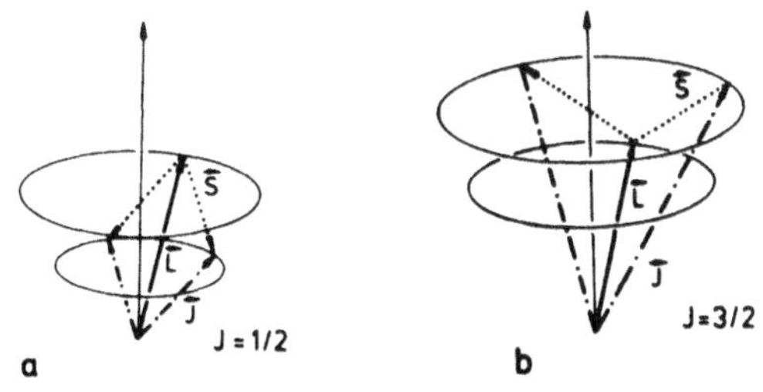

FIGURE 8. Semiclassical vector-coupling model for $L = 1$ and $S = 1/2$. (a) Only spin vectors **S** antiparallel to the given orbital angular momentum **L** can form a $J = 1/2$ state. (b) Only spin vectors **S** parallel to **L** can form a $J = 3/2$ state.[24]

and atoms (we have here discussed only H atoms, but the following reactions apply to all light one-electron atoms) we have

$$
\begin{aligned}
&e(\uparrow) + A(\downarrow) \rightarrow A^*(\downarrow) + e(\uparrow): && |f|^2 \\
&e(\uparrow) + A(\downarrow) \rightarrow A^*(\uparrow) + e(\downarrow): && \text{``forbidden'' (see text)} \\
&e(\uparrow) + A(\uparrow) \rightarrow A^*(\uparrow) + e(\uparrow): && \text{``forbidden''} \\
&e(\downarrow) + A(\downarrow) \rightarrow A^*(\downarrow) + e(\downarrow): && |f - g|^2 \\
&e(\downarrow) + A(\uparrow) \rightarrow A^*(\uparrow) + e(\downarrow): && \text{``forbidden''} \\
&e(\downarrow) + A(\uparrow) \rightarrow A^*(\downarrow) + e(\uparrow): && |g|^2
\end{aligned}
\tag{9}
$$

Other reactions that do not conserve spin are forbidden for the light atoms considered here.

The number of electrons with spin up after collisions is $N_\uparrow \propto (|f|^2 + |g|^2)$, which is obviously different from $N_\downarrow \propto |f - g|^2$. In other words, the electrons are polarized normal to the scattering plane after collision according to the definition (1). Likewise, a spin up-down or left-right scattering asymmetry will be found.

In the picture drawn here the strength of the spin-orbit interaction that causes the fine-structure splitting is irrelevant. This interaction can be very small. We merely need to *resolve* the spin-orbit splitting. When exchange is negligible (i.e., $g = 0$) this polarization will vanish ($N_\uparrow = N_\downarrow$).

If exchange is not negligible, spin effects may be directly observed in double-scattering experiments: One may either use unpolarized beams of electrons and atoms and measure the spin polarization of the scattered electrons, or use a spin-polarized electron beam and measure the left-right asymmetry of the scattered intensity. Similar polarization phenomena are found in Mott scattering. However, in these mechanisms, spin polarization effects are produced in different regions of the target. Whereas spin polarization by Mott scattering is produced at small distances from the target nucleus, the spin polarization caused by exchange and orientation effects depends strongly on the angular momentum configuration of the outermost target electrons after the collision. Moreover, for an unresolved (and small) fine-structure splitting this polarization effect must vanish. So the sign of the polarization must differ after excitation of different members of the fine-structure multiplet. From the vector-coupling model shown in Figure

8 we conclude this immediately. A careful examination of this model also shows that

$$P'(J = 1/2) = -2P'(J = 3/2) \tag{10}$$

Remembering that the cross section for exciting the $J = 1/2$ state is half that for the $J = 3/2$ state, we can verify that the total polarization is indeed zero. Formulas for two-electron atoms corresponding to (10) can be found elsewhere.[2,33]

With increasing nuclear charge of the targets the situation becomes more complex. The assumption of LS coupling for target states may no longer be warranted. Examples are the $np^5(n+1)$s states in the rare gases, where the spin–orbit interaction for the hole in the p shell is strong enough to violate LS coupling significantly even for Ne ($Z = 10$). On the other hand, the influence of the spin–orbit interaction for the continuum electron is still small for Ne.[34] (By "small" we mean that the absolute values of the polarization are less than 5% over the whole angular range.)

For the fine-structure effect discussed so far this violation of the LS-coupling scheme means that general relations like equation (10) will be violated, even though the spin polarization is still caused by an orbital angular momentum orientation that is transferred to the atomic spin via the spin–orbit interaction of the target electrons and simultaneously transferred to the scattered electrons via exchange collisions. In contrast to the H case, the strength of this spin–orbit interaction is important, as can be seen by the degree of the violation of relations like equation (10).

For a heavy target such as Hg ($Z = 80$), the spin–orbit interaction for the continuum electron cannot be neglected, in general, and the situation becomes very complex. At certain energies and angles, however, either the spin–orbit interaction of the continuum electron or exchange can dominate. Only in such exceptional cases can straightforward conclusions about the different spin effects be drawn.

2.3.2. The Effect of Fine-Structure Splitting in Elastic Collisions. So far we discussed the fine-structure effect for inelastic collisions only. Is the fine-structure effect also significant in elastic collision?

It is worthwhile noting that the role of the valence electrons in elastic collisions is basically different from that in inelastic collisions: For inelastic scattering the electron–electron interaction is the principal "transition operator"; the core contributes only in higher order to the scattering amplitudes.[2] For elastic collisions the attractive core is dominant, in general, and the valence electrons mainly determine the long-range behavior of the scattering potential, in particular terms arising from the charge-cloud polarization. There is, however, no reason that an open-shell atom whose ground state is a P state cannot be oriented by an elastic collision. Because we have basically a three-body problem, transfer of orbital angular momentum is possible even in elastic collisions. If the target shows a fine-structure splitting, the lowest fine-structure level must be considered the ground state. Examples are $^2P_{1/2}$ atoms such as Tl or B, although experimental discrimination between scattering from different fine-structure states is infeasible for B.

The possible influence of the orientation of the target electrons on the elastic scattering process may be explained by the following simple picture (Figure 9). A beam of unpolarized $^2P_{1/2}$ atoms can be thought of as consisting of 50% with $M_J = 1/2$ and 50% with $M_J = -1/2$. For atoms with $M_J = 1/2$ (oriented normal to the scattering plane), the orbital angular momentum is oriented mainly counterclockwise in the scattering plane, while the spin is preferentially down. The situation is just the opposite for atoms with $M_J = -1/2$. These different orientations may cause different cross sections owing to an orbit effect: The simple picture shown in Figure 9 may suggest that if the continuum electron and the valence electron move "parallel" (i.e., $M_J = 1/2$ atoms and electrons scattered to the left), then the probability of the two electrons being found near another is smaller than for "opposite orbits" ($M_J = -1/2$ atoms). Thus the repulsive electron–electron Coulomb interaction for scattering from $M_J = 1/2$ atoms is weaker than that for scattering from $M_J = -1/2$ atoms.

So far this different potential is an *orbit* effect. If this orbit effect causes different cross sections for elastic scattering from atoms with $M_J = 1/2$ and from atoms with $M_J = -1/2$, the scattered electrons may be spin polarized by exchange collisions even if the beams are initially unpolarized. Let us assume, for the sake of simplicity, that the cross section for scattering from $M_J = 1/2$ atoms is much larger than that for scattering from $M_J = -1/2$ atoms. The scattering then takes place with target atoms that are *effectively spin polarized*, because for $M_J = 1/2$ atoms the atomic spin is (preferentially) down. Consequently, part of this effective target spin polarization can be transferred to the scattered electrons via exchange collisions: Spin up electrons are either scattered directly without spin–flip (with a probability

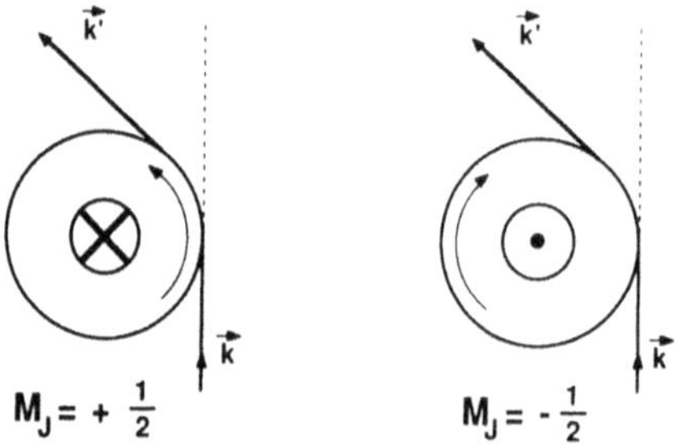

FIGURE 9. Fine-structure effect for elastic collisions of electrons from $^2P_{1/2}$ atoms: The cross section may depend on the relative orientation of the atoms for a given orbit of the continuum electron, because the effective electron–electron repulsion between the continuum electron and the target electron depends on the orientation.

proportional to $|f^+|^2$) or with a spin-flip ($|g^-|^2$), whereas spin down electrons are scattered without spin-flip and a probability proportional to $|f^+ - g^+|^2$. The superscripts on f and g indicate that the collision takes place without (+) a change of atomic orbital angular momentum or with (−) such a change. Collisions involving a change of the orbital momentum orientation of the target atom may be less likely. The number of electrons with spin up after collision, $N_\downarrow \propto |f^+|^2$, is different from that with spin down, $N_\downarrow \propto |f^+ - g^+|^2 + |g^-|^2$, i.e., the scattered electron beam is polarized.

A model calculation for elastic scattering of electrons from B ($2\,^2P_{1/2}$) atoms[35] yields a strong spin polarization, as shown in Figure 10. This result is only a qualitative check on the size of the effect because very crude approximations were used. The same calculation shows that the influence of the spin-orbit interaction of the continuum electron is negligible (less than 0.1%). Experimental study of elastic scattering from $2\,^2P_{1/2}$ B atoms is very difficult because of the high temperatures needed to produce an atomic beam and the small fine-structure separation of the target. However, investigations using atoms such as Tl ($6\,^2P_{1/2}$) have been performed recently (see Section 3.3).

2.4. *Elastic Scattering of Electrons by Cesium Atoms*

Spin effects in elastic collisions of electrons by Cs atoms are governed by exchange, and, unlike the systems discussed in Section 2.2, by the spin-orbit interaction for the continuum electron. No spin-orbit interaction can be effective, however, for the 6*s*-valence electron. A theoretical treatment of this collision system has been given by Burke and Mitchell.[18]

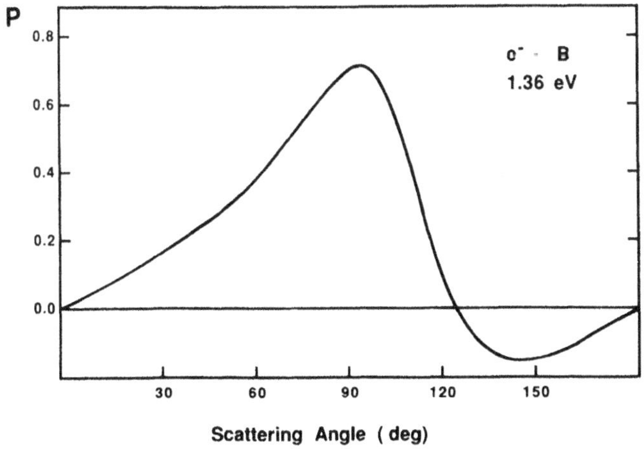

FIGURE 10. Calculated spin polarization for elastic scattering from B ($2\,^2P_{1/2}$) atoms.[35]

Farago pointed out that a left-right asymmetry will be observed when unpolarized electrons are scattered by Cs atoms that are polarized perpendicular to the scattering plane, provided the exchange and spin-orbit interactions act simultaneously.[36] If only one of the two processes is effective, no such asymmetry can be found.

This phenomenon can be illustrated in simple terms. By $\sigma_{\uparrow\downarrow}(\theta)$ we denote the cross section for elastic scattering of electrons to the left in the case of atomic and electron spin initially up and down, respectively. The corresponding cross section for scattering to the right is denoted by $\sigma_{\uparrow\downarrow}(-\theta)$, etc. There are, in general, four different cross sections corresponding to the four different initial spin states, when the quantization axis is chosen normal to the scattering plane:

$$\sigma_{\uparrow\uparrow}(\theta) = \sigma_{\downarrow\downarrow}(-\theta) \tag{11a}$$

$$\sigma_{\downarrow\downarrow}(\theta) = \sigma_{\uparrow\uparrow}(-\theta) \tag{11b}$$

$$\sigma_{\uparrow\downarrow}(\theta) = \sigma_{\downarrow\uparrow}(-\theta) \tag{11c}$$

$$\sigma_{\downarrow\uparrow}(\theta) = \sigma_{\uparrow\downarrow}(-\theta) \tag{11d}$$

The relations between the cross sections for scattering to the left and scattering to the right in equations (11) can be seen by rotating the scattering plane by π around the incident beam direction. This rotation changes the scattering angle from θ to $-\theta$ and the spins in the opposite directions. Because of space isotropy, however, this rotation does not change the probability for the scattering process or its cross section. If, for example, the atoms are polarized with spin up, then the total number of electrons scattered to the left is $N_l \propto \{\sigma_{\uparrow\uparrow}(\theta) + \sigma_{\uparrow\downarrow}(\theta)\}$, and the number scattered to the right is $N_r \propto \{\sigma_{\uparrow\uparrow}(-\theta) + \sigma_{\uparrow\downarrow}(-\theta)\}$. These numbers are different, in general, and thus a left-right asymmetry will be observed.

If, however, only exchange is involved, the cross section will not depend on the relative orientation of the spins to the orbit, and we have $\sigma_{\uparrow\uparrow}(\theta) = \sigma_{\uparrow\uparrow}(-\theta)$ and $\sigma_{\uparrow\downarrow}(\theta) = \sigma_{\uparrow\downarrow}(-\theta)$, giving no asymmetry. But if only the spin-orbit interaction of the continuum electron is important, spin-up and spin-down atoms will no longer scatter differently, i.e., the potential will be the same. In this case, we have $\sigma_{\uparrow\uparrow}(\theta) = \sigma_{\downarrow\uparrow}(\theta)$ and $\sigma_{\downarrow\downarrow}(\theta) = \sigma_{\uparrow\downarrow}(\theta)$. With equations (11)—which are always valid—we eventually have $\sigma_{\uparrow\uparrow}(\theta) = \sigma_{\uparrow\downarrow}(-\theta)$ and $\sigma_{\uparrow\downarrow}(\theta) = \sigma_{\uparrow\uparrow}(-\theta)$, and no left-right asymmetry can be observed.

Although such an experiment was proposed as early as 1974,[36] only recently have data for spin-orbit and exchange effects in elastic scattering of spin-polarized electrons from spin-polarized Na atoms been reported.[37] These authors presented their data as exchange asymmetry-corresponding to the spin antiparallel-parallel asymmetry (19)—and as spin-orbit asymmetry—corresponding to the spin up-down asymmetry (3). As expected for the Na target, these asymmetries at 54.5 eV are 3-4% at most. An experiment using polarized Cs atoms is in progress.[38]

3. ELASTIC COLLISIONS

3.1. Direct Observation of Exchange

The scattering of electrons by light one-electron atoms such as Na ($Z = 11$) may be spin dependent because of exchange collisions. Because of the small nuclear charge, spin-orbit interaction will be negligible. Hence we can distinguish the following basic reactions[39] that conserve spin:

$$e(m_s) + A(m_s) \rightarrow e(m_s) + A(m_s): \qquad f - g \tag{12a}$$

$$e(m_s) + A(-m_s) \rightarrow e(m_s) + A(-m_s): \qquad f \tag{12b}$$

$$e(m_s) + A(-m_s) \rightarrow e(-m_s) + A(m_s): \qquad -g \tag{12c}$$

Here we denote by $m_s = \pm 1/2$ the magnetic spin-quantum numbers for an arbitrary quantization axis and by $f(\theta) = |f(\theta)|e^{i\delta_1}$ and $g(\theta) = |g(\theta)|e^{i\delta_2}$ the complex amplitudes for direct and exchange scattering, respectively. The differential cross section for scattering of unpolarized beams is given by

$$\sigma_0(\theta) = \tfrac{1}{4}|f(\theta) - g(\theta)|^2 + \tfrac{1}{4}|f(\theta)|^2 + \tfrac{1}{4}|g(\theta)|^2 \tag{13}$$

A general description of scattering by polarized beams has been given by Kessler.[5] The differential cross section $\sigma(\theta)$ and polarization $\mathbf{P}'$ of the scattered electrons can be written

$$\sigma(\theta) = \sigma_0(\theta)[1 - A(\theta)\mathbf{P}_e \cdot \mathbf{P}_A] \tag{14}$$

$$\mathbf{P}'_e = \frac{T_e(\theta)\mathbf{P}_e + T_A(\theta)\mathbf{P}_A + S(\theta)(\mathbf{P}_e \times \mathbf{P}_A)}{1 - A(\theta)\mathbf{P}_e \cdot \mathbf{P}_A} \tag{15}$$

where $\mathbf{P}_e$ and $\mathbf{P}_A$ are the polarization vectors of incident electron and atom beam, respectively, and

$$A(\theta) = \frac{|f(\theta)||g(\theta)| \cos(\delta_1 - \delta_2)}{\sigma_0(\theta)} = \frac{|f(\theta) + g(\theta)|^2 - |f(\theta) - g(\theta)|^2}{|f(\theta) + g(\theta)|^2 + 3|f(\theta) - g(\theta)|^2} \tag{16a}$$

$$T_e(\theta) = 1 - \frac{|g(\theta)|^2}{\sigma_0(\theta)} \tag{16b}$$

$$T_A(\theta) = 1 - \frac{|f(\theta)|^2}{\sigma_0(\theta)} \tag{16c}$$

$$S(\theta) = \frac{|f(\theta)||g(\theta)| \sin(\delta_1 - \delta_2)}{\sigma_0(\theta)} \tag{16d}$$

The parameter $A(\theta)$ describes an asymmetry for polarization vectors parallel and antiparallel as shown by equation (19), $T_e(\theta)$ describes the reduction of the initial polarization component due to exchange ($T_e = 1$ if $g = 0$), $T_A(\theta)$ describes the transfer of atomic polarization to electron polarization ($T_A = 0$ if $g = 0$, because then $\sigma_0 = |f|^2$), and $S(\theta)$ describes an additional polarization component normal to the plane defined by the directions of $\mathbf{P}_e$ and $\mathbf{P}_A$. Obviously A, T_e, and T_A are not independent, because

$$A(\theta) = T_A(\theta) + T_e(\theta) - 1 \tag{17}$$

This relationship can be used for a consistency check of experimentally determined parameters.

The scattering problem can be attacked using the following three experimental arrangements.

1. Unpolarized electrons on polarized atoms. With a Mott analyzer the parameter $T_A(\theta)$ can be determined by measuring the polarization component

$$\mathbf{P}'_e = T_A(\theta)\mathbf{P}_A \tag{18}$$

parallel to $\mathbf{P}_A$ which should be chosen to be transverse to the direction of scattered electrons. A direct manifestation of exchange collisions is the finding that $\mathbf{P}'_e \neq 0$.

2. Asymmetry for polarized electrons on polarized atoms. We detect the intensity of scattered electrons for the case where polarization vectors of electrons and atoms are antiparallel ($I_{\uparrow\downarrow}, \mathbf{P}_e \cdot \mathbf{P}_A = -P_eP_A$) and parallel ($I_{\uparrow\uparrow}, \mathbf{P}_e \cdot \mathbf{P}_A = P_eP_A$), obtaining

$$\frac{I_{\uparrow\downarrow} - I_{\uparrow\uparrow}}{I_{\uparrow\downarrow} + I_{\uparrow\uparrow}} = A(\theta)P_eP_A \tag{19}$$

3. Electron polarization after collision of polarized beams. We prepare a polarized atom beam with the polarization vector in the direction of the scattered electrons. The incident electron beam is polarized normal to the scattering plane. Thus the transverse polarization of the scattered electrons,

$$\mathbf{P}_e' = T_e(\theta)\mathbf{P}_e + S(\theta)(\mathbf{P}_e \times \mathbf{P}_A) \tag{20}$$

can be determined in a single measurement when the (transverse) components $T_e\mathbf{P}_e$ and $S(\mathbf{P}_e \times \mathbf{P}_A)$ are measured simultaneously in a Mott analyzer with four counters.[5]

Together with absolute differential cross sections, which must be measured separately, experiments (1)-(3) fulfill the requirements of a complete or "perfect" experiment,[5,40] since in them $|g|$, $|f|$ and $\delta_1 - \delta_2$ are determined.

No experiment of type (3) has been reported yet; however, in their pioneering work the New York group obtained the parameter $T_e(\theta)$ by measuring the change of the polarization of recoil K atoms,[41]

$$\mathbf{P}_A' = T_e\mathbf{P}_A \tag{21}$$

The only experiment of type (1) so far has been performed by Hils *et al.*[42] for polarized K atoms.

In recent experiments of type (2) the asymmetry (19) has been studied for elastic scattering of polarized electrons from polarized H atoms,[43] polarized Li atoms,[44] and polarized Na atoms.[37] A schematic diagram of the apparatus used by Baum *et al.*[44] is shown in Figure 11, and their results

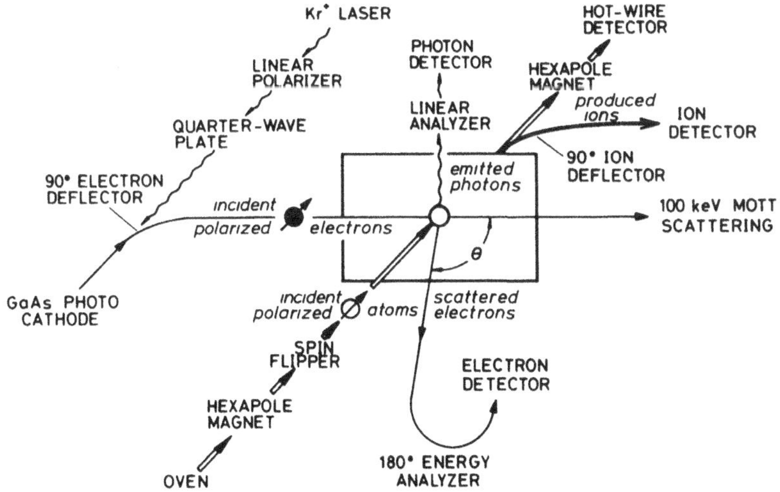

FIGURE 11. Scheme of the experimental setup of the Bielefeld group[44] for measuring angle resolved spin antiparallel-parallel scattering asymmetries.

for elastic electron–Li collisions in Figure 12. For the scattering angles 65°, 90°, and 107.5° the asymmetry was measured between 1 and 30 eV.

In this experiment, the polarized beam was produced by photoemission from a GaAs crystal using circularly polarized light from a Kr-ion laser operating at the 752-nm line. The beam was bent electrostatically through 90°, thereby transforming the polarization from longitudinal to transverse.

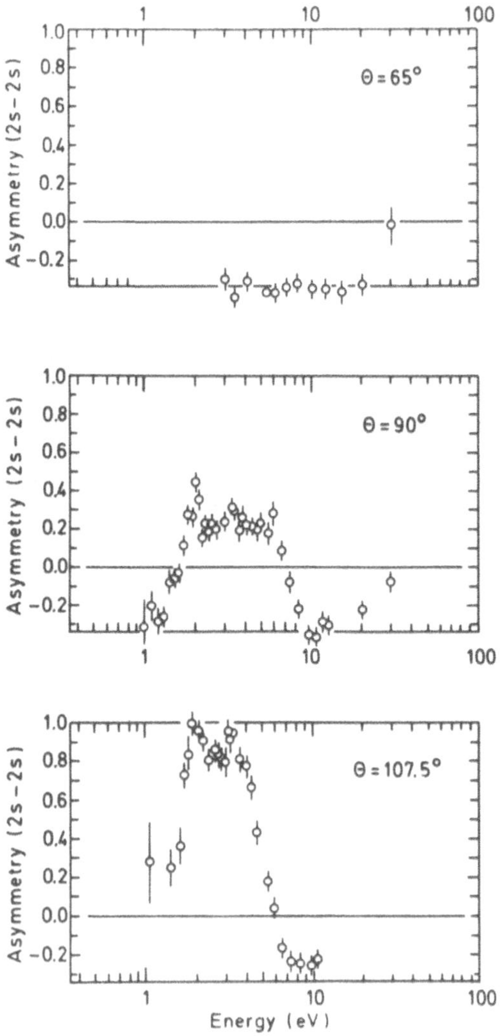

FIGURE 12. Measured spin asymmetries $A(2s\text{-}2s)$ for elastic electron–Li collisions as a function of scattering energy for scattering angles of $\theta = 65°$, 90°, and 107.5°.[44]

A Mott polarimeter was used to determine the electron polarization, which was typically between 0.25 and 0.34. The atomic beam of ^{6}Li was polarized by high-field state selection in a permanent hexapole magnet. An identical magnet served as an analyzer for the atom-beam polarization, which was 0.29. The spin-flipper was used for reversal of the atom-beam polarization. Electron and atomic beam were cross fired and electrons scattered by fixed angles 65°, 90°, or 107.5° were detected after they passed a 180° energy analyzer.

The measured asymmetries (Figure 12) show variations between the extremes of +1 and −1/3 [cf. equation (16a)]. Their behavior can be best described by close-coupling calculations. For details the reader is referred to the original publication.[44]

3.2. Scattering from Spinless Atoms

Spin effects in elastic collisions of electrons from spinless heavy atoms like Hg and Xe have been the subject of a great number of theoretical and experimental investigations during the last 25 years.[5] There are two reasons for this. To a good approximation the spinless target acts as a local potential scatterer. Thus potential scattering techniques can be employed in numerical calculations, at least at energies well above excitation thresholds. Experimentally, only four observables must be measured to obtain the maximum possible information that is accessible about elastic collisions from spinless atoms. These observables are functions of two complex scattering amplitudes: the direct scattering amplitude $f = |f|e^{i\gamma_1}$ and the spin-flip amplitude $g = |g|e^{i\gamma_2}$. (Do not confuse this spin-flip amplitude g with the g used in Section 3.1 for the exchange amplitude):

$$\sigma = |f|^2 + |g|^2 \qquad \text{(differential cross section)} \tag{22a}$$

$$S = -2|f||g|\sin(\gamma_1 - \gamma_2)/\sigma \qquad \text{(Sherman function)} \tag{22b}$$

$$U = 2|f||g|\cos(\gamma_1 - \gamma_2)/\sigma \tag{22c}$$

$$T = (|f|^2 - |g|^2)/\sigma \tag{22d}$$

The amplitudes f and g are defined with respect to a quantization axis parallel to the direction of the incident beam and depend on scattering angle θ, energy E, and the target under study.

If the polarization vector **P** of the incident electron beam is parallel to the scattering plane, the polarization vector **P**′ of scattered electrons is given by

$$\mathbf{P}' = S\hat{n} + T\mathbf{P} + U(\hat{n} \times \mathbf{P}) \tag{23}$$

where $\hat{n}$ is a unit vector normal to the scattering plane. The meaning of the parameters S, T, and U has been stated in Section 2.1. The Sherman function S can be measured in a double scattering experiment with $\mathbf{P} = 0$ [cf. equations (1) and (23)]. The parameters T and U can be measured by determining the components in the scattering plane $T\mathbf{P}$ and $U(\hat{n} \times \mathbf{P})$ of $\mathbf{P}'$ (23) if the incident electrons are polarized with $\mathbf{P}$ parallel to the scattering plane.

Recently, Berger and Kessler[(45)] reported on experiments in which detailed angular distributions of the parameters T and U have been measured for the first time. Their apparatus is shown in Figure 13: In this experiment polarized photoelectrons were released from a p-doped $GaAs_{0.6}P_{0.4}$ (100) crystal by circularly polarized light from a He–Ne laser. The electrons were accelerated to 100 eV and passed through a 90° electrostatic deflector. They were then focused by an electrostatic lens into a deflection system that could be rotated about the axis of the atomic beam. Thereafter, they entered the scattering chamber, where they were scattered by the atomic beam. Electrons that were elastically scattered through an angle θ passed through a filter lens and a Wien filter, were accelerated to 120 keV, and entered a Mott detector for polarization analysis.

The polarization vector of the photoelectrons was parallel to the direction of the laser beam and was nearly unchanged by the electrostatic

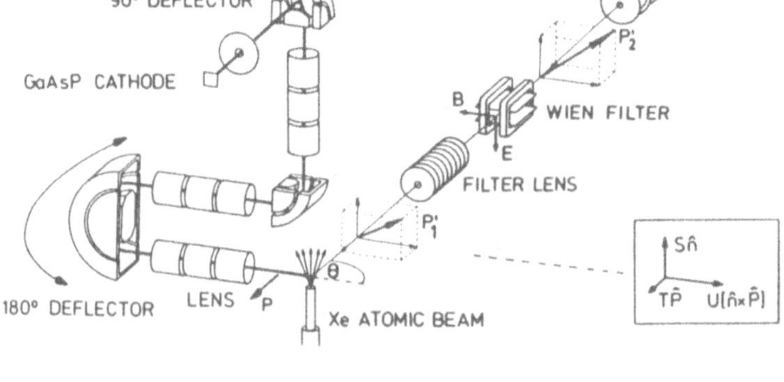

FIGURE 13. Schematic diagram of the apparatus for measuring S, T, and U parameters.[(45)]

deflections, thus being parallel to the detection direction at the target. Typically a primary beam current of 2 μA with a polarization of 0.32 was obtained. The Wien filter rotated the component $T\mathbf{P}$ (cf. Figure 13) into a direction transverse to the scattering plane, so this component could be determined by the left-right scattering asymmetry in the Mott detector, whereas the component $U(\hat{n} \times \mathbf{P})$ was determined simultaneously by the up-down scattering asymmetry.

Results for elastic collisions from Hg at 25 eV and Xe at 60 eV are shown in Figure 14. The overall agreement between experimental and various theoretical data[46-49] is good, though not perfect. Berger and Kessler[45] conclude that it is not sufficient to measure one of the polarization parameters or the differential cross section alone; only the analysis of the complete set of observables gives a stringent test of the theoretical models. Although in the past few years progress has been made in calculating the spin-dependent parameters, there are still certain aspects that need work, such as the absolute scale for the differential cross section[50] and some features of the polarization parameters.

3.3. *Scattering from Heavy Atoms with Unsaturated Angular Momentum*

Examples of heavy atoms with unsaturated angular momentum are Cs ($6s\,^2S$) and Tl ($6p\,^2P_{1/2}$). There is, however, a characteristic difference between these two targets: The angular momentum of the Cs atom is purely

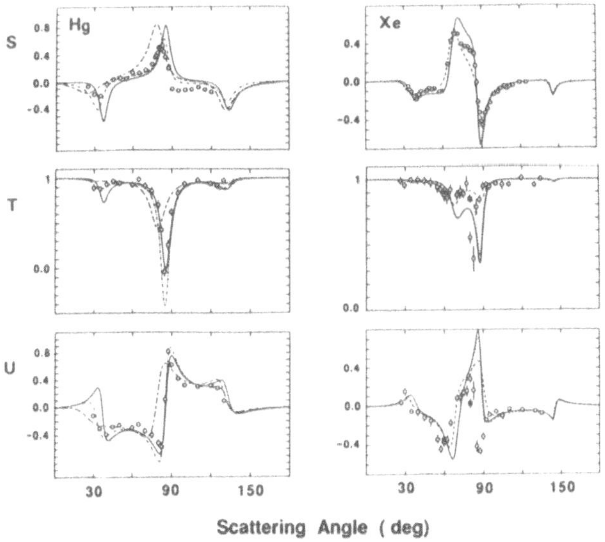

FIGURE 14. Angular distribution of the parameters S, T, and U for Hg at 25 eV and for Xe at 60 eV measured by Berger and Kessler[45] (○). Theory: Haberland *et al.*[48] (—); Walker[46] with exchange (- · -); McEachran and Stauffer[49] (· · ·), Awe *et al.*[47] (- - -).

spin angular momentum, whereas Tl has spin and orbital angular momentum with a fine-structure splitting. The $6p\,^2P_{3/2}$ state of Tl is an excited state separated from the $6p\,^2P_{1/2}$ ground state by about 1 eV. Therefore the Tl atom may show the fine-structure effect in elastic collisions that the Cs atoms cannot show. Although exchange processes may also be involved, the polarization obtained in elastic electron–Cs collisions[51] is caused solely by the spin–orbit interaction of the continuum electron (Mott scattering).

The aim of a recent investigation of Kaussen *et al.*[52] was to study the relative influence of Mott scattering and the fine-structure effect in elastic electron scattering from Hg, Tl, Pb, and Bi atoms. For Hg (ground state: $6s^2\,^1S_0$) or Bi ($6s^26p^3\,^4S_{3/2}$), polarization effects can arise only due to Mott scattering, because the total orbital angular momentum is zero, so no atomic-orientation effect is possible. On the other hand, for Tl ($6s^26p^2P_{1/2}$) and Pb ($6s^26p^2\,^3P_0$) both mechanisms, Mott scattering and fine-structure effect, are possible.

How can we distinguish between these two mechanisms? Theoretical calculations show that for $E \geqslant 25$ eV the polarization after elastic collisions of initially unpolarized electrons with Hg, Pb, and Bi atoms ($Z = 80, 82, 83$) differs only slightly for the different targets[53] if they are approximated by local potentials. If the fine-structure effect is strong, there must be significant differences between the polarization measured when electrons scatter elastically from Tl and Pb atoms (where the fine-structure effect is possible) and that obtained in elastic scattering from Hg and Bi atoms (where the fine-structure effect is not possible). These arguments cannot be reversed, however, because the different configurations of the valence electrons will cause significant differences in the long-range behavior of the scattering potential. Thus at the low energies, where we expect an influence due to the fine-structure effect, these different potentials can also cause differences in the polarization curves.

The experimental arrangement of Kaussen *et al.*[52] was a "standard" double-scattering experiment. Its scheme is contained in Figure 13, except that the source of polarized electrons was replaced by a simple electron gun and no Wien filter was used in the detection system. Thus the polarization $\mathbf{P} = S\hat{n}$ was determined through the left–right scattering asymmetry in the Mott detector.

In Figure 15 is shown the polarization of electrons scattered elastically by Hg, Tl, Pb, and Bi atoms versus scattering angle at 11 eV. The experimental curves differ remarkably for the different targets. At first glance one might think that these differences are due to the fine-structure effect. But the theoretical curves by Haberland and Fritsche[54] shown also in Figure 15 confute this conclusion. In these calculations the different angular-momentum configurations of the valence electrons were not taken into account, but their contribution to an approximate, local, and spherical

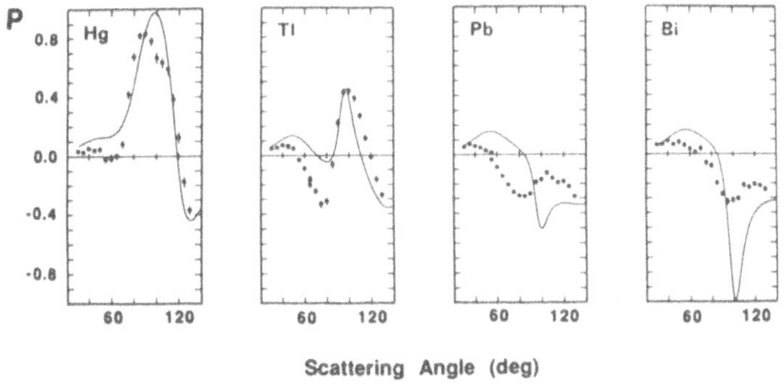

FIGURE 15. Angular dependence of the polarization P of electrons scattered elastically from Hg, Tl, Pb, and Bi at 11 eV. Experimental results from Kaussen *et al.*[52] (●); theoretical curves of Haberland and Fritsche[54] (——).

exchange-correlation potential was calculated. The agreement between experimental and numerical data is not perfect, but the overall behavior of the different polarization curves is surprisingly well described by the calculation.

It is an open question why this calculation suffices to describe the experimental outcome, even though the fine-structure effect should be significant also (as is demonstrated in Figure 10). Further experimental and theoretical work is needed to answer this question.

4. EXCITATION AND SUPERELASTIC COLLISIONS

4.1. *Pure Spin-Exchange Effects in Electron-Impact Excitation*

In this subsection we describe recent experiments in which impact excitation and ionization of polarized light atoms by polarized electrons has been studied. If the spin-orbit interaction is negligible, spin effects can arise only through exchange. Then the formal description given in Section 3.1 for elastic collisions of electrons from light one-electron atoms is also applicable for $^2S \to {}^2S$ electron-impact excitation of these targets. But for $^2S \to {}^2P$ electron-impact excitation processes, we must consider three different possible orientations of the final atomic P state with respect to some reference direction. These are specified by magnetic quantum numbers $m_l = 0, \pm 1$. Accordingly, we have three different sets of amplitudes, instead of the one set given in equations (12); these may be denoted by f_{m_l} and g_{m_l}. From general symmetry consideration we conclude that $f_{-1} = -f_1$ and $g_{-1} = -g_1$.[2] If we even resolve the fine-structure splitting, we will obtain the fine-structure effect and the whole description must be changed. This is discussed in Sections 2.3 and 4.2.

It is difficult to select a particular excitation process if the energy levels of the target are close together. To achieve this, quite different methods have been used: selective quenching,[55] linear-momentum-transfer analysis of recoil atoms,[56] spectral resolution of emitted light,[57,58] and energy resolution in the scattered electron spectrum.[59] In various experiments where the excitation into the continuum (ionization) has been studied,[43,60-63] no energy analysis of final continuum states has been carried out.

In line with our main interest, we will discuss only experiments in which polarized electrons are involved. In earlier experiments a Bielefeld group studied electron-impact excitation of the lowest P levels of Li, Na, and K atoms using the intensity of the fluorescence emission as a measure of the total cross sections for parallel and antiparallel spins.[57] Recently this group reported an angle resolved experiment.[59] Figure 16 shows the observed spin antiparallel-parallel asymmetry A [see equation (19)] for

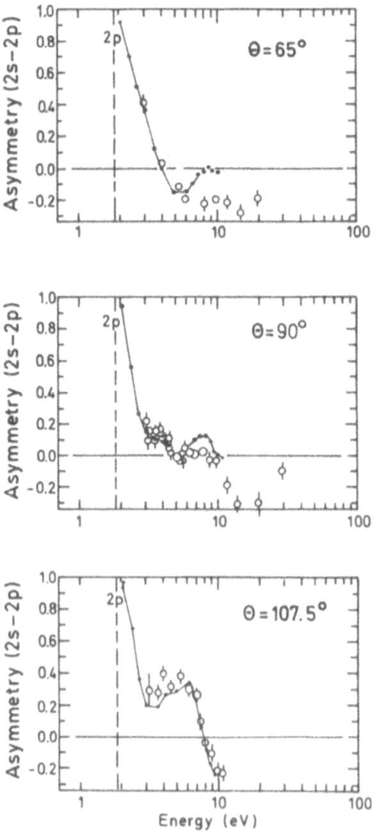

FIGURE 16. Spin antiparallel-parallel asymmetry A measured for electron-impact $2s$–$2p$ excitation of Li atoms by Baum *et al.*[59] —•—, Close-coupling calculation by Moores.[64]

electron impact $2s$–$2p$ excitation of Li atoms. The experimental scheme was the same as in Figure 11. The asymmetry has been measured between 3 and 10 eV. The overall agreement with asymmetries from a close-coupling calculation by Moores[64] is very good, whereas older calculations (not shown in Figure 16) deviate clearly from the experimental data.

Much effort has been focused in recent years on the ionization of polarized one-electron atoms from H through K by polarized electrons.[43,60–63] The cross sections for spin antiparallel and parallel,

$$\sigma_{\uparrow\downarrow} = \int \frac{k_1' k_2'}{k} (|f|^2 + |g|^2)\, d\mathbf{k}_1'\, d\mathbf{k}_2'\, dE_1' \tag{24a}$$

and

$$\sigma_{\uparrow\uparrow} = \int \frac{k_1' k_2'}{k} |f - g|^2\, d\mathbf{k}_1'\, d\mathbf{k}_2'\, dE_1' \tag{24b}$$

have been measured by detecting the ions. Here $k_1' = |\mathbf{k}_1'|$, k_2', and k represent the wave numbers of scattered, ejected, and incident electron, respectively, and E_1' is the energy of the scattered electron. The integrations are necessary, because in the experiments neither a scattering or ejection angle nor an energy analysis has been performed.

Figure 17 shows the results of the Bielefeld group for electron-impact ionization of Li, Na, and K atoms.[63] The apparatus was basically the one in Figure 11, where the ion-counting detector was used to measure the ionization cross sections (24). In Figure 17 the integrated asymmetry $A = (\sigma_{\uparrow\downarrow} - \sigma_{\uparrow\uparrow})/(\sigma_{\uparrow\downarrow} + \sigma_{\uparrow\uparrow})$ is plotted versus the ratio E/I, where E is the energy and I is the threshold energy for the different target atoms. The ionization

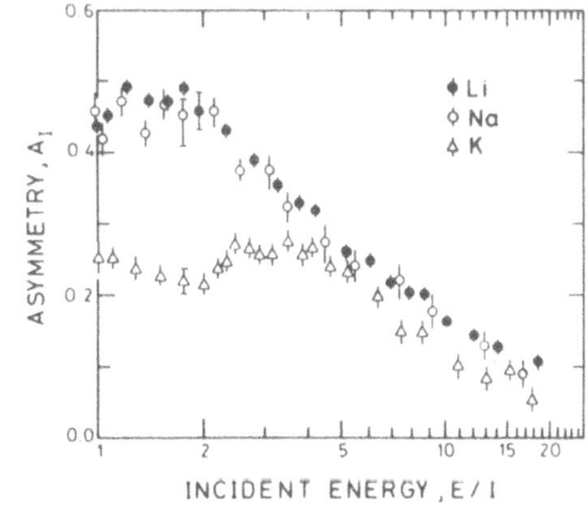

FIGURE 17. Values of the ionization asymmetry A_I as function of the incident energy E, in units of threshold energy I, measured by Baum *et al.*[63] for Li (I = 5.39 eV), Na (I = 5.14 eV), and K (I = 4.34).

asymmetries of Li and Na are nearly identical in absolute values as well as in energy dependence. The data for H atoms of the Yale group[43] follow closely the Li and Na data shown in Figure 17. The atomic core does not seem to be relevant for the observed spin dependence in these cases. For K atoms, however, there is a substantial deviation from this behavior, in agreement with data obtained by Hils *et al.*[60] The Bielefeld group argues that the different behavior for K atoms might be due to the influence of spin–orbit coupling. This point needs to be studied further both experimentally and theoretically.

One of the reasons for the interest in these ionization studies is the problem of the threshold law, i.e., the behavior of the asymmetry when the two escaping electrons have nearly zero kinetic energy. Special attention to the threshold behavior has been paid by the Bielefeld group for Li[63] and the NBS group for Na.[62] No significant dependence of the asymmetry on energy near threshold has been found. A predicted oscillatory behavior[65] could not be confirmed. The dependence of the double escape at threshold on orbital angular momentum, spin, and parity of the two-electron state has been analyzed using Wannier theory.[66–68] The asymmetry gives information about the participating L waves; e.g., $\sigma_{\uparrow\uparrow}$ depends only on odd L values. Thus $A < 1$ shows directly that nonzero L values must contribute even very close to threshold, contrary to the expectation from semiclassical threshold arguments.

Experiments using polarized metastable He($2\,^3S$) atoms to study the spin dependence of the single-ionization process are in progress.[38,59]

4.2. Fine-Structure Effect

The fine-structure effect, in its pure form, alows us to study simultaneously the behavior of atomic orientation and exchange in electron collisions. Theoretical studies have been performed for H atoms,[26] He atoms,[69] and Mg atoms.[70] As a general result, the atomic orientation and the asymmetry functions versus scattering angle exhibit more structure as the target atoms become more complex.[70] This is shown in Figure 18, where the spin-polarization function $S_P(\mathbf{k}, \mathbf{k}')$ [see equation (1)] is plotted for $^1S \rightarrow {}^3P_2$ electron-impact excitation of He, Mg, and Hg. Exchange effects still dominate electron-impact excitation of Hg, even though spin–orbit interaction is not negligible and general relations like equation (10) are only approximately valid. This is discussed in Section 4.2.

Experimentally the fine-structure effect could be studied in its pure form in electron–Na collisions. Because of the low nuclear charge ($Z = 11$), we can assume that spin–orbit interaction is negligible during the $3s$–$3p$ electron-impact excitation process. The fine-structure splitting between the $3p\,^2P_{1/2}$ and $3p\,^2P_{3/2}$ states is only 2 meV, which is too small to be resolved

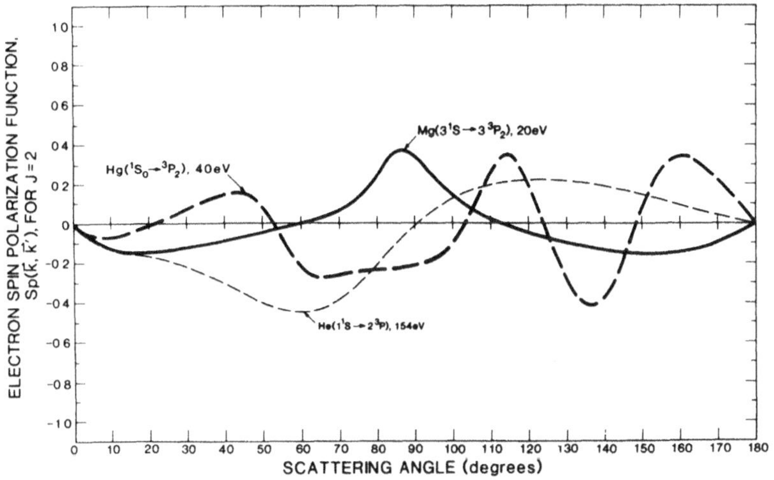

FIGURE 18. Electron spin polarization function $S_P(\mathbf{k}, \mathbf{k}')$ versus scattering angle for ${}^1S \rightarrow {}^3P_2$ electron impact excitation of He, Mg, and Hg for $E/W = 7.2$, where W is the excitation energy. The data are taken from Csanak and Cartwright,[69] Meneses *et al.*,[70] and Bartschat and Madison.[90]

in conventional inelastic scattering experiments. Therefore, the time-reversed superelastic process was studied: Na atoms were prepared in only one of the $3\,{}^2P_{1/2,3/2}$ fine-structure states by laser excitation, and the deexcitation by electron collisions was studied, e.g.,

$$e^-(E - W) + \text{Na}^*(3\,{}^2P_{3/2}) \rightarrow \text{Na}(3\,{}^2S) + e^-(E) \tag{25}$$

where the superelastically scattered electrons gain the excitation energy $W = 2.1$ eV, and E is the kinetic energy of the electron after scattering. The reaction from the right to the left is the inelastic process.

A group at the FOM institute used multimode laser excitation to prepare unpolarized excited $3\,{}^2P_{1/2}$ or $3\,{}^2P_{3/2}$ Na atoms.[71] Unpolarized electrons were scattered superelastically by these excited atoms and the polarization of scattered electrons was measured. Within experimental errors, relation (10) was confirmed for a scattering angle of 20° at 20 eV incident energy.

A group at the NBS in Gaithersburg reported recently a beautiful experiment in which aligned $3\,{}^2P_{3/2}$ Na atoms were prepared prior to scattering.[72] Figure 19a is a scheme of their experimental geometry, and Figures 19b and 19c show results. No full description of the apparatus exists yet, so we use here the description given in the paper[72]: Polarized electrons were produced[8] by photoemission from a negative electron affinity GaAs photocathode and transported through electron optics to the scattering center, where they formed a 2-mm, 10-eV, 1-μA beam with an energy spread of typically 0.1 eV. This beam intersected an atomic Na beam

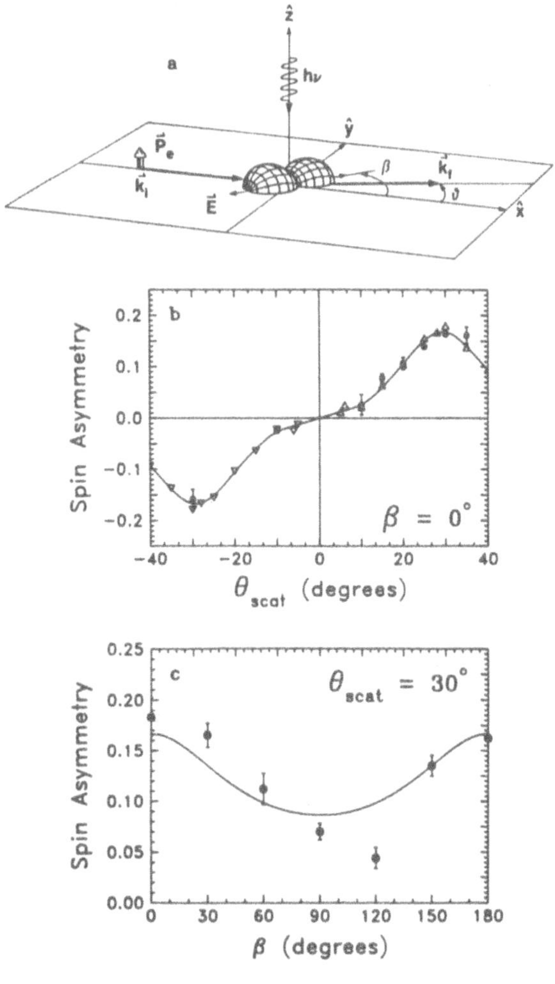

FIGURE 19. (a) Scheme of the experimental geometry of the NBS experiment.[72] (b) Results for the angular dependence of the spin asymmetry for scattering to the left ($+\theta$) and to the right ($-\theta$). (c) Spin asymmetry plotted versus laser polarization angle β.

of density 10^{10} atoms/cm^3 and diameter 4 mm produced by an effusive oven. The scattering region was illuminated with light from a frequency-stabilized, single-frequency ring dye laser locked to the Na $2S_{1/2}(F = 2) \rightarrow 3P_{3/2}(F = 3)$ transition, creating a significant population of Na atoms in the first excited state. The laser light was linearly polarized, and was incident perpendicular to the scattering plane from the $+z$ direction. Superelastically scattered electrons, which have gained the 2.1-eV Na excitation energy, were detected with a channel electron multiplier mounted on a rotatable turntable. The detector included a retarding-field analyzer with resolution (approximately 0.5 eV) sufficient to reject all elastically or inelastically scattered electrons. The electron spin was modulated at 100 Hz by changing the helicity of the circularly polarized photoemission light with a Pockels cell driven by a high-voltage bipolar square wave. Electrons were counted separately for each spin orientation, and after averaging over ten measurements of 200 s each, the spin asymmetry was calculated from $P_e^{-1}(N_\uparrow - N_\downarrow)/(N_\uparrow + N_\downarrow)$, P_e being the magnitude of the incident electron polarization (26%) and N being the number of electrons detected when the incident electrons have spin up ($\uparrow$) or down ($\downarrow$) with respect to the laboratory frame. Typical superelastic counting rates ranged from 300 Hz at small angles to 20 Hz at large angles.

Figure 19b shows the measured angular dependence of the spin asymmetry for positive and negative scattering angles, i.e., for scattering to the left and to the right, respectively. The effect on the spin asymmetry of changing the direction of polarization of the laser with respect to that of the incident electrons is shown in Figure 19c. As there is no orientation of the excited atoms prior to scattering—and as the spin-orbit interaction is negligible—the observed asymmetry must be caused by the mechanism described in Section 2.3. The observed asymmetry depends strongly on the initial preparation of the excited state. Rotation of the electric vector of the linearly polarized light changes the relative phase between the coherently excited substates. This gives rise to the angular dependence of the curve shown in Figure 19c.

More recently this group presented similar data from oriented $3\,^2P_{3/2}$ Na atoms.[73] These data allow an almost "complete" or "perfect" analysis of the 3s-3p electron-impact excitation of Na. Suggestions for future investigations towards this goal (a "quantum mechanical dream") and comparison of presently available experimental and numerical data have been reported.[74]

4.3. *Collisions with Heavy Atoms*

Spin-polarization effects in elastic scattering of electrons from heavy atoms have been the subject of a great number of theoretical and experimental investigations during the last 25 years.[5] There have been, however, only a few experimental attempts to study polarization effects in inelastic collisions[23,75-78] because of the greater difficulties of that type of experiment. No targets other than Hg atoms have been investigated experimentally so far, but theoretical data from R-matrix close-coupling calculations are available for Cs[7] (see Figure 2), Tl, and Pb.[35]

Pure exchange effects for light atoms have been studied so far only by using polarized atoms. Another way to observe exchange effects directly is to use the process described by equation (6). The only inelastic scattering experiment to date that uses this method studied $6\,^1S_0$–$6\,^3P_{1,2}$ excitation of Hg atoms by transversely polarized electrons.[23]

In all other investigations of inelastic electron-Hg collisions, scattering into large angles has been studied by use of "double"-scattering experiments [see equations (1)–(3)]. Because of the high nuclear charge of Hg ($Z = 80$), the spin-orbit interaction of the colliding electrons is strong and acts in two ways:

1. Spin–orbit interaction of the continuum electron in the field of the screened nucleus; and
2. Spin–orbit interaction of the target electrons (fine-structure splitting, breakdown of LS coupling).

Preliminary experiments and calculations also showed sharp structures in the energy dependence of the asymmetry[17,77] at those energies where the cross sections show resonances due to the formation of temporary negative-ion states.[79–84]

To better understand this complex situation, detailed measurements of the scattering asymmetry as a function of scattering angle and energy were carried out.[85,86] Figure 20 shows the experimental setup for a spin up–down asymmetry measurement. Longitudinally polarized electrons (with $P = 0.3$–0.4 and $\Delta E \leqslant 140$ meV) were photoemitted by circularly polarized light of a He–Ne laser from a GaAsP crystal in a UHV chamber. The electrons passed through two electrostatic 90° deflectors and a magnetic coil system that rotated the polarization vector **P** to a direction perpendicular to the x–z scattering plane. The electron beam was finally decelerated from 200 eV, the transportation energy, to the desired scattering energy (5–12 eV). The electron beam then crossed a Hg atomic beam. Electrons scattered by 90° to the right were energy analyzed by a simulated spherical spectrometer.[87]

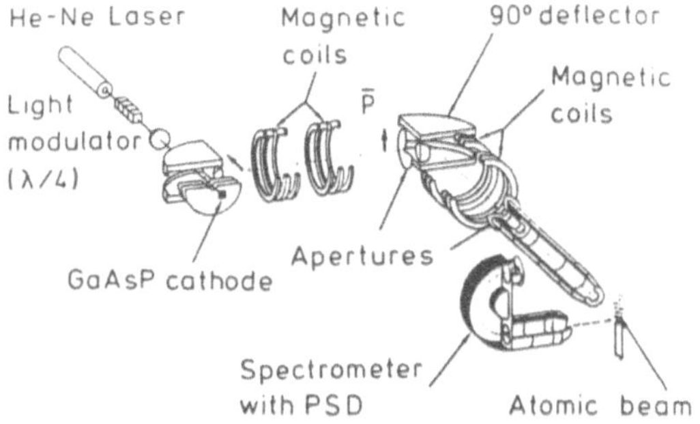

FIGURE 20. Experimental setup for spin up–down asymmetry measurement with position-sensitive electron detector used by Wolcke *et al.*[85] for electron–Hg excitation.

A position sensitive detector at the exit of the electron spectrometer accumulated a large part of the energy-loss spectrum, making feasible simultaneous observation of scattering into different channels.

In this experiment the spin up-down asymmetry could be measured for electrons that have separately excited the fine-structure states $6\,^3P_0$, $6\,^3P_1$, and $6\,^3P_2$. Figure 21 shows measured asymmetry functions S_A [equation (2)] at a scattering angle of 90° plotted versus energy. $S_A(^3P_0)$ could be measured in this experiment for the first time. This function shows remarkable negative values with distinct structure between 8 and 11 eV due to resonances. Values less than -0.8 mean that the cross section for scattering of spin-up electrons to the right—as observed in the experiment—is more than nine times larger than that for spin-down electrons. The data for $S_A(6\,^3P_2)$ agree well with results of a previous experiment,[77] whereas $S_A(6\,^3P_1)$ differs from the earlier data in the energy range between 5.5 and 8 eV. In the previous experiment, the $6\,^3P_0$-excitation channel could not be well separated from the $6\,^3P_1$ channel, which accounts for the observed difference. Figure 21 shows that the relation

$$S_A(^3P_0) = 2S_A(^3P_1) = -2S_A(^3P_2) \tag{26}$$

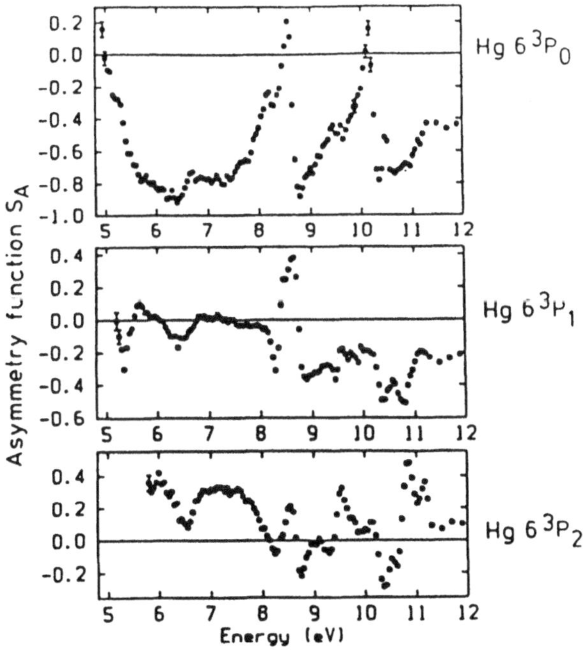

FIGURE 21. Measured scattering asymmetry functions S_A for 90° scattering angle versus scattering energy for impact excitation of $6\,^3P_0$, $6\,^3P_1$, and $6\,^3P_2$ states of Hg by polarized electrons.[85]

which holds if spin–orbit interaction is small,[2] is not obeyed, except for a tendency in the signs.

The general behavior of the asymmetry functions has been studied in greater detail in another recent experiment.[86] Angular distributions of the left–right scattering asymmetry were measured in the energy range between 5.5 and 50 eV and discussed with respect to a simple model[2,33] and numerical calculations.[88,90] The apparatus used in these measurements was essentially that shown in Figure 20, except for a second detector for electrons scattered to the left. The detectors could be independently rotated. Conventional exit slits in the spectrometers allowed the detection of only one scattering channel at a time.

Figure 22 shows measured asymmetry functions for excitation of the $6\,^3P_1$, $6\,^3P_2$, and $6\,^1P_1$ states in Hg at energies of 7.5, 15, and 50 eV. Also shown are DWBA[90] and R-matrix[88] calculations. The experimental data at 7.5 eV tend to obey the relation

$$S_A(6\,^3P_1) = -S_A(6\,^3P_2)$$

but R-matrix calculations show this relationship only for energies up to 6.5 eV.[88,89] The experimental findings indicate that exchange is the

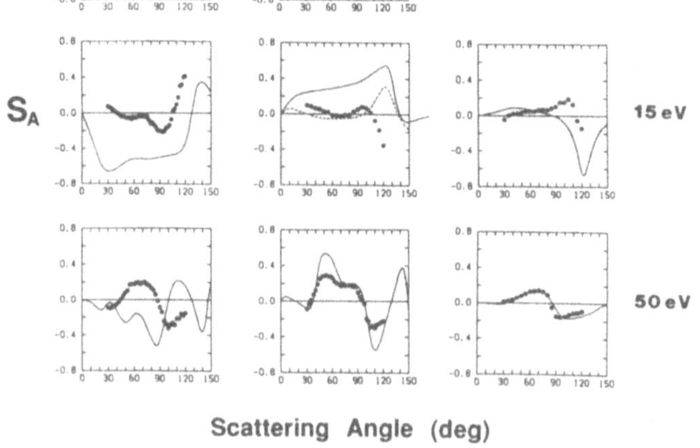

FIGURE 22. Angular distribution of S_A for excitation of the Hg $6\,^3P_1$, $6\,^3P_2$, and $6\,^1P_1$ states at collision energies 7.5, 15, and 50 eV. (ϕ) Experimental data from Borgmann *et al.*[86] - - -, R-matrix calculation[88]; —, DWBA calculations.[90]

dominant spin effect in this energy and angular range. A detailed analysis of the data around 15 eV indicates that exchange is still dominant but that breakdown of LS coupling for the $6\,^3P_1$ and $6\,^1P_1$ states influences significantly the behavior of the corresponding asymmetry functions. Above 30 eV the influence of exchange decreases for excitation of these states, and direct excitation processes dominate. The spin effects observed above 50 eV in the excitation of these states are caused mainly by the spin-orbit interaction of the continuum electron, as has been discussed in earlier investigations.[75-76]

The comparison of the experimental and theoretical data shows fair agreement only in exceptional cases. It is not clear how the approximations implicit in the close-coupling and DWBA methods and approximations in the target wave functions and potentials used in the calculations contribute to the discrepancies.

5. POLARIZED LIGHT FROM COLLISIONALLY EXCITED ATOMS

In their famous paper Percival and Seaton[25] treated the polarization of atomic-line radiation resulting from electron-impact excitation. They considered an experiment in which the atoms were excited by unpolarized electrons but the scattered electrons were not observed. In such an experiment the polarization of line radiation is due to alignment of the excited atoms along the axis of incidence. This alignment is caused by angular-momentum transfer and modified by the fine-structure and hyperfine-structure interactions.

It has been pointed out by Farago and Wykes[91,92] that in such collisions atoms can be oriented by spin-exchange collisions with polarized electrons. The oriented atoms will emit circularly polarized light in the direction of the polarization of the electron spin. The formal description of impact excitation of atoms by polarized electrons has been generalized by Bartschat *et al.* and extended to electron-photon coincidence experiments.[93] In electron-photon coincidence experiments, a subensemble of the electron-atom scattering system is selected by observing photons emitted from atoms that have scattered the electrons into a particular direction.

The angular distribution and the polarization of the emitted light obtained in these experiments directly reflect the charge-cloud (probability) distribution of the selected ensemble immediately before the decay. Figure 23 shows that orientation and alignment, or state multipoles, of the charge-cloud distribution can be determined with respect to a distinguished x–z plane by the Stokes parameters P_1, P_2, P_3, and P_4. The distinguished plane may be the scattering plane or the plane of the initial orientation of the

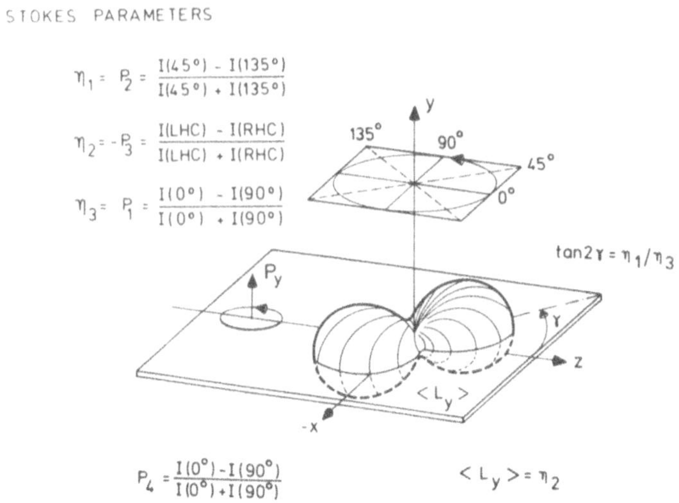

FIGURE 23. Atomic charge-cloud (probability) distribution after electron-impact excitation and related Stokes parameters η_ν and P_ν for subsequent emission of fluorescence light. For definition of the intensities see text.

electrons or atoms. The parameters $P_1 = \eta_3$, $P_2 = \eta_1$, and $P_3 = -\eta_2$ are the polarization components for light that is emitted perpendicular to the plane (see Figure 23), and P_4 is a linear polarization component for light emitted in the plane. $I(\alpha)$ is the fraction of intensity transmitted by a linear polarization filter that is inclined to the z axis by an angle α, and $I(\sigma^+)$ and $I(\sigma^-)$ are the intensities transmitted by filters for left and right circular polarization.

By measurement of the Stokes parameters P_1, P_2, and P_3, the alignment angle γ, the ratio of length to width, and the orientation of the charge cloud distribution can be determined; P_4 is a measure for its relative height at the waist.

Such a charge-cloud distribution is strongly influenced by the collision dynamics and the internal target interaction after the collision. In cases where the internal fine-structure and hyperfine-structure interactions are weak, the collision dynamics can be disentangled from the internal dynamics using angular-momentum algebra.[94] It is then assumed that total orbital, electron-spin, and nuclear-spin angular momenta are at rest during the brief collision and that they are therefore completely decoupled during the collision. The collision is then spin dependent only because of electron-exchange effects.

In cases where the spin–orbit interaction is strong, LS coupling will not hold during the collision. This will weaken the symmetry of the excitation process, giving rise to new parameters.[93,95]

5.1. Integrated Stokes Parameters

Following an idea of Wykes,[92] Eminyan and Lampel were the first to show a transfer of electron-spin polarization to an atomic orientation.[96] This transfer results in circular polarization of the light emitted from the excited states. Wykes proposed to use electron-impact $^1S_0 \rightarrow {}^3S_1$ excitation of a two-electron atom and to observe the optical decay $^3S_1 \rightarrow {}^3P_J$. The resulting circular polarization for photons emitted in the direction of electron polarization P is

$$\eta_2 = -P_3 = k_J P \tag{27}$$

where k_J is a factor that takes account of the fine-structure and hyperfine structure interactions within the target. In the case of pure exchange collisions, k_J is energy independent and can be calculated by angular-momentum coupling. In this case a calibration free determination of electron polarization is possible.

Eminyan and Lampel studied the excitation of Zn ($4s^2\,{}^1S_0 \rightarrow 4s5s\,{}^3S_1$, with an energy loss of 6.65 eV) by electron impact and the subsequent decay into the $4s4p\,{}^3P_J$ states. Recently, such experiments have been performed for electron-impact excitation of Hg.[97,98] The accuracy of such experiments has been considerably improved, so that now data with low statistical errors such as that shown in Figure 24[99] can be measured. The apparatus will be described later (cf. Figure 26). Briefly, Hg atoms in their ground $6s^2\,{}^1S_0$ state were excited to the $6s7s\,{}^3S_1$ state by an electron beam with transverse polarization. The polarization of the light emitted along the direction of electron polarization was analyzed for the transitions $6s7s\,{}^3S_1 \rightarrow 6s6p\,{}^3P_0$ (404.7 nm), $^3P_1$ (435.8 nm), and $^3P_2$ (546.1 nm).

Near threshold (7.73 eV), Wykes' theoretical result has been confirmed: Wyke predicted for the 404.7-nm line a circular polarization $\eta_2 = 0.887P$, and scaling factors 1/2 and −1/2 for the 435.8 and 546.12 nm lines, respectively. The accuracy of the comparison is limited by the absolute calibration

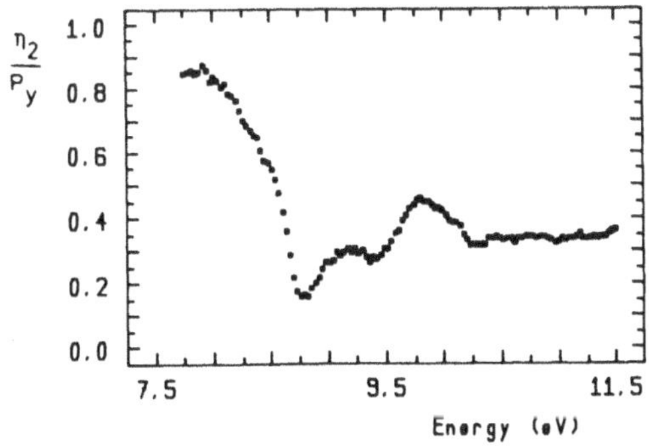

FIGURE 24. Circular light polarization η_2 normalized to the transverse electron polarization P_y by which it is caused. The transition Hg $7\,{}^3S_1 \rightarrow 6\,{}^3P_0$ (404.7 nm) is shown.[99]

of the electron polarization; the uncertainty in this quantity is ±10%. As the electron energy is increased, η_2 decreases, showing a pronounced structure that can be attributed to the resonances above 8 eV found in prior studies.[79-84] Another mechanism that affects the polarization of the emitted light occurs if the electrons have enough energy to excite energy levels above the $7\,^3S_1$ state so that this state becomes additionally populated by decay of the higher levels. But Figure 24 shows that the effect of such cascading on the circular polarization is less dramatic than that of a resonance. The results show also that one can utilize the circular polarization of nonresonant impact radiation for electron-polarization measurements. Because of the complicated dependence of η_2 on the electron energy, the original proposal[92] to use this method without calibration cannot be recommended, except in a narrow energy region above threshold (cf. Figure 24). Compared to certain resonance lines, the intensity of the present lines is fairly low, so that the time needed for the measurements was more than 10 times larger than that for the 253.7-nm line.[82]

Gay[100] proposed using the exchange excitation of the He $3\,^3P_J$ manifold, and measuring circular polarization of the $3\,^3P \rightarrow 2\,^3S$ 388.9-nm multiplet. An experimental analysis of the linear and circular polarization components of the 388.9-nm fluorescence emission will determine the electron polarization without requiring experimental calibration of the apparatus. Gay argues that resonances due to short-lived He^- states should not affect this calibration-free measurement of electron polarization. Preliminary measurements by the Muenster group gave promising results.[101]

The purpose of other investigations with Hg[82] ($6\,^1S_0 \rightarrow 6\,^3P_1$ excitation) and alkalis[102,103] ($ns \rightarrow np$ excitation) was to study the influence of resonances, exchange, and spin-orbit coupling during the collision. The excitation of the first $^2P_{1/2,3/2}$ fine-structure states of K and Cs atoms by polarized electrons was studied by the Mainz group[102,103] by measuring the circular polarization of the fluorescence emission. The agreement with theoretical data by Moores[104] and Nagy *et al.*[105] is fairly good. Figure 25 shows the results of the experiments with K.[103]

5.2. Electron–Photon Coincidence Experiments

Electron–photon coincidence experiments allow one to study the population of magnetic sublevels of atoms excited by electron impact with the electron scattering angle resolved. If the scattering is spin dependent, information is lost if only unpolarized electrons are used in these experiments. Experiments of this type using polarized electrons are feasible now,[106] and several are currently in progress.[107,108] Initial results have been presented recently for electron-impact excitation of Hg.[106] Electrons that have excited the $6\,^3P_1$ state of Hg were detected in coincidence with

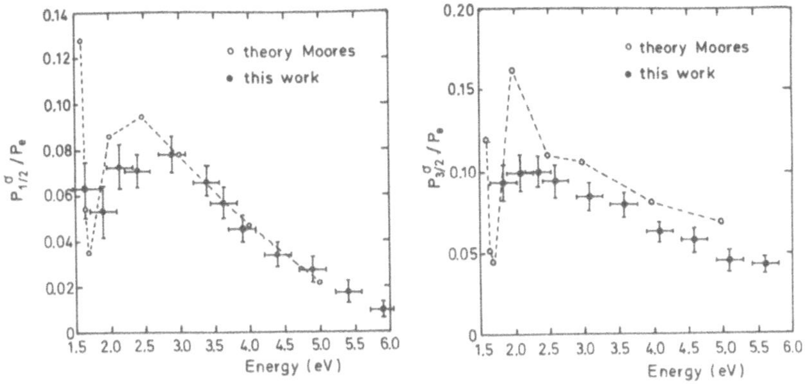

FIGURE 25. Ratio P^σ/P_e as a function of energy for electron-impact excitation of K.[103] P^σ is the measured light polarization and P_e is the polarization of electrons. Theoretical data for K from Moores.[104]

photons ($6\,^3P_1$–$6\,^2S_0$ transition, $\lambda = 253.7$ nm) transmitted by linear and circular polarization filters. The electron-scattering angle in this experiment was about 0°, being an extension of a previous experiment with unpolarized electrons.[109] The initial electron polarization defined a distinguished plane, and photons emitted perpendicular to that plane were observed (cf. Figure 23).

From a theoretical point of view this investigation was of interest for several reasons. Firstly, η_1 and η_2 vanish identically if the incident electrons are unpolarized. That is, the tilt angle γ and the orientation of the charge-cloud distribution are zero in this case (cf. Figure 23). This conclusion follows by adapting the symmetry arguments of Bartschat and Blum[110] to this case. Hence it may be expected that η_1 and η_2 will be sensitive tests of spin-dependent interactions. Furthermore, the Bonham–Ochkur approximation predicts $\eta_1 = 0$ even for polarized electrons.[111] A measurement of η_1 therefore allows one to draw some immediate conclusions concerning the validity of these approximations for forward-scattered electrons. This point is of interest in view of recent discussions.[111]

A scheme of the experimental apparatus is shown in Figure 26. Longitudinally polarized electrons were emitted from a GaAs photocathode that was placed in UHV and irradiated with circularly polarized light from a Kr-ion laser. After the electrons were deflected by 90°, their polarization was rotated by two magnetic coils through 90°, which made their polarization transverse and parallel to the axis of the light analyzer (cf. Figure 26). The electrons then passed through a differential pumping stage, where they were again deflected by 90°. A lens system focused the polarized electron beam onto the Hg target. Some of the Hg atoms were excited by electron impact to the $6\,^3P_1$ state (with 4.9 eV energy loss). Electrons that were scattered in

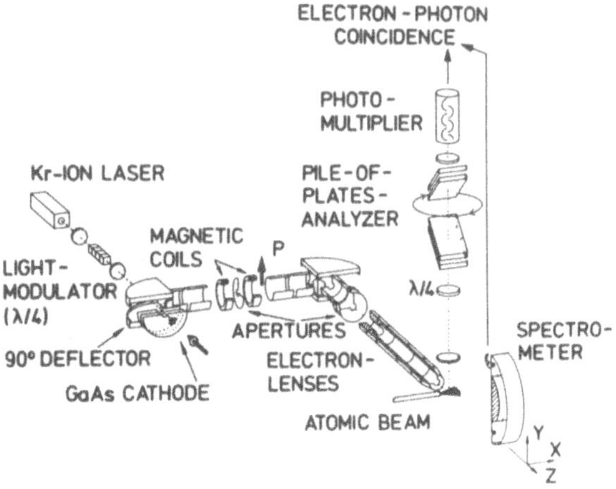

FIGURE 26. Scheme of the experiment for electron–photon coincidence studies with polarized electrons.[106]

the forward direction and that passed through an energy analyzer were detected in coincidence with photons ($6\,^3P_1$–$6\,^1S_0$ transition, 254 nm) that were transmitted by the photon-analyzer system. The photon analyzer detected photons that were emitted in the direction of the electron-spin polarization vector. It consisted of a collimating lens system, a $\lambda/4$ plate, and a pile-of-quartz-plates analyzer to measure either linear or circular polariz-

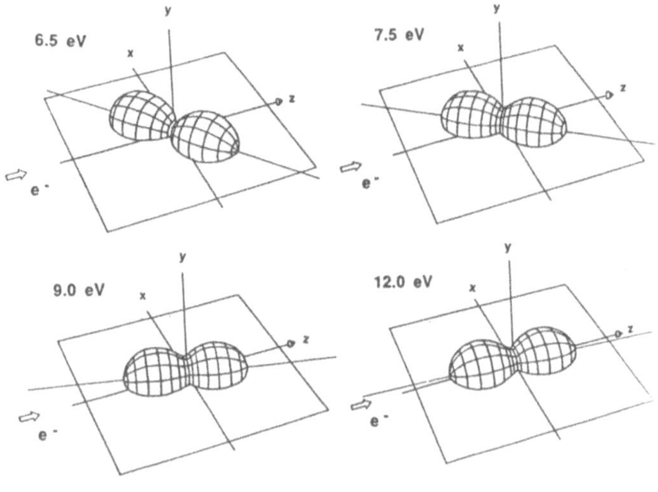

FIGURE 27. Charge-cloud distributions calculated[112] from experimentally determined Stokes parameters in the electron–photon coincidence experiment with polarized electrons by Wolcke *et al.*[106]

ation components. A wavelength filter was placed in front of the photomultiplier. The same apparatus was used for measurements of integrated Stokes parameters[82,97,99,101] without detection of scattered electrons, of course.

Instead of the measured Stokes parameters we show in Figure 27 some plots of the charge-cloud distributions.[112] The tilting angle γ is large below 8 eV, decreasing slowly with increasing energy. The orientation of the atom (not shown here) can be easily understood as the result of exchange collisions, but an obvious and simple picture for the nonzero tilt angle cannot be given. Blum argues that a nonzero tilt angle is the result of the interaction between the oriented atom and the electric field of the scattered electron.[113] Numerical calculations using the R-matrix close-coupling method[17] give results that agree well with experimental data between 6 and 8 eV, but these calculations cannot be extended to energies above 8 eV.

Recently, electron-photon coincidences have been used to study the spin up-down asymmetries for electron excitation of magnetic sublevels of Hg.[114]

6. CONCLUSIONS

Our knowledge of the role that the spin plays in electron collisions has been improved considerably during the last few years owing to the experimental and theoretical progress documented in this review. The quantitative agreement between experimental results and numerical calculations is by no means satisfactory; it is particularly bad in the energy region between the thresholds of the first excited states and three times the ionization thresholds. But qualitative arguments and simple models have been established now that reliably guide our intuition.

A number of problems remain still unsolved, and data are fragmentary: the role of exchange in elastic collisions from open-shell atoms, careful tests of the validity of the Percival-Seaton hypothesis, and tests of threshold laws. More experimental data and theoretical calculations are needed to explain shape and dynamics of the charge-cloud distribution obtained by electron-photon coincidence experiments. A systematic study of how the various optically forbidden types of electron excitations depend on spin effects is lacking, too. Certainly these basic studies should be extended to electron-molecule collisions and to scattering in intense electromagnetic fields.

We have considered only experiments in which the spin polarization of the free electrons was involved. Basic processes like exchange could be studied with bound polarized electrons, i.e., polarized atoms, as well.[41,55,56,58,115,116] Future reviews of electron-spin polarization in collisions should encompass investigations using polarized bound electrons as

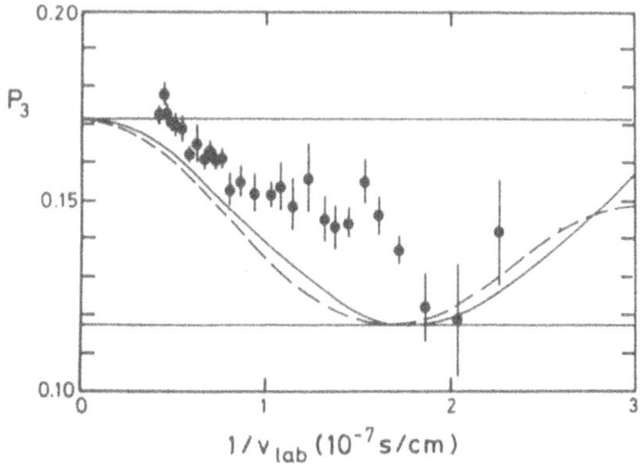

FIGURE 28. Measured and calculated circular polarization P_3 of the fluorescence light versus inverse impact velocity of v_{lab}. The upper and lower straight lines correspond to conservation of the spin orientation ($P_s' = P_s$) and to fulll depolarization ($P_s' = 0$), respectively, during the collision. The two other curves correspond to spin depolarization by electron exchange between the Ne^+2p^5 core and valence electron; spin-orbit coupling in Ne^+2p^5 neglected (—), and approximately taken into account (- - -). Figure taken from Jitschin *et al.*[116]

well, because the basic phenomena are the same: Exchange, or more generally the Pauli principle,[115] and the spin-orbit interaction. As an example of this wider view we show finally a result obtained by a group in Bielefeld.[116] Polarized Na atoms were excited by Ne^+ ions,

$$\mathrm{Ne}^+ + \mathrm{Na}(3s\uparrow) \rightarrow \mathrm{Ne}^+ + \mathrm{Na}(3p\uparrow)$$

$$\searrow$$

$$\mathrm{Na}(3s) + h\nu(\eta_2)$$

and the circular polarization of the fluorescence emission was observed. Depolarization of the spin orientation in the excitation process at slow impact velocities is clearly seen in the results shown in Figure 28. This observation is the first experimental evidence of a spin change caused by electron exchange in an energetic atomic collision.

ACKNOWLEDGMENTS. The author is deeply grateful to Dr. M. Morrison for reading the manuscript and offering valuable criticisms and suggestions. The author is indebted to Johnette Ellis, who skillfully typed and corrected the manuscript.

REFERENCES

1. H. KLEINPOPPEN AND J. F. WILLIAMS, editors, *Coherence and Correlation in Atomic Collisions* (Plenum Press, New York, 1980).
2. G. F. HANNE, Spin Effects in Inelastic Electron-Atom Collisions, *Phys. Rep.* **95**, 95-165 (1983).
3. K. BLUM AND H. KLEINPOPPEN, Spin-Dependent Phenomena in Inelastic Electron-Atom Collisions, *Adv. At. Mol. Phys.* **19**, 187-264 (1983).
4. J. SLEVIN, Coherence in Inelastic Low-Energy Electron Scattering, *Rep. Prog. Phys.* **47**, 461-512 (1984).
5. J. KESSLER, *Polarized Electrons*, 2. ed. (Springer Verlag, Berlin, 1985).
6. M. A. MORRISON, The Physics of Low-Energy Electron-Molecule Collisions: A Guide for the Perplexed and the Uninitiated, *Aust. J. Phys.* **36**, 239-286 (1983).
7. N. S. SCOTT, K. BARTSCHAT, P. G. BURKE, W. B. EISSNER, AND O. NAGY, Low-Energy Scattering of Electrons by Cesium Atoms, *J. Phys. B: At. Mol. Phys.* **17**, L191-195 (1984).
8. D. T. PIERCE, R. J. CELOTTA, G.-C. WANG, W. N. UNERTL, A. GALEJS, C. E. KUYATT, AND S. R. MIELCZAREK, GaAs Spin Polarized Electron Source, *Rev. Sci. Instrum.* **51**, 478-499 (1980).
9. I. V. HERTEL AND W. STOLL, Collision Experiments with Laser Excited Atoms in Crossed Beams, *Adv. At. Mol. Phys.* **13**, 113-228 (1977).
10. C. W. MCLUCAS, H. J. E. WEHR, W. R. MACGILLIVRAY, AND M. C. STANDAGE, Isotope Specific Analysis of the Electron Impact Excited $6\,^1S_0 \to 6\,^1P_1$ transition of mercury using stepwise electron and laser excitation, *J. Phys. B: At. Mol. Phys.* **15**, 1883-1898 (1982).
11. M. EMINYAN, K. B. MACADAM, J. SLEVIN, AND H. KLEINPOPPEN, Electron-Photon Angular Correlations in Electron-Helium Collisions: Measurement of Complex Excitation Amplitudes, Atomic Orientation and Alignment, *J. Phys. B: At. Mol. Phys.* **7**, 1519-1542 (1974).
12. J. UNGURIS, D. T. PIERCE, AND R. J. CELOTTA, Low-Energy Diffuse Scattering Electron-Spin Polarization Analyzer, *Rev. Sci. Instrum.* **57**, 1314-1323 (1986).
13. D. HILS, W. JITSCHIN, AND H. KLEINPOPPEN, Production of a Highly Polarized Sodium Atomic Beam, *Appl. Phys.* **25**, 39-47 (1981).
14. J. J. MCCLELLAND AND M. H. KELLEY, Detailed Look at Aspects of Optical Pumping in Sodium, *Phys. Rev. A* **31**, 3704-3710 (1985).
15. N. T. PADIAL, G. D. MENESES, F. J. DA PAIXAO, G. CSANAK, AND D. C. CARTWRIGHT, Electron-Impact Excitation of the Lowest Four Excited States of Argon, *Phys. Rev. A* **23**, 2194-2212 (1981).
16. K. BARTSCHAT, D. H. MADISON, AND G. F. HANNE, Spin Polarisation and Scattering Asymmetry for Electron Impact Excitation of the ($6s6p$) States of Mercury: DWBA Treatment, *J. Phys. B: At. Mol. Phys.* **18**, 1847-1858 (1985).
17. K. BARTSCHAT AND P. G. BURKE, Low Energy Scattering of Electrons by Heavy Atoms, *Comments At. Mol. Phys.* **16**, 271-291 (1985).
18. P. G. BURKE AND J. F. B. MITCHELL, Spin-Polarization in the Elastic Scattering of Electrons by One-Electron Atoms, *J. Phys. B: At. Mol. Phys.* **7**, 214-228 (1974).
19. R. A. BONHAM, Inelastic Scattering from Atoms at Medium Energies. I. Bound States, *J. Chem. Phys.* **36**, 3260-3269 (1962).
20. V. I. OCHKUR, The Born-Oppenheimer Method in the Theory of Atomic Collisions, *Sov. Phys. JETP* **18**, 503-508 (1964).
21. R. A. BONHAM, Second Born Zero Angle Differential Cross Sections in the High Energy Limit for the Electron Impact Excitation of He: The $2\,^3S \leftarrow 1\,^1S$ and $2\,^3P \leftarrow 1\,^1S$ Transitions, *J. Chem. Phys.* **57**, 1604-1612 (1972).

22. L. A. BURKOVA AND V. I. OCHKUR, Calculations of Inelastic Electron Scattering from Atoms in the Second Born Approximation, *Sov. Phys. JETP* **49**, 38-45 (1979).
23. G. F. HANNE AND J. KESSLER, Study of Exchange Excitation in Mercury by Means of Polarized Electrons. I. Experiment, *J. Phys. B: At. Mol. Phys.* **9**, 791-804 (1976).
24. G. F. HANNE, Orientation and Spin Exchange in Electron-Atom Collisions, *Comments At. Mol. Phys.* **14**, 163-171 (1984).
25. I. C. PERCIVAL AND M. J. SEATON, The Polarization of Atomic Line Radiation Excited by Electron Impact, *Philos. Trans. R. Soc. A* **251**, 113-138 (1958).
26. W. N. SHELTON, Spin Polarisation of Electrons Resulting from Electron Impact Excitation of the $2p$ Fine-Structure States of the Hydrogen Atom, *J. Phys. B: At. Mol. Phys.* **19**, L257-L259 (1986).
27. N. C. STEPH AND D. E. GOLDEN, Alignment and Orientation in the Electron-Impact Excitation of the $2\,^1P$ State of He from 40 to 500 eV, *Phys. Rev. A* **21**, 1848-1855 (1980).
28. H. W. HERMANN AND I. V. HERTEL, Orientation in Atomic Collisions: Correspondence of Classical Model and Partial-Wave Analysis, *J. Phys. B: At. Mol. Phys.* **13**, 4285-4297 (1980).
29. M. KOHMOTO AND U. FANO, Orientation by Collision, *J. Phys. B: At. Mol. Phys.* **14**, L447-L451 (1981).
30. N. ANDERSEN, I. V. HERTEL, AND H. KLEINPOPPEN, Shape and Dynamics of States Excited in Electron-Atom Collisions: A Comment on Orientation and Alignment Parameters by Consideration of Attractive and Repulsive Forces, *J. Phys. B: At. Mol. Phys.* **17**, L901-L908 (1984).
31. N. ANDERSON AND I. V. HERTEL, A Fresh Look at Alignment and Orientation in Collisional Excitation, *Comments At. Mol. Phys.* **19**, 1-34 (1986).
32. D. H. MADISON, G. CSANAK, AND D. C. CARTWRIGHT, The Sign of the Orientation Parameter in Electron-Photon Coincidence Experiments, *J. Phys. B: At. Mol. Phys.* **19**, 3361-3366 (1986).
33. G. F. HANNE, Study of Exchange Excitation in Mercury by Means of Polarized Electrons II. Theory, *J. Phys. B: At. Mol. Phys.* **9**, 805-815 (1976).
34. K. SCHACKERT, Spinpolarisation langsamer Elektronen durch elastische Streuung an Edelgasatomstrahlen, *Z. Physik* **213**, 316-322 (1968).
35. K. BARTSCHAT, Angular Momentum Orientation in Elastic Electron-Atom Scattering, *J. Phys. B: At. Mol. Phys.* **20**, L815-L819 (1987).
36. P. S. FARAGO, On the Detection of Spin-Orbit Interaction in the Elastic Scattering of Electrons from One-Electron Atoms, *J. Phys. B: At. Mol. Phys.* **7**, L28-L31 (1974).
37. J. J. MCCLELLAND, M. H. KELLEY, AND R. J. CELOTTA, Spin-Orbit and Exchange Effects in Elastic Scattering of Spin-Polarized Electrons from Spin-Polarized Na Atoms, *Phys. Rev. Lett.* **58**, 2198-2200 (1987).
38. G. BAUM, W. RAITH, AND H. STEIDL, Stossexperimente mit polarisierten Elektronen, *SFB 216 Finanzierungsantrag 1986-1988*, 95-137 (1985).
39. P. G. BURKE AND H. M. SCHEY, Polarization and Correlation of Electron Spin in Low-Energy Elastic Electron-Hydrogen Collisions, *Phys. Rev.* **126**, 163-168 (1962).
40. B. BEDERSON, The "Perfect" Scattering Experiment. I, *Comments At. Mol. Phys.* **1**, 41-44 (1969).
41. R. E. COLLINS, B. BEDERSON, AND M. GOLDSTEIN, Differential Spin Exchange and the Elastic Scattering of Low-Energy Electrons by Potassium, *Phys. Rev. A* **3**, 1976-1987 (1971).
42. D. HILS, M. V. MCCUSKER, H. KLEINPOPPEN, AND S. J. SMITH, Differential and Direct Differential Elastic Scattering Cross Sections for Electrons and Potassium Atoms, *Phys. Rev. Lett.* **29**, 398-401 (1972).

43. G. D. FLETCHER, M. J. ALGUARD, T. J. GAY, V. W. HUGHES, P. F. WAINWRIGHT, M. S. LUBELL, AND W. RAITH, Experimental Study of Spin-Exchange Effects in Elastic and Ionizing Collisions of Polarized Electrons with Polarized Hydrogen Atoms, *Phys. Rev. A* **31**, 2854-2884 (1985).
44. G. BAUM, M. MOEDE, W. RAITH, AND U. SILLMEN, Measurement of Spin Dependence in Low-Energy Elastic Scattering of Electrons from Lithium Atoms, *Phys. Rev. Lett.* **57**, 1855-1858 (1986).
45. O. BERGER AND J. KESSLER, Elastic Scattering of Polarised Electrons from Mercury and Xenon, *J. Phys. B: At. Mol. Phys.* **19**, 3539-3557 (1986).
46. D. W. WALKER, Relativistic Effects in Low Energy Electron Scattering from Atoms, *Adv. Phys.* **20**, 257-323 (1971) and calculations based on his program.
47. B. AWE, F. KEMPER, F. ROSICKY, AND R. FEDER, Elastic Scattering of Slow Electrons from Xe Atoms, *J. Phys. B: At. Mol. Phys.* **16**, 603-612 (1983).
48. R. HABERLAND, L. FRITSCHE, AND J. NOFFKE, Elastic Scattering of Low-Energy Electrons by Ne, Ar, Kr, and Xe, *Phys. Rev. A* **33**, 2305-2314 (1986).
49. R. P. MCEACHRAN AND A. D. STAUFFER, Spin Polarisation of Electrons Elastically Scattered from Xenon, *J. Phys. B: At. Mol. Phys.* **19**, 3523-3538 (1986).
50. J. KESSLER, Absolute Cross Sections for Electron Scattering in the Light of Complete Experiments—A Fresh Look at an Old Topic, *Comments At. Mol. Phys.* **18**, 279-288 (1986).
51. M. KLEWER, M. J. M. BEERLAGE, AND M. J. VAN DER WIEL, Polarisation and Angular Distribution of Electrons Elastically Scattered from Cesium Atoms at 13-25 eV, *J. Phys. B: At. Mol. Phys.* **12**, L525-L528 (1979).
52. F. KAUSSEN, H. GEESMANN, G. F. HANNE, AND J. KESSLER, Study of Spin Polarisation in Elastic Scattering of Electrons from Hg, Tl, Pb, and Bi Atoms, *J. Phys. B: At. Mol. Phys.* **20**, 151-165 (1987).
53. M. FINK AND A. C. YATES, Tables of Scattering Amplitudes and Spin Polarizations of 25, 50, 75, 150, 175, 200, 300, 400, 600, 800 eV Electrons Scattered Elastically From H, He, C, Ne, Ar, Kr, Rb, Xe, Cs, Au, Hg, Pb, and Bi-I, Technical Report No. 88, Electronics Research Center (Austin), 1970.
54. R. HABERLAND AND L. FRITSCHE, On the Elastic Scattering of Low-Energy Electrons by Hg, Tl, Pb, and Bi atoms, *J. Phys. B: At. Mol. Phys.* **20**, 121-126 (1987).
55. W. LICHTEN AND S. SCHULTZ, Cross Sections for the Excitation of the Metastable $2s$ State of Atomic Hydrogen by Electron Collision, *Phys. Rev.* **116**, 1132-1139 (1959).
56. K. RUBIN, B. BEDERSON, M. GOLDSTEIN, AND R. E. COLLINS, Electron-Alkali-Metal Inelastic Recoil Experiments With Spin Analysis: Experimental Method and the Small-Angle Behavior of the $4^2S_{1/2} \rightarrow 4^2P_{1/2,3/2}$ Excitation of Potassium, *Phys. Rev.* **182**, 201-214 (1969).
57. G. BAUM, M. MOEDE, W. RAITH AND W. SCHRÖDER, Ionization and ns-np Excitation Asymmetry in Polarized Electron-Polarized Alkali Atom Collisions, XIV ICPEAC (Palo Alto) Abstracts of Contributed Papers, 144 (1985).
58. W. JITSCHIN, S. OSIMITSCH, R. REIHL, H. KLEINPOPPEN, AND H. O. LUTZ, Electron Exchange in the Na $3p$ Electron Impact Excitation, *J. Phys. B: At. Mol. Phys.* **17**, 1899-1912 (1984).
59. G. BAUM, W. RAITH, U. SILLMEN, AND H. STEIDL, Measurement of Relative Singlet and Triplet Contributions in Inelastic Collisions of Electrons with Low-Z Atoms, XV ICPEAC (Brighton) Abstracts of Contributed Papers, 153 (1987).
60. D. HILS AND H. KLEINPOPPEN, The Ionisation of Polarised Potassium Atoms by Polarised Electrons, *J. Phys. B: At. Mol. Phys.* **11**, L283-L287 (1978).
61. D. HILS, W. JITSCHIN, AND H. KLEINPOPPEN, The Spin Dependent Ionisation Asymmetry in Sodium, *J. Phys. B: At. Mol. Phys.* **15**, 3347-3357 (1982).

62. M. H. KELLEY, W. T. ROGERS, R. J. CELOTTA, AND S. R. MIELCZAREK, Near-Threshold Measurements of the Spin Dependence of Electron-Impact Ionization, *Phys. Rev. Lett.* **51**, 2191–2193 (1983).
63. G. BAUM, M. MOEDE, W. RAITH AND W. SCHRÖDER, Measurement of Spin Asymmetries in the Electron Impact Ionisation of Alkali Atoms, *J. Phys. B: At. Mol. Phys.* **18**, 531–538 (1985).
64. D. L. MOORES, Scattering of Low-Energy Electrons by Neutral Lithium Atoms: Five-state Close-Coupling Calculations, *J. Phys. B: At. Mol. Phys.* **19**, 1843–1851 (1986).
65. A. TEMKIN, Threshold Law for Electron–Atom Impact Ionization, *Phys. Rev. Lett.* **49**, 365–368 (1982).
66. H. KLAR AND W. SCHLECHT, Threshold Multiple Ionization of Atoms. Energy Dependence for Double and Triple Escape, *J. Phys. B: At. Mol. Phys.* **9**, 1699–1711 (1976).
67. C. H. GREENE AND A. R. P. RAU, Double Escape of Two Electrons at Threshold: Dependence on L, S, and π, *Phys. Rev. Lett.* **48**, 533–537 (1982).
68. A. D. STAUFFER, The Nodal Structure of Two-Electron Wave Functions on the Wannier Ridge, *Phys. Lett.* **91A**, 114–116 (1982).
69. G. CSANAK AND D. C. CARTWRIGHT, The Fine-Structure Effect for Electron Impact Excitation of $n^3P_{0,1,2}$ States of LS-coupled Atoms: I. Helium, *J. Phys. B: At. Mol. Phys.* **20**, L49–L54 (1987).
70. G. D. MENESES, L. E. MACHADO, G. CSANAK, AND D. C. CARTWRIGHT, The Fine-Structure Effect of Electron Impact Excitation of $n^3P_{0,1,2}$ States of LS-Coupled Atoms. II. The Mg atom, *J. Phys. B: At. Mol. Phys.* **20**, L717–L722 (1987).
71. G. F. HANNE, CZ. SZMYTKOWSKI, AND M. VAN DER WIEL, Spin Polarisation of Electrons Scattered Superelastically by Laser-Excited Na Atoms, *J. Phys. B: At. Mol. Phys.* **15**, L109–L113 (1982).
72. J. J. MCCLELLAND, M. H. KELLEY, AND R. J. CELOTTA, Spin Dependence in Superelastic Electron Scattering from Na(3P), *Phys. Rev. Lett.* **55**, 688–691 (1985).
73. J. J. MCCLELLAND, M. H. KELLEY, AND R. J. CELOTTA, Spin-Dependent Superelastic Scattering from Pure Angular Momentum States of Na(3P), *Phys. Rev. Lett.* **56**, 1362–1365 (1986).
74. I. V. HERTEL, M. H. KELLEY, AND J. J. MCCLELLAND, Analysis of Collisional Alignment and Orientation Studied by Scattering of Spin-Polarized Electrons from Laser Excited Atoms, *Z. Phys. D: At. Mol. Clusters* **6**, 163–183 (1987).
75. W. EITEL AND J. KESSLER, Spinpolarisation und differentielle Wirkungsquerschnitte bei der inelastischen Elektron-Atom-Streuung, *Z. Phys.* **241**, 355–368 (1971).
76. G. F. HANNE, K. JOST, AND J. KESSLER, Spinpolarisation und Winkelverteilung der Elektronen nach Anregung verschiedener Zustände des Hg-Atoms, *Z. Phys.* **252**, 141–146 (1972).
77. K. BARTSCHAT, G. F. HANNE, A. WOLCKE, AND J. KESSLER, Resonance Features and Fine-Structure Effect in the Asymmetry of Polarized Electrons Scattered Inelastically from Mercury Atoms, *Phys. Rev. Lett.* **47**, 997–999 (1981).
78. K. FRANZ, G. F. HANNE, K. H. HIDDEMANN, AND J. KESSLER, Electron Spin Polarisation by Inelastic Scattering from Mercury Atoms, *J. Phys. B: At. Mol. Phys.* **15**, L115–L120 (1982).
79. T. W. OTTLEY AND H. KLEINPOPPEN, Resonances near Threshold in the Electron Impact Excitation and Polarization of the $6\,^1S_0$–$6\,^3P_1$ Line of Mercury, *J. Phys. B: At. Mol. Phys.* **8**, 621–627 (1975).
80. O. B. SHPENIK, V. V. SOUTER, A. N. ZAVILOPULO, I. P. ZAPESOCHNYI, AND E. E. KONTROSH, Investigation of Resonances in the Total Cross Sections for Excitation of Mercury Atoms by Slow Monoenergetic Electrons, *Sov. Phys. JETP* **42**, 23–28 (1976).
81. S. M. KAZAKOV, A. I. KOROTKOV, AND O. B. SHPENIK, Resonant Scattering of Low-Energy Electrons by Mercury Atoms, *Sov. Phys. JETP* **51**, 847–851 (1981).

82. A. WOLCKE, K. BARTSCHAT, K. BLUM, H. BORGMANN, G. F. HANNE, AND J. KESSLER, Investigation of Stokes Parameters for Studying Resonances near Threshold for $6\,^1S_0$-6^3P_1 Excitation of Mercury by Polarised Electrons, *J. Phys. B: At. Mol. Phys.* **16**, 639-655 (1983).
83. L. KOCH, T. HEINDORFF, AND E. REICHERT, Resonances in the Electron Impact Excitation of Metastable States of Mercury, *Z. Phys. A* **316**, 127-130 (1984).
84. D. S. NEWMAN, M. ZUBEK, AND G. C. KING, A Study of Resonance Structure in Mercury Using Metastable Excitation by Electron Impact with High Resolution, *J. Phys. B: At. Mol. Phys.* **18**, 985-998 (1985).
85. A. WOLCKE, C. HOELSCHER, E. WEIGOLD, AND G. F. HANNE, A Position-Sensitive Detector for the Simultaneous Measurement of the Spin Asymmetry of Different Electron Impact Excitation Channels, *J. Phys. E: Sci. Instrum.* **20**, 299-304 (1987).
86. H. BORGMANN, J. GOEKE, G. F. HANNE, J. KESSLER, AND A. WOLCKE, Asymmetry of Polarised Electrons Scattered Inelastically from Mercury Atoms, *J. Phys. B: At. Mol. Phys.* **20**, 1619-1633 (1987).
87. K. JOST, Novel Design of a "Spherical" Electron Spectrometer, *J. Phys. E: Sci. Instrum.* **12**, 1006-1012 (1979).
88. K. BARTSCHAT, private communication (1985), extension of calculations presented by Bartschat *et al.*[89]
89. K. BARTSCHAT, K. BLUM, P. G. BURKE, G. F. HANNE, AND N. S. SCOTT, The Fine-Structure Effect in the Low-Energy Scattering of Electrons on Hg and Tl Atoms, *J. Phys. B: At. Mol. Phys.* **17**, 3797-3803 (1984).
90. K. BARTSCHAT AND D. H. MADISON, Inelastic Electron Scattering from Mercury Atoms at Intermediate Energies (8-50 eV), *J. Phys. B: At. Mol. Phys.* **20**, 1609-1618 (1987).
91. P. S. FARAGO AND J. S. WYKES, Optical Detection of Electron Polarization, *J. Phys. B: At. Mol. Phys.* **2**, 747-756 (1969).
92. J. WYKES, A General Optical Detector of the Polarization of Low Energy Electrons, *J. Phys. B: At. Mol. Phys.* **4**, L91-L94 (1971).
93. K. BARTSCHAT, K. BLUM, G. F. HANNE, AND J. KESSLER, Electron-Photon Coincidences with Polarised Electrons, *J. Phys. B: At. Mol. Phys.* **14**, 3761-3776 (1981).
94. K. BLUM, *Density Matrix Theory and Applications* (Plenum Press, New York, 1981).
95. F. J. DA PAIXAO, N. T. PADIAL, G. CSANAK, AND K. BLUM, Spin-Orbit Coupling Effects in Parameters that Describe Electron-Photon Coincidence Experiments, *Phys. Rev. Lett.* **45**, 1164-1167 (1980).
96. M. EMINYAN AND G. LAMPEL, Optical Measurement of Free-Electron Polarization, *Phys. Rev. Lett.* **45**, 1171-1174 (1980).
97. J. GOEKE, G. F. HANNE, J. KESSLER, AND A. WOLCKE, Circular Polarization of Impact Radiation Produced by Polarized Electrons in Mercury, *Phys. Rev. Lett.* **51**, 2273-2275 (1983).
98. H. WOLF AND E. REICHERT, Impact Excitation of the 7^3S_1-State of Mercury by Polarized Electrons, XIII ICPEAC (Berlin), Abstracts of Contributed Papers, 148 (1983).
99. J. GOEKE, Erprobung eines Lichtleitersystems für Laserstrahlung und Untersuchungen zur optischen Kontrolle polarisierter Elektronen, diploma thesis (Münster) 1983.
100. T. J. GAY, A Simple Optical Electron Polarimeter, *J. Phys. B: At. Mol. Phys.* **16**, L553-L556 (1983).
101. J. GOEKE, J. KESSLER, AND G. F. HANNE, Circularly Polarized He Radiation for Electron Polarimetry, *Phys. Rev. Lett.* **59**, 1413-1415 (1987).
102. P. NASS, N. LUDWIG, AND E. REICHERT, Impact Excitation of Potassium and Cesium by Polarized Electrons, XIV ICPEAC (Palo Alto) Abstracts of Contributed Papers, 187 (1985).

103. N. LUDWIG, A. BAUCH, P. NASS, E. REICHERT, AND W. WELKER, Impact Excitation of 4^2P-Potassium by Polarized Electrons, *Z. Phys. D: At. Mol. Clusters* **4**, 177–183 (1986).
104. D. L. MOORES, The Scattering of Electrons by Potassium Atoms, *J. Phys. B: At. Mol. Phys.* **9**, 1329–1349 (1976).
105. O. NAGY, K. BARTSCHAT, K. BLUM, P. G. BURKE, AND N. S. SCOTT, Calculation of Stokes' Parameter for Inelastic Electron–Cesium Scattering, *J. Phys. B: At. Mol. Phys.* **17**, L527–L530 (1984).
106. A. WOLCKE, J. GOEKE, G. F. HANNE, J. KESSLER, W. VOLLMER, K. BARTSCHAT, AND K. BLUM, Electron–Photon Coincidence Studies in Collisions of Polarized Electrons with Mercury Atoms, *Phys. Rev. Lett.* **52**, 1108–1111 (1984).
107. A. VASILAKIS, M. S. LUBELL, K. RUBIN, F. C. TANG, J. SLEVIN, AND M. EMINYAN, The Three-Body Interaction with Long-Range Forces: Spin-Dependent Electron–Hydrogen Collisions, XIV ICPEAC (Palo Alto) Abstract of Contributed Papers, 180 (1985).
108. To our knowledge such experiments are in progress also at the University of W.A., at the Flinders University of S.A., and at the University of Oklahoma.
109. G. F. HANNE, K. WEMHOFF, A. WOLCKE, AND J. KESSLER, Electron–Photon Coincidence Experiment for Measuring the Polarisation of the $6\,^3P_1$–$6\,^1S_0$ Resonance Line of Hg after Electron Excitation with $\Delta M_L = 0$, *J. Phys. B: At. Mol. Phys.* **14**, L507–L510 (1981).
110. K. BARTSCHAT AND K. BLUM, Theory and Physical Importance of Integrated State Multipoles, *Z. Physik A: At. Nuclei* **304**, 85–88 (1982).
111. R. A. BONHAM, Polarisation of the $6\,^3P_1$–$6\,^1S_0$ Line of Mercury, *J. Phys. B: At. Mol. Phys.* **15**, L361–L366 (1982).
112. M. SOHN, private communication (1987).
113. K. BLUM, private communication (1987).
114. J. GOEKE, G. F. HANNE, AND J. KESSLER, Study of Spin up-down Asymmetries for Electron Excitation of Magnetic Sublevels of Hg Using Electron–Photon Coincidences, XV ICPEAC (Brighton) Abstracts of Contributed Papers, 175 (1987).
115. W. JITSCHIN, Spin-Depolarisation in Energetic Atomic Collisions by the Exchange of Electrons, *Z. Phys. D: At. Mol. Clusters* **1**, 135–136 (1986).
116. W. JITSCHIN, S. OSIMITSCH, H. REIHL, D. W. MUELLER, H. KLEINPOPPEN, AND H. O. LUTZ, Spin Exchange in the Excitation of Spin-Polarized Na Atoms by Ne^+-Ion Impact, *Phys. Rev. A* **34**, 3684–3691 (1986).
117. K.-N. HUANG, Spin-Polarization Correlation in Electron–Atom Scatterings, *Chinese J. Phys.* **25**, 156–174 (1987).

CHAPTER 3

ELECTRON SCATTERING FROM CHIRAL MOLECULES

K. BLUM

1. INTRODUCTION: ORIGIN OF OPTICAL ACTIVITY IN LIVING SYSTEMS

It is generally believed that the formation of optically active molecules was a necessary step towards the origin of life. Optically active (or chiral) molecules usually exist in two forms, the L- and D-forms, which are mirror images of each other. In living organisms only one of the two alternative enantiomers is found. Apart from a few exceptions, the amino acids occurring in natural proteins belong to the L-species and sugars to the D-species.

In order to discuss the origin of optical purity one must distinguish between the first generation of optical asymmetry (which might be very small) and its further amplification. There is no generally accepted theory of the first generation of optical isomerism. The simplest possibility is to assume that at the beginning of evolution a racemic mixture of L- and D-structures, capable of self-reproduction, was present. Fluctuations led then to small differences in their densities, which were amplified by autocatalytic processes.

Another explanation was put forward after parity nonconservation had been discovered. It has been established that electrons, emitted in weak interaction decay processes, possess an intrinsic left-chirality (that is, momentum and spin vectors are antiparallel) and positrons a right-handed chirality. It is tempting to look for a connection between this asymmetry and optical isomerism.

Vester and Ulbright[1] proposed that molecular dissymmetry may have originated through preferential destruction of one enantiomer of a racemic

K. BLUM • Institute for Theoretical Physics I, University of Münster, D-4400 Münster, West Germany.

mixture by spin polarized electrons arising from β decay, or circularly polarized bremsstrahlung produced during their deceleration (β radiolysis). However, as yet, no unequivocal enantio-selective radiolysis has been obtained by using polarized electrons produced by particle accelerators, or emitted by radio nucleides.

A new aspect occurred with the discovery of parity-violating weak neutral currents (PNC). PNC interactions can produce an energy difference between a chiral molecule and its optical isomer, which is positive for one structure and negative for its antipode. The energy splitting depends explicitly on spin-dependent interactions, as can be deduced from symmetry requirements.[2] Garay and Hrasko[3] developed a model suggesting that weak interactions can differentiate between L- and D-forms because their orbital electrons have a nonvanishing helicity.

A formal development has been given by Hegstrom *et al.*[4] Estimates based on model calculations for twisted ethylene give a energy shift of the order of 5×10^{-19} eV. *Ab initio* calculations show that this shift is sensitive to the molecular conformation.[5] Of particular interest is that Mason and Tranter[6] found that it is the dominant L-amino acids that are favored by the PNC interaction.

The small energy differences between the two enantiomers produced by the PNC interaction give rise to small differences in their *Arrhenius reaction rates*. On the other hand, radiolysis proceeds via the usual electromagnetic interaction, which does not produce energy shifts but small differences in the *concentration* of optically active antipodes. Both effects are very small. However, it has been shown recently that autocatalytic chemical systems can effectively amplify these small differences and provide a mechanism for producing the observed optical purity of biological macromolecules over the long periods of prebiotic evolution. For a more detailed discussion of these and related theories we refer to the review by Mason.[7]

In the development of all these ideas the spin dependence of the interaction between electrons and chiral molecules plays an essential part. Investigations of these spin-dependent interactions can therefore be important for a deeper understanding of the origin of optical purity.

Information of spin-dependent effects can be obtained by studying elastic electron scattering from chiral molecules. It is the purpose of the present chapter to give a brief review of the theoretical and experimental results obtained so far.

We will not consider the related problem of angular distribution and polarization of photoelectrons emitted from chiral molecules. For details we refer to the theoretical papers by Richie[8] and Cherepkov.[9] First experiments are in progress.[10]

2. SPIN POLARIZATION IN ELASTIC SCATTERING OF ELECTRONS FROM CHIRAL MOLECULES

2.1. *General Conclusions*

Possible chiral effects in the spin polarization of elastically scattered electrons were first discussed by Farago.[11] The arguments are based on the expansion of the spin scattering matrix M, which must be linear in the Pauli matrices and invariant under rotations (see, for example, Kessler's book[12]):

$$M = f1 + f'\boldsymbol{\sigma}\mathbf{n}_1 + g\boldsymbol{\sigma}\mathbf{n}_2 + h\boldsymbol{\sigma}\mathbf{n}_3 \tag{1}$$

where f, f', g, h are scalar functions of energy and scattering angle. Here 1 is the two-dimensional unit matrix, $\boldsymbol{\sigma} = (\sigma_x, \sigma_y, \sigma_z)$ denotes the Pauli matrices, and a coordinate system defined by the following unit vectors has been used:

$$\mathbf{n}_1 = \frac{\mathbf{k}_2 - \mathbf{k}_1}{|\mathbf{k}_2 - \mathbf{k}_1|}, \qquad \mathbf{n}_2 = \frac{\mathbf{k}_1 \times \mathbf{k}_2}{|\mathbf{k}_1 \times \mathbf{k}_2|}, \qquad \mathbf{n}_3 = \mathbf{n}_1 \times \mathbf{n}_2 \tag{2}$$

$\mathbf{k}_1$ and $\mathbf{k}_2$ are the initial and final electron momenta.

The requirement that M should be invariant against time reversal gives the further condition $f' = 0$, and equation (1) reduces to the expression

$$M = f1 + g\boldsymbol{\sigma}\mathbf{n}_2 + h\boldsymbol{\sigma}\mathbf{n}_3 \tag{3}$$

If the target is in an eigenstate of parity and the interaction parity invariant then h vanishes too.

Starting with an unpolarized primary beam of unit intensity the polarization components of the scattered electrons can be calculated from equation (3).[11] The results are given by the following equations:

$$\begin{aligned} P_1 &= \frac{|g + ih|^2 - |g - ih|^2}{2S_0} \\ P_2 &= \frac{|f + g|^2 - |f - g|^2}{2S_0} \\ P_3 &= \frac{|f + h|^2 - |f - h|^2}{2S_0} \end{aligned} \tag{4}$$

where

$$S_0 = |f|^2 + |g|^2 + |h|^2$$

is the scattered intensity. For nonchiral targets $h = 0$, P_1 and P_3 vanish and one is left with the polarization component P_2 perpendicular to the scattering plane. Nonvanishing in-plane components are therefore manifestations of parity-violating phenomena.

There are two more effects that can in principle be produced by chiral targets: (1) the dependence of the attenuation of a longitudinal polarization of the incident electron beam and (2) the rotation of a transverse spin polarization about the beam axis if the incident beam is transversally polarized. The first phenomenon is the electron optic analog of *circular dicroism* and the second one is the analog of the rotation of the plane of polarization of light.[13]

2.2. Characterization of Chiral Molecules

In this section we will generalize Farago's expression (3) for the spin-scattering matrix. The main point is the following. Equation (3) does not clearly distinguish between parity-violating effects caused by the interaction itself and those phenomena that are due to the asymmetric structure of the target. Since the former effects (due to PNC interactions) are too small to have a noticeable influence on low-energy electron scattering, they can safely be ignored. Equation (3) should therefore be reformulated and generalized in such a way that parity-violating effects, caused by the interaction, are *explicitly* ruled out and that in particular in-plane polarization components are directly related to certain structural elements of the chiral molecule.

A complete geometrical characterization of asymmetric molecules would clearly lead to expressions for M that would be too complex to reveal the basic physics. At the present stage, however, it might be of interest to concentrate on very few but essential points. We will therefore proceed in the following way.

We will introduce a very simple model of a chiral molecule that only includes the minimal number of structural elements that are absolutely necessary to characterize optically active molecules. This description will be used to construct an expression for the spin scattering matrix M from symmetry principles. The equations obtained will then give some insight into the physical mechanism for the production of in-plane spin polarization in particular.

In the following we will only present an outline of the arguments. A detailed discussion will be published elsewhere.

We will explain the basic idea by considering a molecule with an asymmetric carbon atom and four substituent groups (for example, H, HO, CH_2OH, CHO). Chirality is exhibited in the respective left-handed (anticlockwise) and right-handed (clockwise) sequence of the H, HO, CH_2OH groups attached to the asymmetric carbon atom, when viewed with the CHO-group remote from the observer.

In order to describe formally these structural relations we first define a unit vector **e**, pointing from the CHO-group to the asymmetric carbon atom. We can further define a "sense of rotation" by considering the relative positions of the substituents H, HO, CH_2OH. We associate with the right-handed sequence, say, a positive "sense of rotation" and a negative one with the left-handed sequence. This "sense of rotation" can as usual be represented by an axial vector **a** which is perpendicular to the plane of the groups H, HO, CH_2OH.

e and **a** together form a "molecular screw." Their scalar product **ea** is a rotational invariant (and therefore independent of the orientation of the molecule) and transforms as a pseudoscalar under space reflection. The sign of **ea** clearly distinguishes the two optical antipodes and we will call **ea** the *chirality* of the molecule. **ea** will be positive (negative) for the clockwise (anticlockwise) sequence.

For simplicity, we will assume that **a** is parallel or antiparallel to **e**, that is, we write

$$\mathbf{a} = (\mathbf{e} \cdot \mathbf{a})\mathbf{e} \tag{5}$$

taking **e** and **a** as unit vectors. $(\mathbf{e} \cdot \mathbf{a})$ denotes the scalar product. Equation (5) preserves the important distinction between polar and axial vectors.

In the following we will use **e** and **a** in order to characterize optically active molecules and neglect all other structural elements necessary for a complete description of a given target molecule. On the other hand, **e** *or* **a** alone is clearly not sufficient to distinguish between the two optical isomers.

2.3. Separation of Dynamical and Geometrical Elements

In order to discuss spin effects in elastic scattering processes between electrons and chiral molecules we first derive an expression for the matrix M in the adiabatic representation. As in Section 2.1 we use the combinations $\mathbf{k}_1 + \mathbf{k}_2$, $\mathbf{k}_1 - \mathbf{k}_2$, and $\mathbf{k}_1 \times \mathbf{k}_2$, where we now take $\mathbf{k}_1$ and $\mathbf{k}_2$ as unit vectors, pointing in the directions of initial and final momenta respectively. In addition, we have now the vectors **e** and **a** at our disposal.

By using explicitly that the interaction shall be invariant under rotations and time *and* space reflections and by considering the transformation properties of the vectors $\mathbf{k}_1$, $\mathbf{k}_2$, **e**, and **a** under the various symmetry operations it can be shown that the spin scattering matrix M must satisfy the following two conditions:

1. M must remain invariant under the simultaneous transformations $k_1 \to -k_1$, $k_2 \to -k_2$, $e \to -e$ (which follows from the invariance of the interaction under space reflection);
2. M must remain invariant under the simultaneous transformations $k_1 \to -k_2$, $k_2 \to -k_1$, $\boldsymbol{\sigma} \to -\boldsymbol{\sigma}$ (which follows from the invariance of the interaction under time inversion). This means that M transforms as a *proper scalar operator.* (We will call an expression a proper scalar if it remains unchanged under rotations and space and time reflections.)

Using the conditions (1) and (2) an expression for M can be constructed. Here, we concentrate on forward scattering ($\mathbf{k}_1 = \mathbf{k}_2 = \mathbf{k}$), which is of particular interest.[11] In this case we obtain from (1) and (2):

$$M = g_1 1 + g_2(\mathbf{ea})(\mathbf{k}\boldsymbol{\sigma}) + g_3(\mathbf{pe})(\mathbf{ea})(\mathbf{e}\boldsymbol{\sigma}) + g_4[\mathbf{k} \times \mathbf{e}]\boldsymbol{\sigma} \tag{6}$$

where the g_i are required to be proper scalar functions of $\mathbf{k}$, $\mathbf{e}$, and $\mathbf{a}$ and where the multiplication sign in the last term denotes the vector product of $\mathbf{k}$ and $\mathbf{e}$.

Equation (6) is the most general expression for M within the framework of our model. The dynamics of the scattering process is contained in the functions g_i and the dependence on the molecular structure is explicitly given by the vectors $\mathbf{e}$ and $\mathbf{a}$. This separation between dynamics and geometry is of importance, as will be shown by the following examples.

Since the functions g_i must transform as proper scalars it follows that they can only depend on even powers of the molecular chirality ($\mathbf{ea}$). Hence, the functions g_i have the same values for both optical antipodes. The expressions for M for the two isomers differ therefore only in the sign of the chirality ($\mathbf{ea}$). [It should be noted that terms like $g(\mathbf{k}\boldsymbol{\sigma})$ would only occur in M if the interaction itself would violate parity invariance. Otherwise g would not transform as a proper scalar.]

The second and third term in equation (6) depend on the molecular chirality $\mathbf{ea}$ and vanish therefore if an average is taken over a racemic mixture. The last term in equation (6) is independent of the chirality and is a function of the molecular orientation (axis $\mathbf{e}$) relative to the beam momentum $\mathbf{k}$. This term gives rise to spin-dependent effects, even for nonchiral molecules, because of interactions between the projectile spin and the electric field along the axis $\mathbf{e}$ (for 2-atomic molecules this would correspond to Hund's coupling case, for example). The corresponding component of the polarization vector is perpendicular to the plane formed by **k** and **e**. No longitudinal polarization can be produced within the framework of our simple model of an optically active molecule. This would be different, however, if more structural elements than **e** and **a** would be included. In fact, it is well known that nonenantiomorphous systems can show optical activity for particular orientations.[14]

2.4. Examples and Physical Interpretations

As a first example we will consider forward scattering of electrons from oriented molecules such that **e** is parallel to **k**. If we choose this direction as quantization axis (z axis), equation (6) reduces to

$$M = g_1 1 + g(\mathbf{ea})\sigma_z \tag{7}$$

with $g = g_1 + g_2$. Initial electrons with "spin up" or "spin down" with respect to **k** are transformed into the final states in the following way:

$$\begin{pmatrix}1\\0\end{pmatrix} \to M\begin{pmatrix}1\\0\end{pmatrix} = [g_1 + g(\mathbf{ea})]\begin{pmatrix}1\\0\end{pmatrix} \tag{8a}$$

$$\begin{pmatrix}0\\1\end{pmatrix} \to M\begin{pmatrix}0\\1\end{pmatrix} = [g_1 - g(\mathbf{ea})]\begin{pmatrix}0\\1\end{pmatrix} \tag{8b}$$

As shown by these relations there is no spin-flip for this particular orientation.

The probability of transmission is

$$W(+) = |g_1 + g(\mathbf{ea})|^2 \tag{9a}$$

for initial electrons with spin up and

$$W(-) = |g_1 - g(\mathbf{ea})|^2 \tag{9b}$$

for incident electrons with "spin down."

For the projectile the left-handed or right-handed chirality derives from the respective antiparallel or parallel relation between its momentum vector and its spin vector. Assuming for the sake of definiteness that g_1 and g are positive, we can express our results (8) and (9) in the following way: *Molecules with* **ea** $= +1$ (**ea** $= -1$) *scatter preferentially electrons with right-handed (left-handed) chirality* (that is, the electronic "spin-momentum screw" must be parallel to the molecular screw).

From equation (9) it follows that an unpolarized incident electron beam will be longitudinally polarized after the scattering. This effect vanishes for a racemic mixture.

The physical nature of this effect can be understood by adapting an argument developed by Hrasko and Garay (see Section 3.1). During the scattering the projectile electron moves along the molecular screw along a helical path with opposite sense for L- and D-enantiomers. As a result a circular current appears that produces a magnetic field parallel or anti-parallel to the molecular axis. This magnetic field couples to the spin of the projectile and the spin will preferentially align itself parallel or anti-parallel to the molecular axis.

As a second example we consider forward scattering from oriented molecules with **e** perpendicular to **k**. In this case the third term in equation (6) does not contribute. Choosing the z- and x-axes of our coordinate system parallel to **k** and **e**, respectively we obtain

$$\begin{pmatrix}1\\0\end{pmatrix} \to M\begin{pmatrix}1\\0\end{pmatrix} = [g_1 + g_2(\mathbf{ea})]\begin{pmatrix}1\\0\end{pmatrix} + ig_4\begin{pmatrix}0\\1\end{pmatrix} \tag{10a}$$

$$\begin{pmatrix}0\\1\end{pmatrix} \to M\begin{pmatrix}0\\1\end{pmatrix} = [g_1 - g_2(\mathbf{ea})]\begin{pmatrix}0\\1\end{pmatrix} - ig_4\begin{pmatrix}1\\0\end{pmatrix} \tag{10b}$$

The first term can be discussed similarly to equations (8). The term proportional to g_4 is independent of the molecular chirality and is caused by some type of spin-axis interaction. From the discussion of the preceding example one might expect that g_2 is small for dynamical reasons.

Finally, for unoriented molecules and unpolarized initial electrons the spin polarization of the forward scattered electrons is given by $g(\mathbf{ea})\mathbf{k}$, where g is a dynamical factor.

3. DYNAMICAL ESTIMATES OF EFFECTS CAUSED BY THE MOLECULAR ASYMMETRY

3.1. Dynamical Models of Chiral Molecules

In order to explain various chiral effects a simple model has been suggested.[16] The starting point is the observation that in optically active molecules quantities, which transform as pseudoscalars (more generally pseudotensors), can exist (see Section 2.2). In particular, the helicity $\langle \mathbf{p}\boldsymbol{\sigma} \rangle$ of the bound electrons and the corresponding helicity density are of this type. If $\langle \mathbf{p}\boldsymbol{\sigma} \rangle \neq 0$ then the electron spin is correlated with the momentum, that is, the electron spin is aligned predominantly parallel to the motion in one of the optical isomers and antiparallel in the other one. In order that $\langle \mathbf{p}\boldsymbol{\sigma} \rangle$ differ from zero, some type of spin-orbit interaction, which couples orbital motion and spin, is clearly required.

As has been pointed out by Hrasko[15] the Hamiltonian of an electron, bound in a chiral molecule, can be of the form

$$H = H_0 + w\mathbf{p}\boldsymbol{\sigma}$$
$$= H_0 + iw\boldsymbol{\sigma}\nabla/h \quad (11)$$

where $\boldsymbol{\sigma}$ denotes the Pauli matrices, $\mathbf{p} = i\nabla/h$ is the momentum operator, and w a pseudoscalar, which depends on structural elements typically for optically active molecules. L and D enantiomers are distinguished by $W = \pm 1$.

The term $w\mathbf{p}\boldsymbol{\sigma}$ describes a special type of spin-orbit coupling, namely, the direct coupling of the electron spin to its momentum, which is only possible in chiral systems.

Hrasko[15] has also given a visual model, called the *helical electron gas model* [with $H_0 = p^2/(2m)$]. In an optically active medium helical structures are always present. The bound electrons move on a helical path inducing a magnetic field in the axis of the helix, which tries to turn the spin of the electron in the direction of the axis. Since the helical structures in the two isomers are of opposite sense, the magnetic field and the spin of the electron will be directed differently in the L and D forms.

This model was used to explain qualitatively some experimental results.[3,16]

Hegstrom developed ideas that in their qualitative aspects resemble the helical electron gas model.[16] Taking into account in first order the perturbation of the molecular orbitals by the usual spin-orbit coupling [$\sim(\mathbf{E}\times\mathbf{p})\boldsymbol{\sigma}$, where $\mathbf{E}$ denotes the molecular electric field] Hegstrom showed in particular that the helicity *density* is different from zero and of order $(\alpha Z)^2$ (atomic units), where α is the fine-structure constant and Z the nuclear charge of the heaviest atom in the molecule (it should be noted that the expectation value $\langle\mathbf{p}\boldsymbol{\sigma}\rangle$ is zero in first order in the spin-orbit interaction). This *bound helical model* was applied to several processes of interest, which will be briefly discussed in the following section.

3.2. Application to Elastic Scattering and Ionization

Rich *et al.*[17] developed this bound helical model and applied it to elastic scattering of electrons from chiral molecules in the first Born approximation. Neglecting spin-orbit coupling between the projectile electron and the molecule and taking electron exchange effects into account, the authors derived an estimate for the helicity $H(s)$ induced in an initially unpolarized electron beam:

$$H(s) \sim \eta(\alpha Z)^2 \quad (12)$$

where η is a structure factor, introduced by Rein,[18] which characterizes the influence of the molecular asymmetry.

So far, estimates for η have only been given for twisted ethylene,[4] the simplest chiral molecule. The result suggests

$$\eta \sim 10^{-2}\text{–}10^{-3} \tag{13}$$

Taking this as a typical value, Rich *et al.*[17] concluded that

$$H(s) \sim 10^{-4}\text{–}10^{-5} \tag{14}$$

(with $Z = 6$ for carbon as heaviest atom). This result suggests that target molecules with heavier atoms in the chiral center should be used. Similar results were also obtained by Walker using a different approach.[26]

The bound helical model was also applied by Hegstrom[16] to discuss the difference in the electron ionization rates for L and D isomers (radiolysis). Electrons produced by β decay have unit helicity and energies $\sim$100 keV. For this case the difference in the ionization rates for amino acids (with $Z = 6$) is of the order $\sim 10^{-8}$. As briefly discussed in the Introduction, this result is of relevance to the question of the origin of optical activity in biological systems.

3.3. The Multiple Scattering Approach

It has been pointed out by Kessler[19] that in electron scattering from optically active molecules polarization components parallel in the scattering plane can be produced by intramolecular scattering. The essence of Kessler's argument is as follows.

Assume that the incident electrons are unpolarized. A single scattering from one of the atoms in the molecules can produce a spin polarization which is perpendicular to the instantaneous scattering plane if the usual spin–orbit coupling between the projectile and the atom acts. A consecutive scattering from a second atom can then produce a rotation of the polarization vector whereby it contains a component in the scattering plane.

Formally, this means that in equation (3) the term $h\boldsymbol{\sigma}\mathbf{n}_3$ is neglected and the usual form

$$M = f1 + g\boldsymbol{\sigma}\mathbf{n}_2 \tag{15}$$

is applied to any single scattering event.

Hayashi[20] has developed these ideas and performed numerical calculations for the case of elastic electron scattering from 1,3-dimethylallene. Scattering from the hydrogen atoms was neglected because of their weak scattering powers. For the functions f and g in equation (15) those for free atoms were adapted and the calculated tables of Fink and Yates were used.[21]

Applying some further approximations Hayashi found that the in-plane polarization is an oscillating function of the scattering angle. Its maximum degree was found to be of order 10^{-6}–10^{-8} at electron energies between 50 and 500 eV. These values are about five orders of magnitude smaller than those of the polarization component perpendicular to the scattering plane.

4. EXPERIMENTAL RESULTS

The first experiment to show the influence of molecular chirality on elastic scattering of electrons was performed by Beerlage *et al.*[22] Campher was chosen as target for several reasons: (1) L and D enantiomers are easily available in high purity and large quantities; (2) at room temperatures its vapor pressure is very high; (3) the vapor can easily be pumped using a liquid air trap; and (4) the vapor is chemically inert.

The energy of the initial electrons was 25 eV. The experiment could establish only an upper limit to the in-plane polarization and remained inconclusive.

Gidley *et al.*[23] investigated asymmetrical positronium formation in optically active molecules using slow spin-polarized electrons. They found a positive effect in one case but none in two others.

Recently Campbell and Farago[24] scattered longitudinally polarized electrons of 5 eV and degree of polarization 0.28 from D- and L-camphor. The transmitted intensity for the two electron helicities defines a transmission asymmetry

$$A = \frac{I_+ - I_-}{I_+ + I_-} \tag{16}$$

First results show that an asymmetry exists and that the asymmetries $A(\mathrm{D})$ and $A(\mathrm{L})$ are of the same order of magnitude for the two isomers ($\sim 10^{-3}$) but of opposite sign. Hence, Campbell and Farago conclude that the results confirm the existence of an electro-optic dichroism.

A limit for the value of the quantity

$$X = \frac{\mathrm{Im}\, h}{\mathrm{Im}\, f} \tag{17}$$

could be extracted from the experiments:

$$X > 2 \times 10^{-3} \tag{18}$$

where h and f are defined in equation (3) and "Im" denotes their imaginary part.

So far there are no theoretical estimates of X. From the results of Rich *et al.*[17] follows

$$\mathrm{Re}(h/f) \sim 10^{-5} \tag{19}$$

The results of Hayashi[20] amount to an estimate

$$\mathrm{Re}(h/f) \sim 5 \times 10^{-7} \tag{20}$$

It should be noted, however, that X could be significantly different from $\mathrm{Re}(h/f)$. Furthermore, as Campbell and Farago pointed out, resonance effects can be important at low-energy electron scattering that have not been taken into account in the theoretical estimates.

5. CONCLUSIONS

It has been shown experimentally[25] that the molecular asymmetry manifests itself in the spin polarization of scattered electrons. However, the results and various theoretical estimates show that these effects are very small if the nuclear charge Z of the atoms in the chiral center is small. What is therefore required are model calculations for molecules containing at least one heavy atom and studies of the influence of Z on the spin polarization of the scattered electrons. Such calculations are now in progress.[27]

Furthermore, more detailed studies of the influence of the molecular structure on asymmetry effects are of great interest. Here, model calculations of electron scattering from oriented molecules should be helpful.

REFERENCES

1. F. Vester, T. L. V. Ulbricht, and H. Krauss, *Naturwissenschaften* **46**, 68 (1959).
2. B. Ya. Zel'dovich, D. B. Saakyan, and I. I. Sobelman, *JETP Lett.* **25**, 94 (1977).
3. A. S. Garay and P. Hrasko, *J. Mol. Evolution* **6**, 77 (1975).
4. R. A. Hegstrom, D. W. Rein, and P. G. H. Sandars, *J. Chem. Phys.* **73**, 2329 (1980).
5. S. F. Mason and G. Tranter, *Chem. Phys. Lett.* **94**, 34 (1983).
6. S. F. Mason and G. Tranter, *Mol. Phys.* **53**, 1091 (1984).
7. S. F. Mason, *Nature* **311**, 19 (1984).
8. B. Ritchie, *Phys. Rev. A* **13**, 1411 (1976).
9. N. A. Cherepkov, *J. Phys. B* **16**, 1543 (1983).
10. G. Schönhense, private communication.
11. P. S. Farago, *J. Phys. B* **13**, L567 (1980).

12. J. KESSLER, *Polarised Electrons* (Springer-Verlag, Berlin, 1985).
13. P. S. FARAGO, *J. Phys. B* **14**, L743 (1981).
14. M. V. HOBDEN, *Nature* **216**, 678 (1967).
15. P. HRASKO, Reports of the central research institute of Physics, Budapest, KFKI 73-40.
16. R. A. HEGSTROM, *Nature* **297**, 643 (1982).
17. A. RICH, J. VAN HOUSE, AND R. A. HEGSTROM, *Phys. Rev. Lett.* **48**, 1341 (1983).
18. D. W. REIN, *J. Mol. Evolution* **4**, 15 (1974).
19. J. KESSLER, *J. Phys. B* **15**, L101 (1982).
20. S. HAYASHI, *J. Phys. C* **19**, 2601 (1986).
21. M. FINK AND A. C. YATES, *At. Data* **1**, 385 (1970).
22. M. J. M. BEERLAGE, P. S. FARAGO, AND M. J. VAN DER WIEL, *J. Phys. B* **14**, 3245 (1981).
23. D. W. GIDLEY, A. RICH, J. VAN HOUSE, AND P. W. ZITZEWITZ, *Nature* **297**, 639 (1982).
24. D. M. CAMPBELL AND P. S. FARAGO, *Nature* **318**, 52 (1985).
25. D. W. WALKER, *J. Phys. B* **15**, L289 (1982).
26. D. THOMPSON, private communication.

CHAPTER 4

Applications of Lasers in Atomic Collision Physics

W. R. MacGillivray and M. C. Standage

1. INTRODUCTION

The introduction of lasers to the field of atomic collisions has provided a powerful new tool with which to investigate collision processes. Laser techniques have now been applied to the study of superelastic scattering and stepwise excitation processes, free–free transitions, photon recoil processes, Rydberg atom collisions, and spin-polarization effects. This chapter is restricted to consideration of two of these applications, namely, stepwise excitation and superelastic scattering studies of electron atom scattering.

Stepwise excitation techniques have been developed in which a combination of electron and laser beams are used to excite target atoms. Measurements of the intensity and polarization of fluorescence emitted from stepwise excited atoms have provided new techniques for studying the electron impact excitation of both optically allowed and forbidden atomic transitions. Such experiments may be categorized according to whether the laser excitation occurs in the first or second excitation steps. In the former case, information may be obtained about inelastic scattering processes from excited atomic states, while the latter case provides new methods of investigating inelastic scattering processes from the ground states. Coincidence techniques may also be applied to the investigation of such scattering processes using detection of stepwise excited photons in delayed coincidence with electrons inelastically scattered from target atoms. The narrow bandwidth of laser radiation enables the fine and hyperfine structure of many atomic transitions to be resolved in the laser excited step, thereby allowing the role of such structure in atomic collision processes to be studied.

W. R. MacGillivray and M. C. Standage • School of Science, Griffith University, Nathan, Queensland, Australia 4111.

The preceding discussion indicates very briefly the diversity of applications of lasers to atomic collision studies. In many respects, their use in such applications is still at an early stage of development, and it can be expected that as the advantages that laser techniques can bring to atomic collision studies become more fully realized, and as new laser technologies are developed, the significance and range of application of lasers in this field will steadily grow.

In this chapter, we attempt to highlight novel features of various laser techniques that either extend existing conventional techniques or provide new techniques capable of obtaining previously inaccessible information. A review of the theory associated with superelastic and stepwise excitation techniques is presented in Section 2. Case studies are discussed to illustrate with concrete examples theoretical aspects of the techniques. Section 3 contains a review of experimental work in stepwise excitation and superelastic scattering.

2. THEORY

The theory of both stepwise and superelastic experiments may be described by closely related treatments. Figures 1a and 1b show two stepwise excitation schemes I and II, while Figure 1c shows a superelastic scattering Scheme III.

In Scheme I, electron excitation from the ground state $|g\rangle$ to a first excited state $|e\rangle$ is followed by laser excitation to a higher excited state $|i\rangle$. The intensity and polarization of light emitted as the atom decays from state $|i\rangle$ to $|f\rangle$ provides information about the preceding stepwise excitation processes. Scheme II reverses the order of the excitation with laser excitation from $|g\rangle$ to $|e\rangle$ followed by electron excitation between the excited states $|e\rangle$ and $|i\rangle$. As for Scheme I, the fluorescence emission provides information

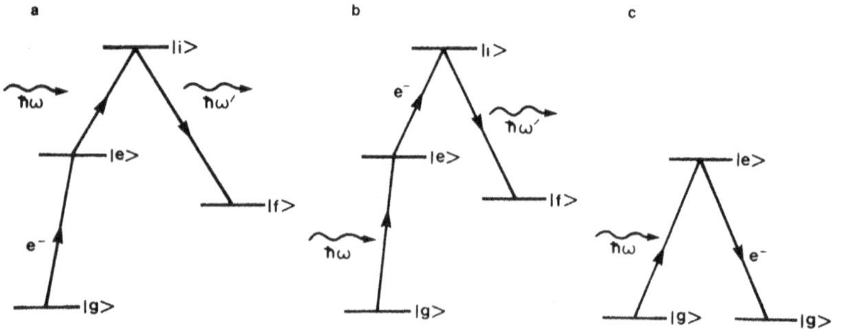

FIGURE 1. (a) Electron-laser stepwise excitation scheme I. (b) Electron-laser stepwise excitation scheme II. (c) Superelastic scattering excitation scheme III.

on the excitation processes. Scheme III involves laser excitation from $|g\rangle$ to $|e\rangle$ followed by electron deexcitation back to $|g\rangle$. Schemes I and III enable the electron scattering parameters for ground state to excited state transitions to be determined, while Scheme II enables inelastic processes to be investigated between electrons and laser excited atoms.

The electron scattering processes in all cases considered here may be represented by a scattering operator T such that the initial density operator ρ_i is related to the final density operator ρ by[1]

$$\rho = T\rho_i T^+ \tag{1}$$

For Type I schemes in which, prior to the electron collision, the target atom is in the ground state, the initial density operator ρ_i can be represented as

$$\rho_i = \frac{1}{2(2J_g + 1)} \sum_{J_g m_g m_i} |J_g m_g\rangle\langle J_g m_g| |m_i\rangle\langle m_i| \tag{2}$$

where it has been assumed that the incident electrons and ground state atoms are uncorrelated. $|J_g m_g\rangle$ is a substate of the ground state of the target atom and $|m_i\rangle$ is the incident electron spin state. If electron spin is neglected (see Hanne 1983[2] for a review of electron spin effects), the density matrix elements formed after the collision has taken place are given by

$$\begin{aligned} \rho^e_{mn} &= \frac{1}{(2J_g + 1)} \sum_{J_g m_g} \langle Jm|T|J_g m_g\rangle\langle J_g m_g|T^+|Jn\rangle \\ &= \langle f_m f_n^* \rangle \end{aligned} \tag{3}$$

where $|Jm\rangle$ is a substate of the excited state $|e\rangle$. Various authors (see Ref. 1) have discussed the symmetry properties of the density matrix elements ρ^e_{mn}. Briefly, the Hermitian character of ρ^e implies that

$$\rho^e_{mn} = \rho^{e*}_{nm} \tag{4}$$

Symmetry of the electron scattering process with respect to reversal of the direction of the incident electron beam, when electron spin is neglected, implies

$$\rho^e_{mm} = \rho^e_{-m-m} \tag{5}$$

For axial symmetry electron scattering processes, which apply for example when the scattered electrons are not detected

$$\rho^e_{mn} = 0 \qquad \text{if } m \neq n \tag{6}$$

In the case of planar symmetry scattering processes applicable to electron–photon coincidence and superelastic scattering experiments the off-diagonal elements are in general nonzero and are related by

$$\rho^e_{mn} = (-1)^{m+n} \rho^e_{-m-n} \tag{7}$$

The laser excitation processes in all three schemes may be divided into two cases; weak and strong optical excitation. For the weak excitation case, it is assumed that only one laser photon interacts with an atom during the optical excitation process. Consideration of the laser excitation step for a Type I excitation scheme shows that the electron excited density matrix elements ρ^e_{mn} are related to the upper state density matrix elements $\rho^i_{m'n'}$ by[3]

$$\rho^i_{m'n'} = \frac{1}{\Gamma_i \Gamma_e} \sum_{mn} \langle m'|\mathbf{e} \cdot \mathbf{P}|m\rangle\langle n|\mathbf{e}^* \cdot \mathbf{P}|n'\rangle \rho^e_{mn} \tag{8}$$

where $\mathbf{e}$ is the polarization vector of the laser light and Γ_i and Γ_e are decay rates for states $|i\rangle$ and $|e\rangle$, respectively, which are assumed to have degenerate substates.

For strong optical excitation, a number of laser photons may interact with an atom during an excited state lifetime. Optical density matrix equations are required to describe the optical processes which take place. Coherences are formed between the lower and upper states of the optically driven transition, and state coherences also arise between degenerate and nearly degenerate substates of either the lower or upper state. The magnitudes of the populations and coherences represented, respectively, by diagonal and off-diagonal density matrix elements are determined by a number of factors such as the laser intensity and tuning relative to the resonance frequency (or frequencies of the transition if the levels are non-degenerate), the laser polarization, the spectroscopic structure of the transition and the atomic decay rates. Space does not permit a detailed account of optical density matrix theory to be given here (see MacGillivray and Standage 1987[4] and references therein for a review of optical density matrix applications to stepwise excitation techniques); however, the formal relationship between upper, $\rho^u_{m'n'}$ and lower, ρ^L_{mn} state density matrix elements of the optically driven transition may be represented by the following equation for all three excitation schemes:

$$\rho^u_{m'n'} = \sum_{mn} \Omega_{m'n'mn} \rho^L_{mn} \tag{9}$$

where the matrix elements $\rho^{u(L)}_{m'n'}$ may represent either populations or state coherences. The terms $\Omega_{m'n'mn}$ represent either decay constants or Rabi frequencies associated with the density matrix elements.

For Type I and II schemes, the intensity and polarization of photons is detected for spontaneous emission decays from state $|i\rangle$ to $|f\rangle$. The fluorescent intensity is found by tracing over the density operator and an emission operator F which represents the spontaneous emission process from state $|i\rangle$ to $|f\rangle$. F has matrix elements given by

$$F_{n'm'} = \sum_q \langle n'|\mathbf{f} \cdot \mathbf{P}|q\rangle\langle q|\mathbf{f}^* \cdot \mathbf{P}|m'\rangle \tag{10}$$

$\mathbf{f}$ is the polarization vector of the analyzer used to measure the polarization of the fluorescent light, $\mathbf{P}$ is the electric dipole moment operator, and $|q\rangle$ are substates of the final state $|f\rangle$. The fluorescent intensity is given by

$$I = \mathrm{Tr}(\rho F) \tag{11}$$

where the trace runs over substates of the upper excited state. Expansion of equation (11) gives

$$I = \sum_{m'n'} \rho_{m'n'} F_{n'm'} \tag{12}$$

For Type I schemes a combination of equations (9) and (12) gives

$$I = \sum_{m'n'mn} F_{n'm'} \Omega_{m'n'mn} \rho^e_{mn} \tag{13}$$

where I is the fluorescent intensity detected from the $|i\rangle \rightarrow |f\rangle$ transition and ρ^e_{mn} are density matrix elements of the electron impact excited state $|e\rangle$.

For Type II schemes, the first excited state is prepared by laser excitation from the ground state. The density operator for this process may be represented by

$$O^e = \sum_{mn} O^e_{mn}|m\rangle\langle n| \tag{14}$$

where O^e_{mn} are density matrix elements resulting from the laser excitation process. The density matrix elements $\rho^i_{m'n'}$ of the upper excited states $|i\rangle$ are formed by the electron impact process. A combination of equations (1) and (14) gives

$$\begin{aligned}\rho^i_{m'n'} &= \langle m'|TO^eT^+|n'\rangle \\ &= \sum_{mn} O^e_{mn}\langle m'|T|m\rangle\langle n|T^+|n'\rangle \\ &= \sum_{mn} O^e_{mn} f_{m'm} f^*_{n'n}\end{aligned} \tag{15}$$

where $f_{m'n}$ are amplitudes representing electron excitation between substates of the excited states $|e\rangle$ and $|i\rangle$. The fluorescent intensity emitted by such stepwise excited atoms is given by

$$I = \sum_{m'n'mn} F_{n'm'} O^{e}_{mn} f_{m'm} f^{*}_{n'n} \tag{16}$$

For Type III schemes, following optical pumping of the target atoms from the ground state, deexcitation of the atom takes place by electron impact in a process that is the time-reversed process for electron excitation. In this case the differential scattering cross section σ is measured of superelastically scattered electrons that have gained energy equal to the transition energy between states $|g\rangle$ and $|e\rangle$. The theoretical form of the signal is given by

$$\begin{aligned} \sigma &= \mathrm{Tr}(O^{e}\rho^{d}) \\ &= \sum_{mn} O^{e}_{mn}\rho^{d}_{nm} \end{aligned} \tag{17}$$

where the matrix elements O^{e}_{mn} are given by equation (14) and ρ^{d} is an operator representing the deexcitation process. The matrix elements of ρ^{d} are complex conjugates of matrix elements of the time-reversed operator ρ^{e} which are given by equation (3). Equation (17) may be written as

$$\sigma = \sum_{mn} O^{e}_{mn}\rho^{e*}_{nm} \tag{18}$$

If the electron impact excitation process has axial symmetry, such as occurs when the incident electron beam has no net spin polarization or when no scattering plane is defined, then for both Type I and Type II schemes only diagonal terms are retained in equations (13) and (15), which correspond to partial total cross sections. If a scattering plane is defined by detection of the inelastically scattered electrons at an angle to the incident electron beam, then in Type I and II schemes, detection of these electrons in coincidence with photons emitted from the stepwise excited atoms will yield both diagonal and off-diagonal terms which give amplitude and phase information about the electron excitation processes. Type III superelastic scattering schemes allow the same type of information to be obtained as from coincidence experiments carried out on the electron impact excitation of the transition $|g\rangle \rightarrow |e\rangle$.

The use of laser techniques in the case of Type I and Type II excitation schemes enables the precise isolation of transitions in the target atoms which can allow the role of hyperfine and fine structure in the collision processes to be studied, or where desired, often eliminated by suitable selection of transitions. Type I and II schemes also enable the electron impact excitation of metastable states or optically forbidden transitions to be studied via optical means. The use of laser techniques often enables more information to be obtained than is possible using conventional techniques. An example of this is the use of a Type I stepwise excitation scheme to obtain partial total cross sections for the magnetic sublevels of the $6\,^3P_2$ metastable state of mercury.[5] Where stepwise techniques yield the same information as conventional techniques it is often the case that this information can be obtained with fewer measurements using stepwise excitation.

The expressions so far derived for the theoretical form of the signals do not as yet take into account specific details, such as the energy level structure of the target atom or the excitation and emission geometry for particular experimental configurations. Several typical cases are considered here for a Type I stepwise excitation scheme and a Type III superelastic scattering scheme.

2.1. *Case 1*

A commonly used excitation-emission geometry is shown in Figure 2 together with a typical Type I energy level diagram. Atoms in a $J = 0$ ground state are first excited by electron impact to a $J = 1$ excited state and then by laser excitation to a higher $J = 2$ state. Fluorescence is detected from the stepwise excited atoms as they spontaneously decay via a transition back to a $J = 1$ state.

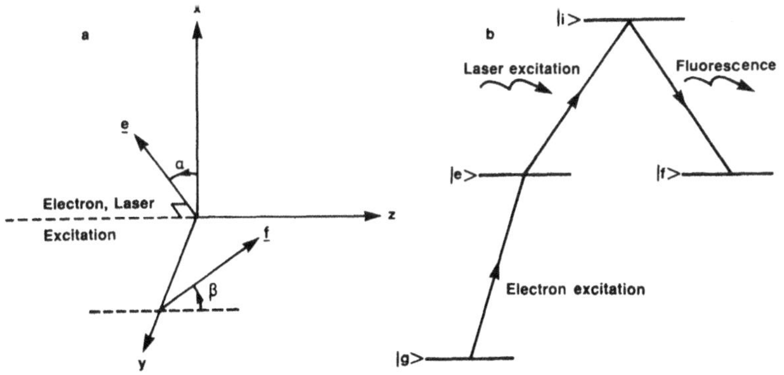

FIGURE 2. (a) Excitation-emission geometry. Collinear electron and laser beams incident along the z axis with photon detection along the y axis. (b) Type I stepwise excitation scheme.

The excitation geometry consists of collinear electron and laser beams propagating along the z axis which is chosen as the collision frame quantization axis. The laser beam may be linearly polarized at angle α or circularly polarized. Detection of photons is along the y axis through a linear or circular analyzer. For planar scattering symmetry calculations the x-z plane is taken as the plane of symmetry.

To apply the theoretical expression for the stepwise excited fluorescent intensity given by equation (12) to this case, it is usual to start by choosing a coordinate system in which the emission operator F is diagonal. For a linear polarizer, the quantization axis of the emission frame is chosen parallel to the polarization vector $\mathbf{f}$, whereas for a circular polarizer the quantization axis is parallel to the direction of propagation of the fluorescence,

For linear analysis of the fluorescence, equation (12) reduces to

$$I = \sum_{m'} F_{m'm'}\rho_{m'm'} \tag{19}$$

where $F_{m'm'}$ are diagonal elements of the emission operator. Evaluation of $F_{m'm'}$ for this case gives the following expression for equation (19):

$$I \propto 3\rho_{1'1'} + 4\rho_{0'0'} + 3\rho_{\bar{1}'\bar{1}'} \tag{20}$$

This expression has been determined in the emission frame and must be related back to the excitation frame to obtain theoretical expressions that can be compared with experimental measurements. To do this, the density matrix elements are expanded in terms of state multipoles which may be rotated into the excitation frame by a rotation operator. Once expressed in the excitation frame those state multipoles are then related back to the density matrix elements. Following the notation of Blum,[1] state multipoles $\langle T^{+}_{kQ}\rangle'$ are related to the density matrix elements $\rho_{m'n'}$ by the expression

$$\langle T^{+}_{kQ}\rangle' = \sum_{m'n'} (-1)^{-J'+n'}\langle J'J'm' - n'|kQ\rangle\rho_{m'n'} \tag{21}$$

where $0 \leqslant k \leqslant 2J'$, $-k \leqslant Q \leqslant k$ and $\langle J'J'm' - n'|kQ\rangle$ is a Clebsch-Gordan coefficient. The inverse relationship is

$$\rho_{m'n'} = \sum_{kQ} \langle T^{+}_{kQ}\rangle'(-1)^{J'-n'}\langle J'J'm' - n'|kQ\rangle \tag{22}$$

The relationship between the state multipole $\langle T^{+}_{kQ}\rangle'$ and state multipoles $\langle T^{+}_{kq}\rangle$ in the excitation frame is given by

$$\langle T^{+}_{kQ}\rangle' = \sum_{q} \langle T^{+}_{kq}\rangle D^{k}_{Qq}(\alpha, \beta, \gamma) \tag{23}$$

where $D^k_{Qq}(\alpha, \beta, \gamma)$ are matrix elements of the rotation operator and the Euler angles (α, β, γ) rotate the emission reference frame into the excitation reference frame.

The expression for the stepwise excited fluorescent intensity may be expressed in terms of state multipoles using equation (22), to give

$$I \propto \sqrt{20}\langle T^+_{00}\rangle' - \sqrt{14}\langle T^+_{20}\rangle' \tag{24}$$

The scalar multipole moment $\langle T^+_{00}\rangle'$ is invariant to rotation. Rotation of the quadrupole multipole moment $\langle T^+_{20}\rangle'$ may be performed using equation (23). The appropriate Euler angles for this rotation in the case of linear analysis of the stepwise excited fluorescence are $(0, -\beta, 0)$. Equation (24) becomes

$$\begin{aligned} I \propto \sqrt{20}\langle T^+_{00}\rangle - \sqrt{7/2}(3\cos^2\beta - 1)\langle T^+_{20}\rangle \\ - \sqrt{21/4}\sin^2\beta(\langle T^+_{22}\rangle + \langle T^+_{2-2}\rangle) \\ + \sqrt{21}\sin\beta\cos\beta(\langle T^+_{21}\rangle - \langle T^+_{2-1}\rangle) \end{aligned} \tag{25}$$

The state multipoles of equation (25) may now be expressed in terms of the density matrix elements $\rho_{m'n'}$ of the upper excited state $|i\rangle$ using equation (21). To express the fluorescent intensity in terms of the electron excited state, the upper density matrix elements must be related to the electron excited state density matrix elements. It is assumed here that the laser excitation process, which connects state $|e\rangle$ and $|i\rangle$, can be regarded as a weak optical process. Equation (8) may then be used to evaluate the density matrix elements of state $|i\rangle$ in terms of the electron excited density matrix elements of state $|e\rangle$. For laser radiation linearly polarized at an angle α, equation (25) becomes

$$\begin{aligned} I \propto (15\sin^2\beta + 4)\rho_{11} + \tfrac{9}{2}(\cos^2\beta + 1)\rho_{00} - 6\sin^2\beta\rho_{1-1} \\ - \left(\frac{15}{\sqrt{2}}\right)\cos\beta\sin\beta(\rho_{10} + \rho_{01}) \\ + \cos 2\alpha\Bigg[6\sin^2\beta\rho_{11} + \tfrac{9}{2}\sin^2\beta\,\rho_{00} + \left(\frac{3}{\sqrt{2}}\right)\cos\beta\sin\beta\,(\rho_{10} + \rho_{01}) \\ -(3\cos^2\beta + 1)\rho_{1-1}\Bigg] \end{aligned} \tag{26}$$

where the symmetry relations equations (4)-(7) have been used to simplify the expression.

For circular analysis of the fluorescence, a quantization axis along the y axis is chosen for the emission frame. In this case equation (12) becomes, on evaluation of the diagonal emission operator matrix elements and expansion of the upper excited density matrix elements into state multipoles,

$$I_{\substack{\text{LHC}\\ \text{RHC}}} \propto \sqrt{20}\langle T_{00}^{+}\rangle' + \left(\frac{7}{\sqrt{14}}\right)\langle T_{20}^{+}\rangle' \mp \left(\frac{15}{\sqrt{10}}\right)\langle T_{10}^{+}\rangle' \tag{27}$$

The appropriate Euler angles are $(\pi/2, -\pi/2, -\pi/2)$ for this case. Following the same approach as used for the linear analysis case gives, for linearly polarized laser excitation

$$\begin{aligned} I_{\substack{\text{LHC}\\ \text{RHC}}} \propto {} & \tfrac{23}{2}\rho_{11} + \tfrac{27}{4}\rho_{00} - 3\rho_{1-1} \mp \frac{27i}{2\sqrt{2}}(\rho_{01} - \rho_{10}) \\ & + \cos 2\alpha\left[\tfrac{9}{4}\rho_{00} + 3\rho_{11} \mp \frac{9i}{2\sqrt{2}}(\rho_{01} - \rho_{10}) - \tfrac{5}{2}\rho_{1-1}\right] \end{aligned} \tag{28}$$

The electron density matrix elements may be parametrized in terms of four parameters[6] such that

$$\begin{aligned} \lambda &= \rho_{00}/\sigma, \qquad \cos\chi = \text{Re}\,\rho_{10}/(\rho_{00}\rho_{11})^{1/2} \\ \sin\phi &= \text{Im}\,\rho_{10}/(\rho_{00}\rho_{11})^{1/2}, \qquad \cos\delta = \rho_{1-1}/\rho_{11} \end{aligned} \tag{29}$$

Other parametrizations have also been recently introduced by Hermann and Hertel.[7]

The polarization state of the fluorescence may be fully characterized by elements of the photon density matrix, or equivalently, the Stokes parameters. The notation of Standage and Kleinpoppen[8] is followed here and four Stokes parameters I, P_1, P_2, P_3 are defined as follows. I is the total intensity and the three parameters associated with linear and circular analysis of the stepwise excited fluorescence are

$$\begin{aligned} P_1 &= (I_0 - I_{90})/I \\ P_2 &= (I_{45} - I_{135})/I \\ P_3 &= (I_{\text{RHC}} - I_{\text{LHC}})/I \end{aligned} \tag{30}$$

where $I = I_0 + I_{90} = I_{45} + I_{135} = I_{RHC} + I_{LHC}$. From equations (26), (28), and (29), the following expressions for the Stokes parameters are obtained:

$$
\begin{aligned}
IP_1 &= 24\lambda - 15 - 3(1-\lambda)\cos\delta\cos 2\alpha + 6(1-\lambda)\cos\delta - 3(2+\lambda)\cos 2\alpha \\
IP_2 &= -6(5-\cos 2\alpha)[\lambda(1-\lambda)]^{1/2}\cos\chi \\
IP_3 &= 18(3+\cos 2\alpha)[\lambda(1-\lambda)]^{1/2}\sin\phi
\end{aligned}
\tag{31}
$$

$$I = 4\lambda + 23 - 5(1-\lambda)\cos\delta\cos 2\alpha - 6(1-\lambda)\cos\delta + 3(2+\lambda)\cos 2\alpha$$

The Stokes parameters P_1 and P_2 depend upon the three parameters λ, $\cos\delta$, and $\cos\chi$, which may be determined by measurements of P_1 and P_2 for different laser polarizations. The parameter $\sin\phi$ may be obtained by measurement of P_3. It should be noted that for conventional coincidence experiments to yield the same information would require photon polarization measurements to be made at two different photon detector positions.[9]

If coincidence techniques are not used, axial symmetry applies and off-diagonal density matrix elements are zero. For linearly polarized laser radiation and linear analysis of the fluorescence, equation (26) reduces to

$$I \propto \tfrac{9}{2}\rho_{00} + 19\rho_{11} + (\tfrac{9}{2}\rho_{00} - 15\rho_{11})\cos^2\beta + (\tfrac{9}{2}\rho_{00} + 6\rho_{11})\cos 2\alpha \sin^2\beta \tag{32}$$

where ρ_{00} and ρ_{11} are now treated as partial total cross sections for the $m = 0$ and $m = \pm 1$ substates. The line polarization for the stepwise excited fluorescence is given by

$$
\begin{aligned}
P &= (I_0 - I_{90})/(I_0 + I_{90}) \\
P &= \frac{9(1-\cos 2\alpha)\rho_{00}/\rho_{11} - 6(5 + 2\cos 2\alpha)}{9(3+\cos 2\alpha)\rho_{00}/\rho_{11} + 2(23 + 6\cos 2\alpha)}
\end{aligned}
\tag{33}
$$

Measurements of the polarization P for the stepwise fluorescence emitted in transition $|i\rangle - |f\rangle$ allows the ratio of the partial total cross section ρ_{00}/ρ_{11} to be determined for the electron impact excited $|g\rangle - |e\rangle$ transition.

2.2. Case 2

An important application of stepwise excitation techniques is to the study of the electron impact excitation of metastable states since it offers an entirely new means of investigating such processes. Here, referring once more to Figure 2a, it is assumed that the same geometry applies, but that the state $|e\rangle$ is a $J = 2$ state and transitions to the $J = 0$ ground state are, as a consequence, optically forbidden. States $|i\rangle$ and $|f\rangle$ are, respectively, $J = 1$ and $J = 0$ states. As in Case 1, it is assumed that the laser light is linearly polarized and that the stepwise excited fluorescence is subject to linear polarization analysis. Only axial symmetry will be considered for the electron excitation step. For a $J = 1$ upper state, the expansion of equation (12) gives

$$I \propto \frac{1}{\sqrt{3}}\langle T^{+}_{00}\rangle' - \frac{1}{3}\sqrt{2}\langle T^{+}_{20}\rangle' \tag{34}$$

Rotation to the excitation frame gives

$$\begin{aligned} I \propto\ & \frac{1}{\sqrt{3}}\langle T^{+}_{00}\rangle - \frac{1}{\sqrt{6}}(3\cos^2\beta - 1)\langle T^{+}_{20}\rangle \\ & + \sqrt{\tfrac{3}{8}}\sin^2\beta[\langle T^{+}_{22}\rangle + \langle T^{+}_{2-2}\rangle] \\ & - \sqrt{\tfrac{3}{8}}\sin\beta\cos\beta[\langle T^{+}_{21}\rangle - \langle T^{+}_{2-1}\rangle] \end{aligned} \tag{35}$$

Evaluation of equation (35) in terms of the electron density matrix elements of the $J = 2$ state $|e\rangle$ using equation (8), where Γ_e is now effectively the transit time of the atom through the laser beam, gives

$$I \propto 3\sin^2\beta\,\rho_{22} + 3\cos^2\beta\,\rho_{11} + \tfrac{1}{2}\sin^2\beta(1 + \cos 2\alpha)\rho_{00} \tag{36}$$

where all off-diagonal terms have been suppressed and ρ_{00}, ρ_{11}, ρ_{22} are partial total cross sections for the magnetic sublevels of the $J = 2$ state. The line polarization for the stepwise excited fluorescence emitted on the $|i\rangle$-$|f\rangle$ transition is given by

$$P = \frac{6\rho_{11}/\rho_{00} - 6\rho_{22}/\rho_{00} - (1 + \cos 2\alpha)}{6\rho_{11}/\rho_{00} + 6\rho_{22}/\rho_{00} + (1 + \cos 2\alpha)} \tag{37}$$

Measurements of the line polarization for two values of the angle α (e.g., $\alpha = 0°, 90°$) enables the ratios of the partial total cross sections ρ_{11}/ρ_{00} and ρ_{22}/ρ_{00} to be determined. The total cross section may then be evaluated with the aid of equation (36). These results demonstrate the advantages provided by stepwise excitation techniques in such cases. The two-step photon process, laser excitation and fluorescence emission, enables the individual partial total cross sections to be obtained for a $J = 2$ state in a situation where other techniques for investigating metastable excitation by electron impact yield only total cross sections.

2.3. *Case 3*

In this example, a superelastic scattering process is studied for electrons scattered from excited sodium atoms which are optically pumped by a single-mode laser tuned to the $3\,^2S_{1/2}$–$3\,^2P_{3/2}$ (589 nm) D_2 transition. This problem has been considered in detail by Hertel and co-workers,[6] although here we utilize results obtained from a full quantum electrodynamical calculation of the D_2 transition, which is based on the theory of Ackerhalt and Eberly[10] and Whitley and Stroud,[11] to investigate the behavior of the theoretical superelastic scattering signal as a function of laser intensity.

The geometry chosen in this case (see Figure 3a) is one that has been frequently used in superelastic scattering experiments. The laser light, either linearly polarized, at an angle β to the z axis, or circularly polarized, propagates along the y axis. The superelastically scattered electrons travel along the z axis. The laser is assumed to be tuned so that the $3\,^2S_{1/2}(F = 2)$–$3\,^2P_{3/2}(F' = 3, 2, 1)$ transitions are excited.

The expression for the superelastic differential scattering cross section is given by equation (18). The calculation proceeds by choosing an optical excitation frame in which only block diagonal terms of the optical density operator O^e are nonzero. The choice of quantization axis for this frame is the same as for the emission frame in the previous two cases considered above. Power broadening and optical pumping effects are clearly of great importance in this case. The close spacing of the hyperfine structure states of the $3\,^2P_{3/2}$ level (see Figure 3b) means that even at quite modest laser intensities (10–50 mW mm^{-2}) power broadening effects are significant.

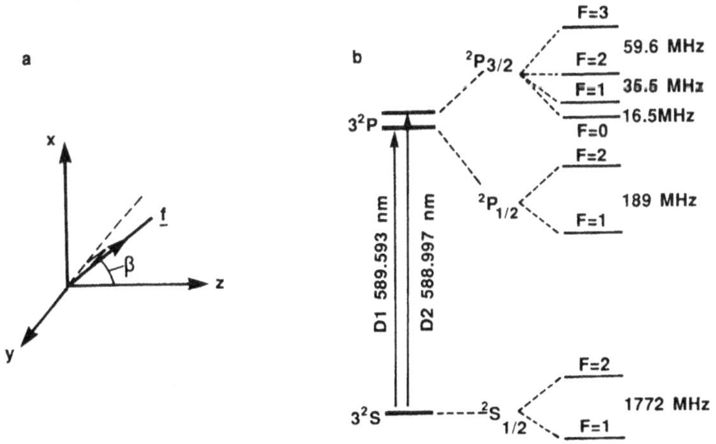

FIGURE 3. (a) Superelastic scattering geometry. (b) Sodium-D transition energy level diagram.

Coherences are formed between hyperfine substates excited by the laser and have to be included in the calculation along with the excited state populations. As a result, block diagonal density matrix elements of O^e include coherences between excited hyperfine substates of different F, but same M_F quantum number.

For a light atom, such as sodium, it is expected that LS coupling holds. In the absence of electron-impact induced electron spin-flips (see Refs. 7 and 12 for some experimental confirmation that the spin-flip cross section for Na is small for scattering angles less than 15°), the density matrix elements for the electron deexcitation process may be expressed in terms of the orbital angular momentum quantum numbers L and M_L such that

$$\langle FM_F|\rho^d|F'M_{F'}\rangle = \sum_{\substack{M_JM_{J'}\\M_I}} \langle IJM_IM_J|FM_F\rangle\langle IJM_IM_{J'}|F'M_{F'}\rangle\langle JM_J|\rho^d|JM_{J'}\rangle$$

where (38)

$$\langle JM_J|\rho^d|JM_{J'}\rangle = \sum_{\substack{M_LM_{L'}\\M_S}} \langle SLM_SM_L|JM_J\rangle\langle SLM_SM_{L'}|JM_{J'}\rangle\langle LM_L|\rho^d|LM_{L'}\rangle$$

Evaluation of equation (18) is performed using an approach that is similar to the stepwise excitation cases already considered in this chapter. Calculation of the matrix elements of the laser excitation operator O^e is carried out using a full quantum electrodynamical approach.[13] Previous treatments of the optical pumping processes in sodium have used either a rate equation approach[6] or a semiclassical density matrix calculation[14] that included optical coherences between the ground and excited states, but neglected the state coherences between the excited hyperfine substates.

We consider first the case where optical transitions occur only between the $3^2S_{1/2}(F=2)$–$3\,^2P_{3/2}(F'=3)$ states. For linearly polarized laser excitation, the quantization axis of the optical excitation frame is chosen parallel to the polarization vector of the laser light. The expression for the superelastic signal in this frame is given by

$$\sigma = \sum_{M_F} f_{M_FM_F}\rho^{e*}_{M_FM_F} \tag{39}$$

where $f_{M_FM_F}$ are optically excited populations of the $F'=3$ excited state. With the aid of equation (38) the electron density matrix elements $\rho^e_{M_FM_F}$ may be rewritten in terms of density matrix elements for $L=1$ to give

$$\sigma = (\tfrac{2}{3}f_2 + \tfrac{7}{15}f_1 + \tfrac{1}{5}f_0)(\rho^*_{11} + \rho^*_{1-1}) + (\tfrac{2}{3}f_2 + \tfrac{16}{15}f_1 + \tfrac{3}{5}f_0)\rho^*_{00} \tag{40}$$

where, for convenience $f_{M_FM_F} \equiv f_{M_F}$.

The density matrix elements ρ_{11}^*, ρ_{00}^*, ρ_{1-1}^* have to be rotated into the emission frame, which is performed using multipoles in exactly the same manner as for the stepwise excitation cases previously considered. For linearly polarized laser excitation the appropriate Euler angles are $(0, -\beta, 0)$. The resulting expression is given by

$$\sigma = 2\rho_{11}(\alpha \cos^2\beta + \gamma \sin^2\beta) + \rho_{00}(\alpha \sin^2\beta + \gamma \cos^2\beta)$$
$$-(\gamma - \alpha)2\sqrt{2}\cos\beta \sin\beta \, \mathrm{Re}\, \rho_{10} \tag{41}$$

where $\alpha = \frac{2}{3}f_2 + \frac{7}{15}f_1 + \frac{1}{5}f_0$ and $\gamma = \frac{2}{3}f_2 + \frac{16}{15}f_1 + \frac{3}{5}f_0$. The symmetry relationships between the electron density matrix elements, given by equations (4)-(7), have also been used in obtaining equation (41).

Measurement of the superelastic scattering differential cross section as a function of laser polarization angle at $\beta = 0°, 90°$ and $\beta = 45°, 135°$ gives the following quantities, which are analogous to the Stokes parameters P_1 and P_2 of equation (30):

$$\frac{\sigma(\beta = 0°) - \sigma(\beta = 90°)}{\sigma(\beta = 0°) + \sigma(\beta = 90°)} = K(2\lambda - 1)$$
$$\frac{\sigma(\beta = 45°) - \sigma(\beta = 135°)}{\sigma(\beta = 45°) + \sigma(\beta = 135°)} = -2K[\lambda(1-\lambda)]^{1/2}\cos\chi \tag{42}$$

where $K = (\gamma - \alpha)/(\gamma + \alpha)$ and λ and $\cos\chi$ are given by equation (29).

If circularly polarized laser radiation is used the result obtained is

$$\frac{\sigma(\mathrm{RHC}) - \sigma(\mathrm{LHC})}{\sigma(\mathrm{RHC}) + \sigma(\mathrm{LHC})} = +2K'[\lambda(1-\lambda)]^{1/2}\sin\chi \tag{43}$$

where $K' = (15f'_3 + 10f'_2 + 5f'_1 - 5f'_{-1})/(15f'_3 + 10f'_2 + 7f'_1 + 6f'_0 + 7f'_{-1})$ and f'_{M_F} are populations for the sublevels of the $F' = 3$ excited state. The definition of circular polarization is the normally accepted one that left-hand circularly polarized light (LHC) carries unit angular momentum parallel to the direction of propagation.

Equations (42) and (43) demonstrate that the superelastic scattering technique gives information on both the diagonal and off-diagonal density matrix elements of the electron density matrix operator.

If all the excited hyperfine states are included, then equations (42) and (43) are unchanged except for terms K and K' which now include all possible populations and coherences. Figure 4a shows the behavior of the factor K as a function of laser intensity. Details of the calculation of the terms f_{M_F} may be found in Ref. 13. In this calculation, an average interaction time of 2 μs has been assumed for the sodium atoms with the laser light.

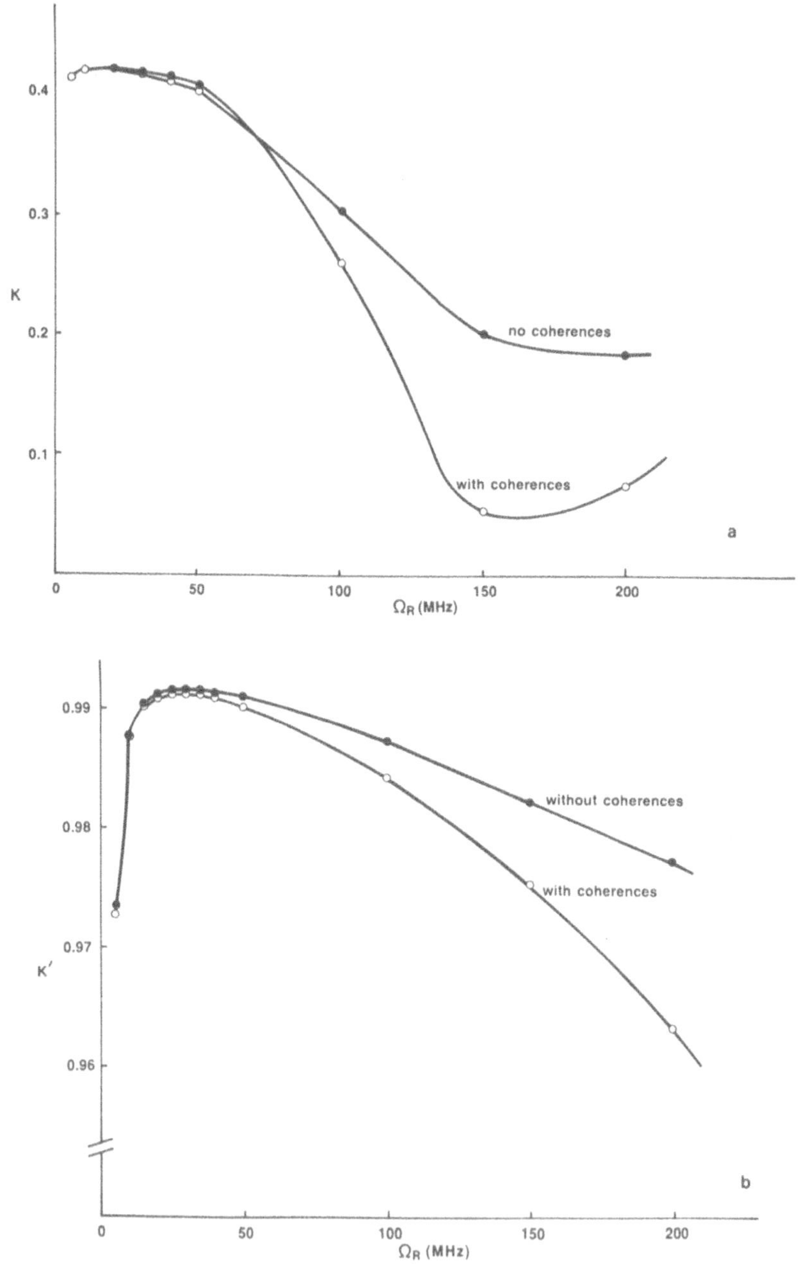

FIGURE 4. (a) Behavior of parameter K for linearly polarized laser excitation as a function of maximum Rabi frequency. (b) Same as for (a), except calculation was performed for circularly polarized laser excitation.

No Doppler effect has been included. The behavior of K is strongly dependent on laser intensity and the excited state coherences play a significant role except at a laser intensity that is sufficiently low that the power broadening induced by the laser is less than the hyperfine structure splittings. Figure 4b shows the behavior of K' as a function of laser intensity. The magnitude of K' changes little over a wide range of laser intensities and the excited state coherences play only a minor role. For low laser intensity, these results converge to those obtained in Ref. 6.

3. EXPERIMENT

The general configuration of a laser-assisted electron–atom collision is shown in Figure 5. A beam of monoenergetic electrons collides with atoms in a region where, in most cases, the laser radiation is injected. The direction of the incident electron beam defines the z axis of the coordinate system. The x-z plane is defined as that plane containing the detected, scattered electron. The angle of scatter is denoted as θ_e. Detection of the scattered electron does not always occur in Type I and Type II experiments, in which case the selection of the x axis may be determined by the laser beam direction or it may be arbitrarily chosen. In Figure 5, the directions of the laser beam, $h\nu$, and the scattered fluorescence, $h\nu'$, are described by the polar angles ϕ_L, θ_L and ϕ_γ, θ_γ, respectively. It is not necessary for determining electron–atom collision parameters to detect the fluorescence in Type III experiments.

Atoms may be available in the interaction region in either vapor or beam form. For excitation by narrow bandwidth lasers, a well collimated atomic beam at right angles to the direction of laser propagation minimizes the Doppler effect.

3.1. *Type I Experiments*

Thus far, the Type I scheme has been used to investigate electron collisions with neon,[15-19] calcium,[20] mercury,[5, 21-23] and helium[19] atoms. Each of the Type I experiments will be reviewed briefly in this section.

Electron excitation cross section measurements of the $2p\,^5 3s$ levels of Ne ($1s_j$; $j = 2, \ldots, 5$ in the Paschen notation) and their Zeeman substates have been carried out in a series of experiments.[15-18] Excitation from the $2p^6(^1S_0)$ ground state was by inelastic collisions between unpolarized electrons in a collimated, monoenergetic beam (± 0.5 eV) and Ne atoms in a low-pressure gas. The second excitation step to an appropriate $2p\,^5 3p$ level was by resonant absorption from a multimode dye laser pumped by an argon ion laser. With reference to Figure 5, the laser beam was incident

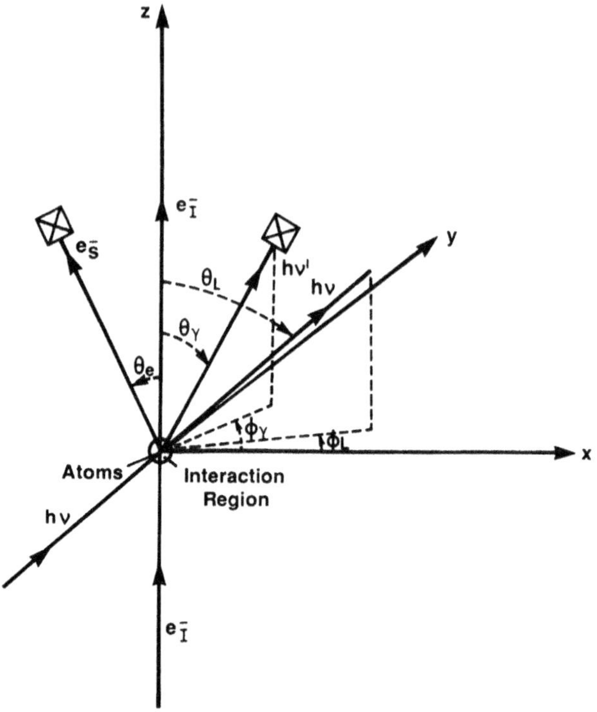

FIGURE 5. Schematic of experimental geometries.

along the x axis ($\phi_L = 0$, $\theta_L = \pi/2$). Tuning of the laser was achieved by observing the optogalvanic effect in a Ne hollow-cathode discharge. Fluorescence from a $2p\,^5 3p \rightarrow 2p\,^5 3s$ relaxation channel (not the same as the excitation transition) was detected along the y axis ($\phi_\gamma = \pi/2$, $\theta_\gamma = \pi/2$). The detection signal was processed and fed to an x–y recorder where it was displayed as a function of electron energy. In the case of experiments to determine the partial cross sections, a polarizer was inserted to analyze the fluorescence.

The stepwise fluorescence was obtained by subtracting the signal with the laser off from that when the laser was on. The measured cross section consisted of the desired direct excitation cross section, Q_a, and a contribution from cascades. Since Q_a for the metastable levels $1s_5$, $1s_3$ ($^3P_2$, $^3P_0$) is negligible at high electron energies, the measured cross section was normalized at 100 eV to previous measurements for the cascade terms.[24] Q_a was then deduced. Radiation trapping was used to increase the effective lifetimes of the levels $1s_2$, $1s_4$($^1P_1$, $^3P_1$) to make the contribution of the cross section of the laser excited level negligible. Since the electron collision cross sections of these optically allowed transitions do not tend to zero rapidly, the normalization procedure was to equate the experimental value at 300 eV with the sum of previously measured cascade terms plus a scaled Born approximation for Q_a. Q_a as a function of electron energy for each level was deduced.

The peak absolute values of the direct excitation cross section for the metastable levels $1s_3$ and $1s_5$ were 2.8×10^{-19} cm^2 and 1.6×10^{-18} cm^2, respectively at 20 eV, while peak values for the $1s_2$ and $1s_4$ levels were 1.0×10^{-17} cm^2 and 8.8×10^{-19} cm^2 at 65 eV. Uncertainties were of the order of 25%.

Evidence of resonance structure at 16.9 eV in the metastable level cross sections was attributed to the formation of Ne^- in the $2p\,^5 3s3p$ levels and subsequent autodetachment. The branching ratio of the resonance into the two metastable levels, $^3P_2$ to $^3P_0$, was estimated at 2:1.

For the measurement of electron excitation partial cross sections, four intensity measurements, $I(\boldsymbol{\varepsilon}_F, \boldsymbol{\varepsilon}_L)$, where $\boldsymbol{\varepsilon}_F(\boldsymbol{\varepsilon}_L)$ is the direction of polarization of the analyzer (laser), were made at an incident electron energy of 18.2 eV. This electron energy is below threshold for higher-lying levels and so the contribution to the cross sections from cascades was minimized. The intensity measurements were $I(\mathbf{k}, \mathbf{k})$, $I(\mathbf{i}, \mathbf{k})$, $I(\mathbf{k}, \mathbf{j})$, and $I(\mathbf{i}, \mathbf{j})$. The sum of the partial cross sections was normalized to the measured total excitation cross section. As expected, the partial cross sections Q_0 and Q_1 for the $m = 0$ and $m \pm 1$ substates of the radiation trapped levels $1s_2$ and $1s_4$ were equal. Also found to be approximately equal were the $1s_3$; Q_0 and the $1s_5$; Q_0, Q_1, and Q_2 cross sections.

Further investigation of the resonance structure of the electron impact excitation functions of the $1s_3$ and $1s_5$ metastable levels has been reported.[(19)] Electrons with 250 meV resolution emanated from a gun and a hemispherical monochromator. About 0.5 μA was produced in a pulse which was fired into the peak density of a burst of gas in a supersonic jet from a pulsed atomic beam traveling in the negative y direction. Once the electron pulse had died away (fall time about 50 ns half maximum) a pulse from a dye laser pumped by a nitrogen laser was injected into the interaction region, coaxially with the electron beam. Fluorescence along the x axis ($\phi_\gamma = 0$, $\theta_\gamma = \pi/2$) was analyzed and collected. The photon counts were detected in coincidence with the laser pulses (40 Hz) and stored in a multichannel analyzer. Helmholtz coils reduced the residual magnetic field in the interaction region to below 10 mG.

The electron excited levels were excited by the laser radiation to the $2p_2$ ($J = 1$) level from which the fluorescence from the $2p_2 \rightarrow 1s_2$ transition (659.9 nm) was detected. For total cross-section measurements, no analysis of the fluorescence was made. The laser light was linearly polarized in the x-y plane at an angle $\phi = 3\pi/4$. Under these experimental conditions, the detected apparent cross section for the $1s_3(^3P_0)$ is deduced, using standard measurement theory,[3,23] to be proportional to Q_0 while for the $1s_5(^3P_2)$ level, the measured cross section is proportional to $Q_0 + 2Q_1 + 6Q_2$.

The energy of the cross section data was normalized to the resonance feature near threshold, referred to as the *b* peak, at 16.906 eV, while the cross sections were normalized to the data reviewed above[18] at 18.2 eV.

By considering the widths of the respective *b* features and using the previously measured ratios of cross sections,[18] the branching ratio of the decay of the *b* resonance into the two levels $^3P_0$, $^3P_2$ was estimated to be $0.3 \pm 60\%$ compared with the 0.5 estimate described above.[16] A theoretical estimate of the branching ratio using a simplified model yielded a ratio of 0.37.

By a suitable selection of laser and analyzer polarization orientations, the $|M_J| = 1$ partial cross section (Q_1) of the $^3P_2$ level was obtained. Resonance structure similar to that of the total cross sections was observed.

Helium was also used as a target atom in these studies. 501.6-nm laser radiation further excited the $1s2s$ electron-excited metastable level to the $1s3p$ level. Fluorescence from the upper level to ground state (52.7 nm) was detected using a channeltron. The total cross section was normalized to calculations[25] at 21.5 eV while the energy scale was calibrated using a resonance feature at 20.955 eV. The energy resolution of the electrons was such that several resonance features were observed.

Absolute total electron impact excitation cross sections of the $4^3P_{0,1,2}$ metastable levels of Ca have been measured using a Type I scheme.[20] A beam of electrons of energies up to 20 eV with a 1 eV energy spread and a Ca atomic beam in the x direction intersected. About 4 cm beyond this interaction region, the atomic beam was crossed by single-frequency dye laser radiation propagating in the $-z$ direction. The laser was selectively tuned to resonance with each of the fine structure transitions $4\,^3P_{0,1,2} \to 5\,^3S_1$ in turn. The fluorescence from the $5\,^3S_1 \to 4\,^3P_2$ (616.2-nm) relaxation channel was detected along the y axis by a photomultiplier tube isolated by an interference filter. Discrimination from background light was attained by pulsing the electron gun with a rectangular pulse at 500 Hz and measuring the output of the photomultiplier tube synchronously.

By measuring all of the experimental parameters such as atomic level populations, total electron beam current, atomic beam temperature etc., absolute total electron cross sections for each level was obtained. The peak values, by 10^{-16} cm^2, deduced for the cross sections were 1.3 (1.08), 3.3 (3.28), and 5.9 (5.52) for the $4\,^3P_0$, $4\,^3P_1$, and $4\,^3P_2$ levels respectively. The values in parentheses were calculated using the Born approximation and Bates–Damgaard functions. The experimental uncertainty was quoted at ±30%.

Two cases of the theoretical development in the first section of this work, Case 1 and Case 2, were chosen because of their relevance to

Type I scheme experiments with atomic mercury.[5,21,22] The geometry for these experiments was the same as that in Figure 2a except that the direction of laser propagation was along the $-z$ axis. Expressions developed for conditions of axial symmetry, such as equations (33) and (37) for line polarization, are unaltered by this change of direction. In these experiments, the radiation came from a narrow bandwidth (1-10 MHz), single-mode, cw ring dye laser pumped by an argon ion laser. Thus the selection of a particular isotope in naturally occurring Hg was possible by tuning the laser to an appropriate hyperfine transition.

The electron gun produced a beam of unpolarized electrons with about 0.5 eV spread in the range 4-30 eV with currents of 1-5 μA. The atoms effused through a collimator with a ratio of about 3:1. The atomic beam traveled in the y-z plane at an angle of approximately $\pi/6$ with the y axis. The laser, electron, and atomic beams intersected in the interaction region where the residual magnetic fields were eliminated by a set of Helmholtz coils.

Fluorescence from the selected channel was detected in the y direction ($\phi_\gamma = \pi/2$, $\theta_\gamma = \pi/2$) by a cooled photomultiplier tube isolated by a narrow-band interference filter. For line polarization measurements, a polarizer analyzed the fluorescence. Photon counting was achieved using standard digital electronics.

Relative total and partial total cross sections were obtained for the $6\,{}^1P_1(I = 0)$, $6\,{}^1P_1(F = \frac{3}{2},\ I = \frac{1}{2})$[21,22] and $6\,{}^3P_2(I = 0)$[5] electron excited levels. For the $6\,{}^1P_1$ case, the electron excited level was further excited to the $6\,{}^1D_2$ level ($F = \frac{5}{2}$ for the $I = \frac{1}{2}$ isotope) by 579-nm radiation. 313-nm photons from the $6\,{}^1D_2$-$6\,{}^3P_1$ relaxation channel were detected. The scheme for the $6\,{}^3P_2$ electron excited metastable level was laser excitation to the $7\,{}^3S_1$ level (546.1 nm) and detection of 404.7-nm photons from the $7\,{}^3S_1$-$6\,{}^3P_0$ transition. To eliminate the contribution to the fluorescence from the direct excitation of the upper levels, fluorescence photon counts were recorded for each setting for a fixed time with the laser radiation present and not present. For line polarization measurements, values of I_0 and I_{90} are required. To account for small instrumental asymmetries, measurements were taken for the analyzing polarizer set at $\beta = 0°$, 90°, 180°, and 270°. For the $6\,{}^3P_2$ measurements, the experiment was automated with the aid of computer-driven slave motors which controlled a laser beam chopper and the setting of the analyzing linear polarizer. The frequency of the laser was also computer controlled.

For measurement of the line polarization $P_{(e)}$ of the electron excited transition to the $6\,{}^1P_1(I = 0)$ level, where

$$P_{(e)} - \frac{\rho_{00}/\rho_{11} - 1}{\rho_{00}/\rho_{11} + 1}, \tag{44}$$

the ratio of the partial cross sections ρ_{00} and ρ_{11} was obtained from measurement of the line polarization of the stepwise excited fluorescence. Circularly polarized laser radiation was used, for which case it can be shown that the fluorescence line polarization is given by[3]

$$P = \frac{9\rho_{00}/\rho_{11} - 30}{18\rho_{00}/\rho_{11} + 46} \tag{45}$$

$P_{(e)}$ was obtained as a function of electron energy and compared with theoretical calculations[26] and data[27] obtained from direct excitation of the naturally occurring isotope mixture. Except at energies near threshold, the experimental results agree quite well and the trend at threshold is to smaller polarizations in both cases, in contrast to the theoretical results.

Line polarization measurements were also made on the $I = \frac{1}{2}$ isotope scheme. Assuming the Percival–Seaton hypothesis[28] holds, the line polarization for the stepwise fluorescence transition $6\,^1D_2(F = \frac{5}{2})$–$6\,^3P_1$ for the same settings as the $I = 0$ scheme is given by

$$P = \frac{3\rho_{00}/\rho_{11} - 10}{11\rho_{00}/\rho_{11} + 20} \tag{46}$$

where ρ_{00} and ρ_{11} are the $I = 0$ partial cross sections for the electron excited level. A comparison of the ρ_{00}/ρ_{11} data as a function of electron energy for each isotope is shown in Table I. Near threshold, there is a marked difference

TABLE I

Measured Ratios for Partial Cross Sections of Hg $6\,^1P_1$
90% Confidence Limits are Indicated

Incident electron energy (eV)	$\rho_{00}/\rho_{11}(I = 0)$	$\rho_{00}/\rho_{11}(I = \frac{1}{2})$
8.0	$5.60 \pm^{0.62}_{0.52}$	$3.05 \pm^{0.42}_{0.36}$
8.5	$3.90 \pm^{0.70}_{0.54}$	$2.43 \pm^{0.15}_{0.14}$
9.0	$3.64 \pm^{0.26}_{0.23}$	$3.36 \pm^{0.23}_{0.21}$
9.5	$3.38 \pm^{0.34}_{0.30}$	$3.54 \pm^{0.58}_{0.45}$
10.0	$3.51 \pm^{0.67}_{0.51}$	$3.52 \pm^{0.66}_{0.51}$
10.5	$4.01 \pm^{0.44}_{0.37}$	$3.42 \pm^{0.50}_{0.41}$

between the ratios, with values of each lying well outside the 90% confidence limit of the other. Thus a breakdown of the Percival–Seaton hypothesis is indicated at these energies. At higher energies, the values of the ratios correspond quite closely. Suggested reasons for the breakdown include spin–orbit interactions, negative ion formation, and long-range interactions.

Relative total and partial cross sections were measured for the electron impact excitation of the $6\,^3P_2$ level. The $I = 0$ isotope was investigated. Of particular interest were the ratios ρ_{11}/ρ_{00} and ρ_{22}/ρ_{00} deduced from the line polarization measurements. Using linearly polarized laser light with $\alpha = 0$ and $\pi/2$ (cf. Figure 2a), these ratios were obtained from equation (37) and are shown in Figure 6 plotted as a function of electron energy. Threshold for this transition is 5.46 eV. Also plotted on these graphs are close coupling theoretical calculations[29] and the experimental results of another study.[23] As will be described in more detail below, the results for the second study are those for the naturally occurring isotope mixture, although the data

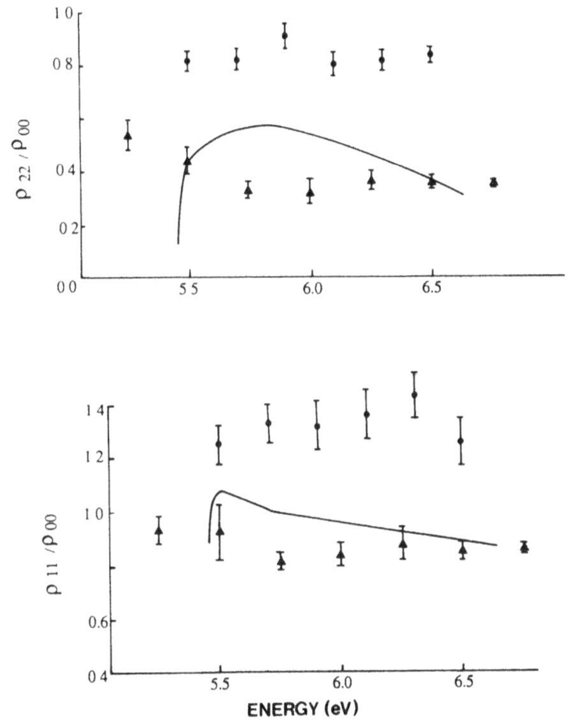

FIGURE 6. Ratios of the partial total cross sections ρ_{22}/ρ_{00} and ρ_{11}/ρ_{00}, for the $6\,^3P_2$ metastable state of mercury isotope 200. ▲, Experimental data, Ref. 5; ●, experimental data, Ref. 23; —, close-coupling theoretical calculations, Ref. 29.

were adjusted for this. The threshold behavior predicted by the theory is not evident in the experimental data. This is most probably due to the energy spread of the electrons of 0.5 eV for the Griffith experiment[5] and 0.18 eV for the Münster experiment.[23] A data point at 5.25 eV for the latter experiment could well have revealed signs of the decrease.

In the Münster study, a laser producing 15-ns pulses with an effective bandwidth of 30 GHz was used. Hence isotope selectivity was not possible. The direction of propagation of the laser beam, with reference to Figure 5, was given by $\phi_L = 0$, $\theta_L = 2\pi/3$. A mercury atomic beam traveled along the y axis. The electron gun emitted a beam whose energy was controlled in the range 4–8 eV with a spread of 180 meV at a current of 10^{-7} A. All beams intersected at the origin. Hemholtz coils reduced the residual magnetic field in the interaction region to less than 1% of the Earth's field.

The fluorescence photons were detected in the direction $\phi_\gamma = 0$, $\theta_\gamma = \pi/3$ by a photomultiplier tube after discrimination by an appropriate interference filter and, if required, after analysis by a linear polarizer. For the study of the $6\,^3P_2$ metastable level, the excitation/emission scheme used was

$$6\,^1S_0 \xrightarrow{e^-} 6\,^3P_2 \xrightarrow{h\nu} 7\,^3S_1 \xrightarrow{h\nu'} 6\,^3P_0$$

The total cross section of the $6\,^3P_0$ level was also measured, in which the roles of the two metastable levels were interchanged. For energy calibration, 254-nm radiation from the $6\,^3P_1 \rightarrow 6\,^1S_0$ transition was recorded by another photon detection system. The electron energy was calibrated to within ±0.05 eV against the 4.9- and 5.5-eV peaks in this cross section. For the $6\,^3P_2$ level, different settings of the laser (linear) polarization angle and analysis of the fluorescence permitted the estimate of relative partial and total cross sections. Ratios, at threshold, of the partial total cross sections to the total cross sections deduced from various sources are shown in Table II.

TABLE II

Ratios of Partial Cross Sections at Threshold of Hg $6\,^3P_2$

Ratios	Griffith[5]	Close coupling[29]	Munster[23]	Threshold theory
ρ_{00}/σ	0.27 ± 0.05	0.335	0.20 ± 0.03	0.4
ρ_{11}/σ	0.25 ± 0.05	0.285	0.24 ± 0.06	0.3
ρ_{22}/σ	0.13 ± 0.03	0.048	0.16 ± 0.03	0.0

The discrepancies between the threshold and the close-coupling calculations are thought to be due to a $6s6p^2\ ^2D_{5/2}$ resonance at 5.5 eV included in the latter.

3.2. Type II and Type III Experiments

As shown in Figures 1b and 1c, the first step in either a Type II or Type III scheme experiment is atomic excitation by laser radiation. Since with subsequent electron collision it is possible to either further excite the atom (Type II) or deexcite the atom (Type III), the experimental requirements for each type are similar. Hence it has been natural to undertake experiments in which both Type II and Type III schemes have been studied. Thus we shall review here experiments of both types together. We commence with a discussion of the Type II only experiments and follow with descriptions of the combined experiments. Finally we describe Type III only studies.

In the first Type II experiment,(30) circularly polarized (σ^+) light was used to optically pump sodium atoms and prepare them in the two-state system $3\ ^2S_{1/2}$, $(F = 2, M_F = 2) \rightarrow 3\ ^2P_{3/2}$, $(F' = 3, M'_F = 3)$. The upper state is a pure $M_L = 1$, $M_S = \frac{1}{2}$ state. A low-energy electron beam, propagating in the opposite direction to the laser beam, intersected the atoms, traveling in the y direction (cf. Figure 5), in the interaction region. Atoms excited to the $3D$ level were studied by detecting, along the x axis, fluorescence from the $3D$–$3P$ transition (819 nm).

The electron gun, using an indirectly heated oxide cathode as its source, produced a total current passing through the interaction region of typically 0.1–0.5 μA at 100 eV. The average energy of the electron beam was calibrated by recording the $3S \rightarrow 3P$ apparent excitation function. The energy resolution of the electrons was measured as 0.29 eV.

The atoms were pumped by the light of a frequency and intensity stabilized cw dye laser of about 1 MHz linewidth and 100 mW power. After pumping, the typical population distribution was 20% in the $3\ ^2P_{3/2}(F' = 3, M'_F = 3)$ state, 25% in the $3\ ^2S_{1/2}(F = 2)$ state and 55% in the $3\ ^2S_{1/2}(F = 1)$ level. The atoms which leak to the $3\ ^2S_{1/2}(F = 1)$ level are not reexcited by the laser.

Great care was taken with the positioning of the electron and laser beams to ensure that the whole electron beam traversed a region with a uniform density of excited atoms. Large background signals, up to two orders of magnitude larger than the $3D$–$3P$ fluorescence due to electron excitation, necessitated experimental runs of long duration (typically 24 h). All backgrounds were measured and subtracted to obtain the signal.

Fluorescence data was recorded with both the laser on and off so that the contribution of the direct electron excitation to the $3D$ level could be accounted for. The measurements were then normalized at high electron energies to a cross section calculated by Born and Born–Ochkur approximation theories where the dominant cascade terms were included. Cross sections for the specific $3P(M_L = 1, M_S = \frac{1}{2}) \rightarrow 3D$ transition in the direction of the detector and the pure $3P \rightarrow 3D$ transition were plotted as a function of electron energy. As well, the $3S \rightarrow 3D$ cross section was obtained.

The threshold behavior of the $3P(M_L = 1, M_S = \frac{1}{2}) \rightarrow 3D$ and $3S \rightarrow 3D$ cross sections was also investigated. It was found that in both cases these cross sections are essentially step functions at threshold, rising to 75% of their peak values within 0.1 eV. This is much more rapid than the onset for the $3S \rightarrow 3P$ cross section and other spin-allowed excitation cross sections and dipole allowed transitions. It was concluded that the threshold behavior might well be dominated by the upper level of the excitation, the $3D$ level in this case.

In an experiment with a different emphasis, cross sections for collisions between laser prepared Na(nd) Rydberg levels and 25-eV electrons have been measured.[31] Atoms were prepared in the appropriate level by stepwise excitation via the $3\,^2P_{3/2}$ level by two nitrogen laser-pumped dye lasers propagating collinearly and at right angles to the atomic beam. An electron beam was directed into the interaction region in the opposite direction to the lasers. Magnetic fields were eliminated by the use of μ metal.

The collision products were analyzed using selective field ionization (SFI). For this method, voltage ramps of equal value but opposite polarity were applied to grids on either side of the interaction region. Electrons produced from the ionization of atoms were detected by an electron multiplier whose output was displayed on a multichannel analyzer after time-to-amplitude conversion. The TAC was initiated by the start of the voltage ramp and stopped by the first detected electron pulse. The distribution of the Rydberg atom population could be inferred from the SFI spectrum since atoms in different Rydberg states ionize at different ramp voltages.

Spectra were collected with the electron gun on and off to obtain the contributions from the electron collisions. The dominant collision process was the state change $nd \rightarrow (n+1)p$. By measuring the number of state changed atoms in the interaction time compared with the initial parent population as well as the electron density and the relative collision velocity, the cross sections for a number of Rydberg levels were calculated. Values of the cross sections increased with n and ranged from 6.6×10^{-10} cm^2 for $n = 36$–3.4×10^{-9} cm^2 for $n = 50$.

In a series of Type II and Type III experiments employing atomic sodium,[6,32–34] one group has measured a number of parameters including relative differential cross sections, alignment and orientation parameters, ratios of partial cross sections, and scattering amplitude phase differences. The standard method of crossed electron, atom, and laser beams was used.

The atom beam was produced in a laval nozzle of about 700°C with a collimation ratio of about 50:1 leading to a residual Doppler width of 30 MHz. The atom density in the interaction region was of the order of 10^9 cm^{-3}. For the initial experiments, the energy resolution of the electron gun was 250 meV, but, with the addition of a hemispherical monochromator, this was improved to 60 meV at 10 nA. The scattered electrons were detected by an electron multiplier after passage through two hemispherical analyzers with 60 meV resolution. Up to 70 mW of single-mode radiation was available from an argon ion laser pumped cw dye laser with a jitter bandwidth of 50 MHz. Tuning of the laser was accomplished by monitoring the $3\,^2P_{3/2} \rightarrow 3\,^2S_{1/2}$ resonance fluorescence.

For the initial experiments the laser radiation was incident in the scattering plane (x-z plane, $\phi_L = 0$) and the atomic beam was perpendicular to this plane. With the laser tuned to excite only the $F = 3$ level of the $3\,^2P_{3/2}$ level, the fraction of the excited atoms was between 4% and 10% for π excitation and 6% to 15% for σ excitation, the efficiency varying with the angle of incidence.

Energy loss (gain) spectra were recorded for the scattered electrons for various θ_e and θ_L and differential cross sections relative to the $3\,^2P \rightarrow 3\,^2S$ cross section were deduced for the dominant processes $3\,^2P \rightarrow 3\,^2S$, $3\,^2P \rightarrow 4\,^2S$ and $3\,^2P \rightarrow 3\,^2D$. The light polarizations used were circular, linear with the vector parallel, and perpendicular to the scattering plane. Differential cross sections as a function of the polarization angle of linearly polarized light for varying scattering geometry and incident electron energies were plotted.

For linearly polarized light, due to reflection symmetry, no odd multipoles are produced (that is, no orientation is produced in the atomic excited state). Thus, theoretically, the scattered electron intensity, for the $^2P \rightarrow {}^2S$ transition, takes on a simple cosine dependence on the angle of polarization of the incident light. Experimental data at various angles of light incidence for 10 eV incident electrons and the $3\,^2P \rightarrow 3\,^2S$ deexcitation scheme confirmed this behavior. The ratio r, of scattered electron intensity for light polarized parallel to the scattering plane to the intensity for light polarized perpendicular to the plane was plotted for the $3\,^2P \rightarrow 3\,^2S$ and the $3\,^2P \rightarrow 4\,^2S$ transitions against the incident angle of the light for various incident electron energies. From least-squares fits to these anisotropy lobes, values for the even state multipoles for the optical preparation of the atoms and the expectation values of the even collision state multipoles up to rank 2 were deduced. Values for the optical alignment varied between -0.82 and -0.92 compared with a theoretical value of -0.96. Nonideal optical pumping due to a number of factors was used to explain the discrepancy. Plots of the collision state multipoles versus scattering angle were made as a function of incident electron energy for the two $^2P \rightarrow {}^2S$ transitions. For the deexcitation case, the experimental data were compared to calculations from the Born approximation, close-coupling theory (CC), and the distorted-wave polarized-orbital method (DWPO), while only Born approximation calculations were available for the excitation case.

The two parameters λ and χ defined in equation (29) in terms of the density matrix elements ρ_{00}, ρ_{11}, and Re ρ_{01}, were deduced and plotted against the electron scattering angle for both transitions. The density matrix elements are related directly to the expectation values of the collision state multipoles. The data were compared with available calculations.

The Born approximation gave surprisingly good agreement for the multipoles and λ even at these low energies. On the other hand, the experimental values of cos χ differed markedly from the Born approximation values of 1 for the deexcitation process and -1 for the excitation process. Both the CC and DWPO methods demonstrated agreement under certain but not all conditions.

Measurement of the orientation (first rank state multipole) was achieved by a change in the geometry of the experiment. Rather than the incident laser beam lying in the scattering plane, it was adjusted to an orientation of $\phi_L = 90°$, $\theta_L = 55°$–$90°$ or $0°$, while the atomic beam now propagated along the x axis. The scattered electron intensity was measured for σ^+ and σ^- polarization, I^+ and I^-, respectively. The difference between the two contains only odd multipoles, while the sum contains only even. With $\theta_L = 90°$, the asymmetry can be expressed as

$$\frac{I^+ - I^-}{I^+ + I^-} \propto \langle L_y \rangle \tag{47}$$

where $\langle L_y \rangle$ measures the momentum transfer perpendicular to the scattering plane.

Firstly, the form of the asymmetry as a function of θ_L was confirmed at $\theta_e = 15°$ and incident electron energies of 10 and 20 eV. Then, with $\theta_L = 90°$, experimental data for $\langle L_y \rangle$ as a function of θ_e were compared with two-state CC and DWPO method calculations for electron energies at 3, 10, and 20 eV. Also at 3 eV, a four-state OC calculation was available. Best agreement was obtained with this latter theory. This theory also gave excellent agreement between measured and calculated data for the phase parameter, ϕ, defined by

$$\sin \phi = \frac{\langle L_y \rangle}{2[\lambda(1-\lambda)]^{1/2}} \tag{48}$$

where the above measured values of λ were used. For negligible exchange, ϕ is equivalent to χ.

Type II and III scheme experiments using barium have formed a study that has produced unusual results.[35,36] In the initial work, differential electron scattering cross sections for excitation and deexcitation from the ^1P, ^3P, ^1D, and ^3D levels were measured. Linearly polarized light from a 30-GHz bandwidth, multimode, cw dye laser, incident in the scattering plane, excited the $6s^2\,{}^1S \rightarrow 6s6p\,{}^1P$ transition. The atoms were in a beam traveling at right angles to the scattering plane. Cascade processes then populated the ^3P, ^1D, and ^3D levels. The experiment was carried out under conditions of radiation trapping so as to populate the substates of the ^1P and hence those of the other levels equally.

The electron gun and analyzer both incorporated a tandem double-hemispherical energy selector yielding an energy resolution of 80 meV. Incident electron energies of 30 and 100 eV were used.

Energy loss (gain) spectra were obtained for the laser on and off. By using the relative scattering intensities ($\theta_e = 5°$), known absolute electron impact excitation cross sections for ^1P transitions and the principle of detailed balancing, absolute differential cross sections were obtained for the observed inelastic processes for the singlet levels. Also recorded were scattering intensities for the triplet–triplet processes relative to the $6s6p\,{}^1P \rightarrow 6s^2\,{}^1S$ intensity.

For the study of electron superelastic scattering from the excited $6s6p\,{}^1P$ state of ^{138}Ba($I = 0$), a single-mode laser with a bandwidth of a few megahertz was used. With the laser incident in the scattering plane with $\theta_L = 45°$ or 90°, the scattered intensity, I^S, was recorded as a function of laser polarization angle ($\psi = 0$ for polarization vector parallel to the scattering plane) for a number of scattering angles, θ_e. All plots exhibited the simple cosinelike form as was detected for the sodium experiments discussed above[33]; however, for scattering angles $|\theta_e| < 5°$ the curves were observed to shift in phase. At $\theta_e = 0$, the scattered intensity for Ba was a maximum for $\psi = 90°$ whereas a minimum for Na had been reported for the same experimental parameter values. For scattering angles $\theta_e \simeq 0°$ and $|\theta_e| \geqslant 5°$, the curves were symmetric in the sense that $I^S(\psi) = I^S(-\psi)$, but for $|\theta_e|$ between 1° and 5° this symmetry was not observed, that is, they were phase shifted.

This phase shift was plotted as a function of scattering angle for both incident electron energies and both incident laser angles. Despite a careful analysis, no experimental cause of this asymmetry was detected and its existence, as yet, remains unexplained.

The remaining Type III experiments reviewed have either used incident electrons that were spin polarized or analyzed the spin of the scattered electrons.

In the first,[37] unpolarized electrons and laser radiation intersected a beam of Na atoms at right angles. The laser power was 500 mW with beam diameter between 3 and 7 mm^2 and its bandwidth was 40 GHz. Coupled with a large residual Doppler width of 200 MHz and a number density of 5×10^{12} cm^{-3} for the atoms, equal populations were produced in the magnetic substates of the excited level. The laser was tuned to excite either the $3\,^2P_{3/2}$ or $3\,^2P_{1/2}$ level. The electrons superelastically scattered at angle θ_e entered an analyzer system that consisted of electron lenses, a 90° spherical deflector, and a Mott chamber where the spin polarization perpendicular to the scattering plane was measured. The 90° deflector also served as a monochromator tuned to the electrons superelastically scattered from the appropriate excited state.

The measured spin polarizations were $P'_{1/2} = 0.095 \pm 0.048$ and $P'_{3/2} = -0.068 \pm 0.044$ with quoted 68% confidence limits. The theoretically predicted relationship was $P'_{3/2} = -\frac{1}{2}P'_{1/2}$ so that the experimental results confirmed the sign change but more accurate measurements are required. The scattering angle was 30° and the incident electrons had an energy of 20 eV.

The final experiments reviewed[38,39] are ones in which incident spin polarized electrons are superelastically scattered from laser excited $3\,^2P_{3/2}$ sodium atoms. Polarized electrons were produced by photoemission from a GaAs photo cathode with a polarization P_e of about 25%. Up to 1 μA of current was produced in the energy range of 1–12 eV with resolution of typically 0.1 eV. The electron beam intersected a beam of sodium of density 10^{10} cm^{-3} traveling in the scattering plane. Light from a frequency stabilized dye laser, tuned to the $3\,^2S_{1/2}(F = 2) \rightarrow 3\,^2P_{3/2}(F' = 3)$ hyperfine transition was incident on the interaction region at right angles to the scattering plane. The light was either circularly polarized or linearly polarized at an angle β to the scattering plane. An electron detector could be rotated in the scattering plane. It consisted of a channel electron multiplier and an analyzer that was set to reject all but appropriate superelastically scattered electrons. The spin of the incident electrons was modulated, up or down compared with the scattering plane, and scattered electrons counted separately for each spin orientation. For linearly polarized laser light, the spin asymmetry of the scattered electrons was calculated from $P_e^{-1}(N_\downarrow - N_\uparrow)/(N_\downarrow + N_\uparrow)$, where N was the number of electrons detected when the incident electrons had either spin up ($\uparrow$) or spin down ($\downarrow$).

The spin asymmetry was plotted as a function of scattering angle for $|\theta_e| < 40°$ and $\beta = 0$. The resulting curve was antisymmetric. This asymmetry was explained by noting that the collisions were dominated by the more heavily populated $M_J = \pm\frac{3}{2}$ states. Also, the electrons scattered to a particular side (positive or negative θ_e) interact mostly with a particular M_L value; here ± 1. Thus choosing a given scattering angle automatically selects the dominant subset of the excited atom. The spin asymmetry changes sign for the exchange process from each of the $M_L = \pm 1$ subsets and so explains the antisymmetry of the observations.

The spin asymmetry was also plotted as a function of linear polarization angle, β, at a fixed scattering angle $\theta_e = -30°$ and demonstrated a simple cosine oscillation with a period of π. This variation was explained in terms of the variation as a function of β of the phase between the states $M_J = \frac{3}{2}$ and $M_J = \frac{1}{2}$.

For circularly polarized light, as noted above,[30] the atom could be prepared in pure M_L and M_S states ($M_L = +1$ or -1 and $M_S = +\frac{1}{2}$ or $-\frac{1}{2}$ for σ^+ and σ^- light respectively). Thus superelastic scattering intensities could be measured for the incident electrons polarized parallel to the target electron spin, giving a purely triplet scattering, or antiparallel to the target spin, giving a scattering consisting of equal contributions from singlet and triplet states. For each M_L state, the asymmetry was defined as

$$A = \frac{1}{P_e} \frac{I_{\uparrow\downarrow} - I_{\uparrow\uparrow}}{I_{\uparrow\downarrow} + I_{\uparrow\uparrow}} \tag{49}$$

where I was the scattering intensity for parallel ($\uparrow\uparrow$) and antiparallel ($\uparrow\downarrow$) spins. For each state, it was shown that, theoretically, the asymmetry could range from $+1$ for purely singlet scattering to $-\frac{1}{3}$ for purely triplet scattering. The two asymmetries $A_+(M_L = +1$ target) and $A_-(M_L = -1$ target) were measured as functions of incident electron energy for fixed scattering angle, $\theta_e = 30°$, and as functions of scattering angle at fixed incident electron energies of 2 and 9.26 eV. It was found that the measured asymmetry was the same for positive and negative scattering angles if both the incident electron spin and the helicity of the exciting laser were reversed. It was also found that for $\theta_e = 30°$, A_- changed from almost completely singlet at low energies to equal amounts of triplet and singlet for high energies, whereas A_+ changed from slightly singlet to almost completely triplet. At 2 eV incident energy, A_- was almost purely singlet at $\theta_e = 35°$, whereas at 9.26 eV, $\theta_e \simeq 30°$, A_- changed, within 5°, from being all triplet to singlet dominated. A_+ demonstrated a relatively smooth behavior. No explanations were presented for these characteristics.

4. CONCLUSIONS

In these concluding remarks possible future developments for stepwise and superelastic scattering techniques are considered. The applicability of these techniques has been limited by the availability of suitable tunable lasers capable of strongly exciting relevant atomic transitions. This limitation is perhaps most severe for Type II stepwise and superelastic scattering experiments in which laser excitation is required from the ground state, since many of these transitions occur at ultraviolet and vacuum-ultraviolet

(VUV) wavelengths. Recent developments in frequency doubling techniques for both cw and pulsed lasers and the development of excimer pumped dye lasers may provide a means of overcoming some of these problems.

Type I stepwise excitation techniques provide a means of investigating inelastic collision processes in which VUV transitions are excited. The investigations already performed on the Hg $6\,^1S_0$–$6\,^1P_1$ (185 nm) transition demonstrate their feasibility and a number of suitable Type I excitation schemes exist in atoms such as He, Ne, Cd, Ca, Ar. The application of Type I schemes to the study of metastable excitations provides new experimental methods and previously inaccessible data. Although at the time of writing, no stepwise coincidence experiments have been reported in the literature, the authors have recently observed stepwise coincidence signals for the Type I excitation scheme $6\,^1S_0(e^-) \rightarrow 6\,^1P_1$ (579 nm) $\rightarrow$ $6\,^1D_2$(313 nm) $\rightarrow 6\,^3P_1$ discussed in Section 2. It would also appear feasible to apply stepwise excitation coincidence techniques to metastable states. These developments represent a considerable widening of the range of applicability of coincidence techniques in atomic collision physics.

Type II stepwise excitation schemes provide a means of investigating inelastic scattering processes between excited states of the target atoms and again coincidence techniques could be applied to yield more detailed information about the collision processes. Superelastic scattering studies could also be extended to study scattering processes involving electron induced transitions between excited states by the use of stepwise laser excitation to excite higher-lying atomic states.

REFERENCES

1. K. BLUM, *Density Matrix Theory and Applications* (Plenum Press, New York, 1981).
2. G. F. HANNE, *Phys. Rep.* **95**, 95 (1983).
3. C. J. WEBB, W. R. MACGILLIVRAY, AND M. C. STANDAGE, *J. Phys. B: At. Mol. Phys.* **17**, 1675 (1984).
4. W. R. MACGILLIVRAY AND M. C. STANDAGE, *Phys. Rep.* (1988).
5. C. J. WEBB, W. R. MACGILLIVRAY, AND M. C. STANDAGE, *J. Phys. B: At. Mol. Phys.* **18**, L259 (1985).
6. I. V. HERTEL AND W. STOLL, *Adv. At. Phys.* **13**, 113 (1977).
7. H. W. HERMANN AND I. V. HERTEL, *Comments At. Mol. Phys.* **12**, 61 (1982); **12**, 127 (1982).
8. M. C. STANDAGE AND H. KLEINPOPPEN, *Phys. Rev. Lett.* **36**, 577 (1976).
9. A. A. ZAIDE, S. M. KHALID, I. MCGREGOR, AND H. KLEINPOPPEN, *J. Phys. B: At. Mol. Phys.* **14**, L503 (1981).
10. J. R. ACHERHALT AND J. H. EBERLY, *Phys. Rev. D* **10**, 3350 (1974).
11. R. M. WHITLEY AND C. R. STROUD, *Phys. Rev. A* **14**, 1498 (1976).
12. J. R. RILEY, P. J. O. TEUBNER, AND M. J. BRUNGER, *J. Phys. B: At. Mol. Phys.* **19**, 129 (1986).
13. P. M. FARRELL, W. R. MACGILLIVRAY, AND M. C. STANDAGE, in preparation.

14. J. J. McCLELLAND AND M. H. KELLEY, *Phys. Rev. A* **31**, 3704 (1985).
15. M. H. PHILLIPS, L. W. ANDERSON, AND C. C. LIN, *Phys. Rev. A* **23**, 2751 (1981).
16. M. H. PHILLIPS, L. W. ANDERSON, C. C. LIN, AND R. E. MEIRS, *Phys. Lett.* **82A**, 404 (1981).
17. R. E. MEIRS, J. E. GASTINEAU, M. H. PHILLIPS, L. W. ANDERSON, AND C. C. LIN, *Phys. Rev. A* **25**, 1185 (1982).
18. J. O. PHELPS, M. H. PHILLIPS, L. W. ANDERSON, AND C. C. LIN, *J. Phys. B: At. Mol. Phys.* **16**, 3825 (1983).
19. P. W. ZETNER, W. B. WESTERWELD, G. C. KING, AND J. W. McCONKEY, *J. Phys. B: At. Mol. Phys.* **19**, 4205 (1986).
20. V. E. DOBRYSHIN, V. I. RAKHOVSKII, AND V. M. SHUSTRYAKOV, *Opt. Spectrosc.* **52**, 364 (1982).
21. C. W. McLUCAS, W. R. MACGILLIVRAY, AND M. C. STANDAGE, *Phys. Rev. Lett.* **48**, 88 (1982).
22. C. W. McLUCAS, H. J. E. WEHR, W. R. MACGILLIVRAY, AND M. C. STANDAGE, *J. Phys. B: At. Mol. Phys.* **15**, 1883 (1982).
23. G. F. HANNE, V. NICKICH, AND M. SOHN, *J. Phys. B: At. Mol. Phys.* **18**, 2037 (1985).
24. F. A. SHARPTON, R. M. ST. JOHN, C. C. LIN, AND F. E. FAJAN, *Phys. Rev. A* **2**, 1305 (1970).
25. W. C. FON, K. A. BERRINGTON, P. G. BURKE, AND A. E. KINGSTON, *J. Phys. B: At. Mol. Phys.* **14**, 2921 (1981).
26. J. C. McCONNELL AND B. L. MOISEIWITSCH, *J. Phys. B: At. Mol. Phys.* **1**, 406 (1968).
27. T. W. OTTLEY, D. R. DENNE, AND H. KLEINPOPPEN, *J. Phys. B: At. Mol. Phys.* **7**, L179 (1974).
28. I. C. PERCIVAL AND M. J. SEATON, *Phil. Trans. R. Soc. London, Ser. A* **251**, 113 (1958).
29. K. BARTSCHAT, private communication (1983).
30. B. STUMPF AND A. GALLAGHER, *Phys. Rev. A* **32**, 3344 (1985).
31. G. W. FOLTZ, E. J. BEITING, T. H. MEYS, K. A. SMITH, F. B. DUNNING, AND R. F. STEBBINGS, *Phys. Rev. A* **25**, 187 (1982).
32. I. V. HERTEL AND W. STOLL, *J. Phys. B: At. Mol. Phys.* **7**, 583 (1974).
33. H. W. HERMANN, I. V. HERTEL, W. REILAND, A. STAMATOVIC, AND W. STOLL, *J. Phys. B: At. Mol. Phys.* **10**, 251 (1977).
34. H. W. HERMANN, I. V. HERTEL, AND M. H. KELLEY, *J. Phys. B: At. Mol. Phys.* **13**, 3465 (1980).
35. D. F. REGISTER, S. TRAJMAR, S. W. JENSEN, AND R. T. POE, *Phys. Rev. Lett.* **41**, 749 (1978).
36. D. F. REGISTER, S. TRAJMAR, G. CSANAK, S. W. JENSEN, M. A. FINEMAN, AND R. T. POE, *Phys. Rev. A* **28**, 151 (1983).
37. G. F. HANNE, CZ SZMYTKOWSKI, AND M. VAN DER WIEL, *J. Phys. B: At. Mol. Phys.* **15**, L109 (1982).
38. J. J. McCLELLAND, M. H. KELLEY, AND R. J. CELOTTA, *Phys. Rev. Lett.* **55**, 688 (1985).
39. J. J. McCLELLAND, M. H. KELLEY, AND R. J. CELOTTA, *Phys. Rev. Lett.* **56**, 1362 (1986).

CHAPTER 5

Excitation and Ionization in Atomic Collisions

Rainer Hippler

1. INTRODUCTION

The encounter of an energetic atomic particle with an atom may result in a number of different effects, which may be classified as elastic, inelastic, superelastic, and radiative.[1] In elastic collisions no energy transfer takes place between the kinetic energy of the projectile and the internal energy of either target or projectile. In contrast, in inelastic collisions the projectile transfers part of its kinetic energy into internal energy of one or both collision partners, thereby exciting or ionizing one or more (bound) electrons. Superelastic collisions can occur when, for example, the target atom is in an excited state prior to the collision. Then it is possible that the internal energy is transferred into kinetic energy. Finally, we mention radiative collisions, for example, free-free transitions, where the projectile interacts with the electromagnetic field during the collision, thereby emitting (bremsstrahlung) or absorbing (inverse bremsstrahlung) photons from the field.

The interaction of energetic atomic particles with free atoms has been the subject of experimental investigations since the early days of this century. For example, by means of X-ray spectroscopy it has been revealed that electron vacancies may be created in inner atomic shells[2]; the work of Franck and Hertz[3] has demonstrated the existence of discrete excitation energies in collisions of electrons with atoms. First theoretical descriptions of ionization processes may be traced back to Bohr,[4] who already in 1913 developed a theory for the stopping power of fast particles in matter. A

Rainer Hippler • Fakultät für Physik, Universität Bielefeld, D-4800 Bielefeld, West Germany quantum-mechanical description of excitation and ionization processes was given by Bethe[5] in 1930 making use of the Born approximation. As further steps one may notice the work of Massey and Mohr,[6] of Burhop,[7] the relativistic calculations of Arthurs and Moiseiwitsch[8] and of Perlman,[9] and the work of Bang and Hansteen,[10] who gave the first semiclassical description of inner-shell ionization processes by light ion impact. In these calculations, the collision is treated as between a charged structureless projectile (electron, proton) interacting via the Coulombic force with a single target electron and causing a (small) perturbation of the atomic states of the target only. A perturbation theoretical treatment of the collision process is then adequate. This is generally justified if the collision itself can be classified as "fast," where "fast" or, respectively, "slow" refers to the relative velocity of the projectile with respect to the (classical) velocity of the (bound) atomic electron under consideration. For example, the collision of an electron with atomic hydrogen is considered fast already for incident energies of a few hundred electron volts. In contrast, the excitation of an uranium K-shell electron in symmetric U-U collisions is "slow" even for incident energies as large as some 100 MeV.

In "slow" ion-atom collisions, the two collision partners transiently form a quasimolecule.[11] This quasimolecular model of atomic collisions provides an intuitive approach to account for the strong distortion of atomic states during the interaction; the model has been successfully used for a quantitative description particularly of inner-shell excitation in heavy ion-atom collisions (see, e.g., Refs. 12-14). Transitions between quasimolecular states are induced by the nuclear motion; large transition amplitudes can occur between states that become energetically degenerate or near-degenerate during the collision.

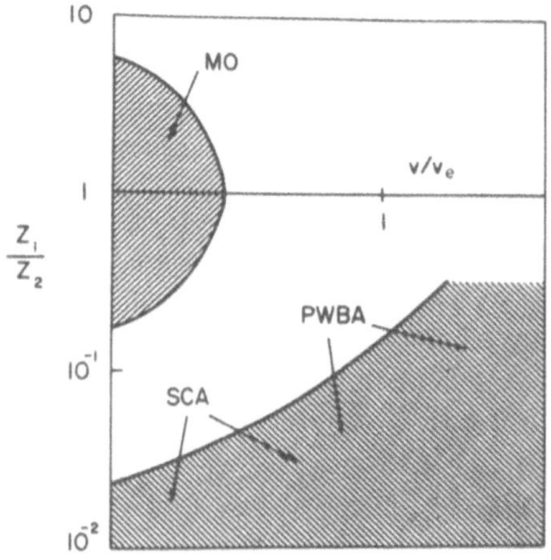

FIGURE 1. Regions of validity for different approximation schemes as a function of Z_1/Z_2 (Z_1 and Z_2 are the effective nuclear charges of projectile and target, respectively), and relative velocity v/v_e (v and v_e are the projectile and target electron velocities, respectively). Shaded regions are for "slow" [quasimolecular (MO) model] and "fast" [plane-wave Born (PWBA), semiclassical approximation (SCA)] collisions (from Ref. 15).

The regions of validity of the above approximations are illustrated in Figure 1. Perturbation theoretical approaches like plane wave Born approximation (PWBA) or the semiclassical approximation (SCA) are valid as long as the perturbation of projectile and target wave functions remains small during the interaction. This is the case for sufficiently fast collisions and/or if the charge of the projectile is small compared to the effective charge of the target atom (so-called asymmetric collisions). The quasimolecular model is applicable if the projectile velocity is small compared to the orbiting velocity of the bound electron under consideration, and for symmetric or near-symmetric collision systems.(15)

2. IMPACT IONIZATION

Much of our present day's knowledge about excitation and ionization processes by charged particle impact was obtained by means of photon (visible and ultraviolet photons, X rays) spectroscopy. Figure 2 shows a neon spectrum in the vacuum ultraviolet (VUV) regime that was produced by 800-keV Ne^+ impact.(16) The spectrum was recorded in the wavelength region 40–150 nm; it displays characteristic lines from various atomic and ionic states of neon. For example, the lines observed at 46.1/46.2 nm result from ionization of Ne − 2*s* electrons, whereas the group around 49 nm stems from simultaneous removal of 2*s* and 2*p* electrons.

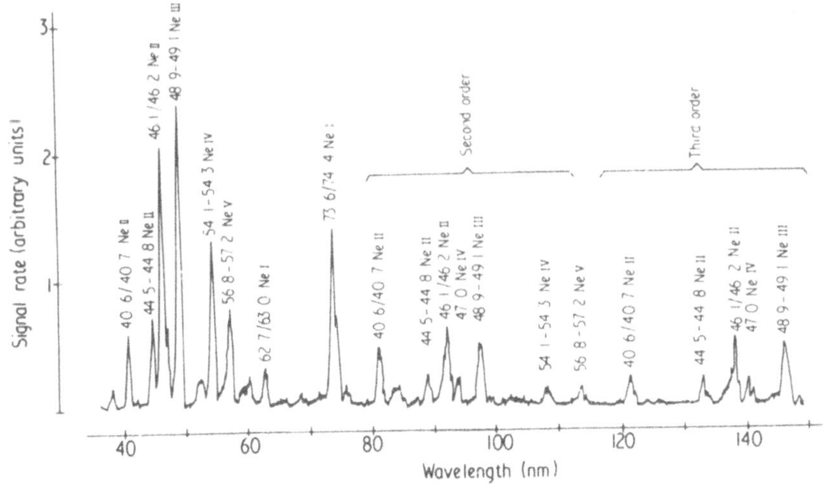

FIGURE 2. Wavelength spectrum in the region from 35 to 150 nm, produced by 800-keV Ne^+-Ne collisions (from Ref. 16).

In first-order perturbation theory, the transition amplitude f for ionization by charged particle impact may be written as[15,17]

$$f = \langle \psi_f | V | \psi_i \rangle \tag{1}$$

where ψ_i and ψ_f are initial and final state wave functions, respectively. $V = Z_1 Z_2 e^2/r_1 - Z_1 e^2/r_{12}$ is the Coulomb-interaction of the projectile (charge $Z_1 e$) with target nucleus (charge $Z_2 e$) and target electron. $\mathbf{r}_1$ and $\mathbf{r}_2$ are projectile and target electron coordinate, respectively, and $r_{12} = |\mathbf{r}_1 - \mathbf{r}_2|$. In the plane wave Born approximation (PWBA) the wave functions ψ are approximated by product wave functions

$$\psi = e^{i\mathbf{k}\mathbf{r}_1} \phi(\mathbf{r}_2) \tag{2}$$

where the projectile is represented by a plane wave and $\phi(\mathbf{r}_2)$ is an eigenfunction of the unperturbed atom. Then, the transition amplitude is given as

$$f_{\mathrm{PWBA}} = -Z_1 e^2/(2\pi^2 \hbar^2 q^2) \langle \phi_f(\mathbf{r}_2) | e^{i\mathbf{q}\mathbf{r}_2} | \phi_i(\mathbf{r}_2) \rangle \tag{3}$$

where $\mathbf{q} = \mathbf{k}_i - \mathbf{k}_f$ is the momentum transfer and $\mathbf{k}_i$ and $\mathbf{k}_f$ are initial and final projectile momentum, respectively. The total ionization cross section $\sigma \propto \int |f|^2 \, d\mathbf{q} \, d\varepsilon$ is obtained by integration of $|f|^2$ between the limits which follow from the conservation of energy ε and momentum during the collision. In Bethe approximation the exponential in equation (3) is expanded as $e^{i\mathbf{q}\mathbf{r}} = 1 + i\mathbf{q}\mathbf{r} + \cdots$, retaining only the first two terms. The integral cross section in Bethe approximation is obtained as[18]

$$\sigma = \frac{4\pi a_0^2}{T/R} M_i^2 \ln(4c_i T/R) \tag{4}$$

where $T = \frac{1}{2}mv^2$, m is the electron mass, a_0 is Bohr's radius, R is the Rydberg energy, $M_i^2 = \int |\langle f | e\mathbf{r} | i \rangle|^2 \, d\varepsilon$ is the square of the matrix element for ionization, integrated over the final energy of the ejected electron, and $1/c_i$ is a certain mean of the energy transfer and of the order of the binding energy. The ionization cross section in the nonrelativistic regime thus falls off with $\ln(T/R)/T$; it is for projectiles with equal velocity independent of the projectile's mass and the sign of the projectile's charge.

Results of PWBA calculations for K-shell and L-shell ionization by electron impact are for sufficiently large incident energies in good agreement with experimental results, particularly if electron exchange is also accounted for.[17,19,20] As an example, Figure 3 compares experimental and theoretical cross sections for ionization of atomic hydrogen by electron impact. This cross section shows a maximum around four times the threshold energy

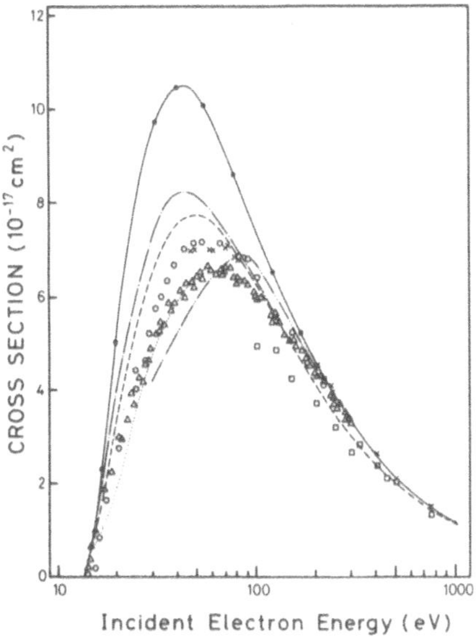

FIGURE 3. Total cross sections for ionization of atomic hydrogen by electron impact. Experimental results[22] are compared with PWBA (solid line), Born-Ochkur (dashed line), Born-exchange[21] (dash-dotted line), pseudo-state (dotted line), and optical model (dash-two-dotted line) calculations (from Ref. 20).

(13.6 eV). PWBA cross sections are in good agreement with experiment for incident energies of about ten times the threshold value. Below, large discrepancies are observed, which in part may be attributed to electron exchange, which is neglected in PWBA calculations. Hence, better agreement is obtained for calculations which take electron exchange into account.[21]

At low incident energies, the observed order-of-magnitude discrepancies between experiment and PWBA calculations (Figure 4) are caused by the inadequate description of the projectile using plane waves.[23] A significant improvement is achieved within the Coulomb-Born-exchange (CBE) approximation using Coulomb waves for the projectile.[24] It should be noted that the experimental cross section displayed in Figure 4 increases somewhat more steeply than expected from the CBE calculation, however. The experimental result is close to an energy dependence $\sigma \propto (E_0 - I)^n$, where E_0 and I are projectile energy and ionization energy, respectively, and with $n = 1.127$ as predicted by Wannier[25] (see also Ref. 26). For Ar-K ionization this threshold behavior was found to hold over an energy range of about 50 eV. In contrast, Xe-$L3$ ionization was found to increase linearly with excess energy $E_0 - I$ in the range 10–200 eV.[23] The results for Ar-K and Xe-$L3$ are both understandable within the frame of the Wannier theory, if the screening of the nucleus by core electrons is taken into account.[27] Estimates of the energy range over which the Wannier theory holds give values of 70 and 10 eV for Ar-K and Xe-$L3$, respectively, and are thus in agreement with the experimental observation.

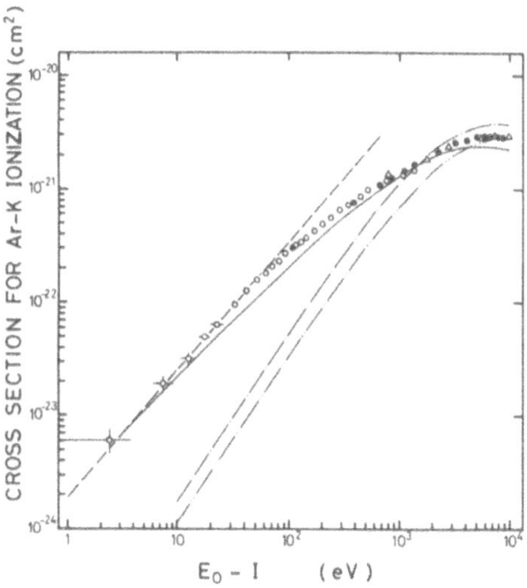

FIGURE 4. Total cross sections for K-shell ionization of argon by electron impact versus excess energy $E_0 - I$. Experimental data[23] are compared with -·-, PWBA[20]; -··-, PWBA-Ochkur[20]; and ——, Coulomb-Born-exchange[24] calculations. The dashed curve corresponds to $\sigma \propto (E_0 - I)^{1.127}$ (see text) (from Ref. 23).

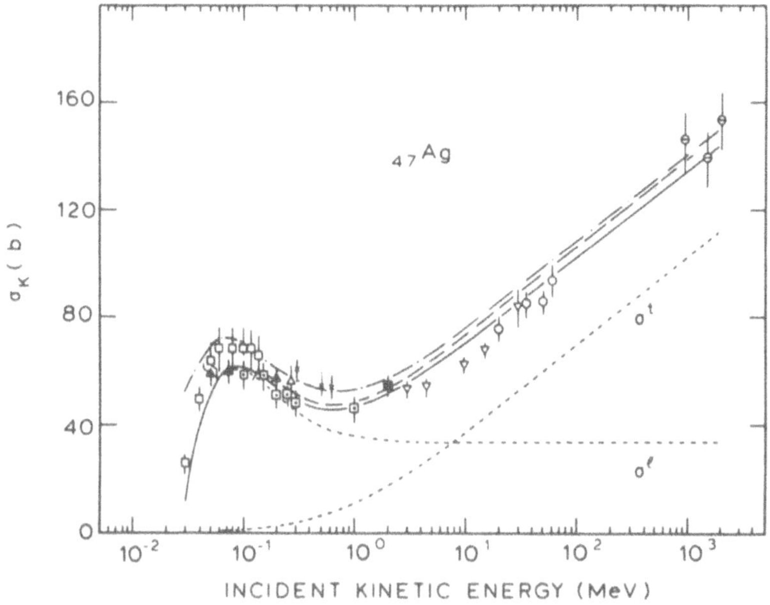

FIGURE 5. Total cross sections for K-shell ionization of silver by electron impact. Experimental results are compared with theoretical calculations of Kolbenstvedt[29] (dash-dotted line), and PWBA calculations of Scofield[30] (dashed line). Also given are theoretical calculations of Eschwey and Manakos[31] (solid line) and the decomposition into a longitudinal, σ^l, and a transverse, σ^t, part (from Ref. 28).

At relativistic velocities the interaction of the projectile with target electrons may be divided into two noninterfering contributions from longitudinal and transverse fields.(15) The longitudinal component results from the pure Coulomb field; it is dominant at nonrelativistic velocities and gives a constant cross section at relativistic energies.(20) The transverse component stems from the magnetic interaction; it becomes large at energies above 1 MeV and is responsible for the logarithmic rise of the cross section.(17,19) In Figure 5 the different components of the total ionization cross section are illustrated for ionization of silver atoms with relativistic electrons.(28)

3. ELECTRON CAPTURE

Ionization of target atoms can be achieved also via capture processes. In this case an electron is transferred from the target to the projectile, and free electrons are not created. In Figure 6 total cross sections for electron

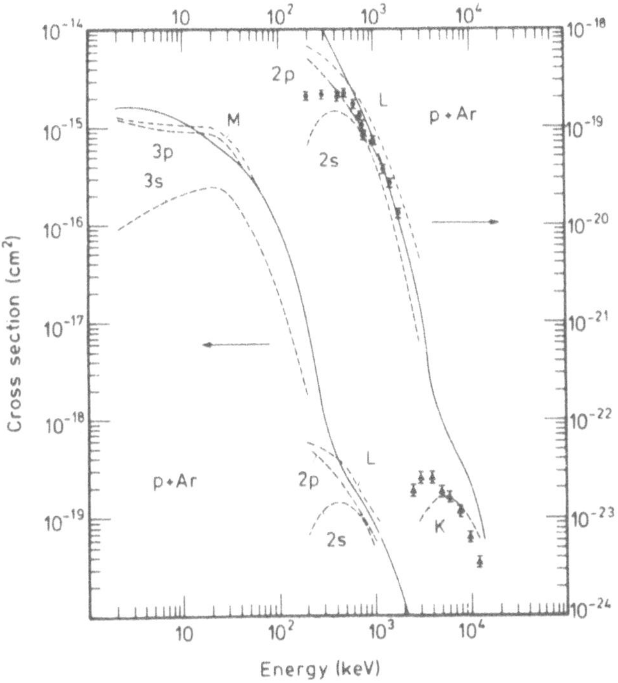

FIGURE 6. Total and subshell electron capture cross sections of argon by incident protons. Experimental total (solid line, Ref. 33), ●, L-shell(34); and ▲, K-shell(35) capture cross sections are compared with calculations of Lin and Tunell(32) for capture from indicated subshells (from Ref. 32).

capture by protons in collisions with argon atoms are displayed. These show the significance of contributions from the different shells of the argon atom. At low projectile energies up to about 300 keV the capture process is dominated by contributions from the argon M shell. In the regime from about 400 keV to 10 MeV, electron capture from the argon L shell provides the main contribution, and above about 10 MeV most electrons are captured from the K shell. The experimental results here are in reasonable agreement with theoretical calculations by Lin and Tunell,[32] which were performed within the atomic-expansion method, expanding the electronic eigenfunctions in terms of traveling atomic eigenfunctions of the two collision partners.

The simplest quantum-mechanical ansatz for electron capture, the first Born or Oppenheimer-Brinkmann-Kramers (OBK) approximation predicts, for $1s$–$1s$ capture and at sufficiently large incident velocities v, a velocity dependence of the capture cross sections $\sigma_c \propto v^{-12}$ (e.g., Refs. 36 and 37). A somewhat weaker velocity dependence is predicted by higher-order theories. For example, the high-velocity limit of the second Born approximation is given by[38,36]

$$\sigma_{B2} \underset{v\to\infty}{\to} [0.295 + 5\pi v/2^{11}(Z_p + Z_t)]\sigma_{\mathrm{OBK}} \tag{5}$$

This prediction is in agreement with the observation that at intermediate velocities theoretical cross sections calculated within OBK differ from experiment by factors of typically about 3.

The reason for the weaker velocity ($\propto v^{-11}$) dependence of higher-order capture theories compared to the simple OBK theory is easily understood. Since a free electron cannot be captured in a single collision, the dominant contribution to the capture process at high velocities comes from a double scattering mechanism. This double scattering mechanism, first recognized by Thomas,[39] consists of two binary collisions. In the first collision, the projectile interacts with the target electron, which recoils with velocity v through an angle of 60°. In a second collision, the recoiling electron is scattered off the target nucleus by 120° (Figure 7a). The electron escapes the target ion in the same direction and with the same velocity as the projectile, i.e., with almost zero momentum with respect to the projectile. In order to conserve energy and momentum during the collision the projectile undergoes a small but observable scattering; the corresponding scattering angle, often called the Thomas angle $\theta_T \approx (m_e/m_p) \sin 60°$ amounts to about $\theta_T = 0.475$ mrad for proton impact.[36,40] Here m_e and m_p are the mass of electron and projectile, respectively. The appearance of a structure attributed to the above double-scattering mechanism has been observed experimentally by Horsdal-Pedersen *et al.*[41] in H^+–He and by Vogt *et al.*[42] in H^+–H

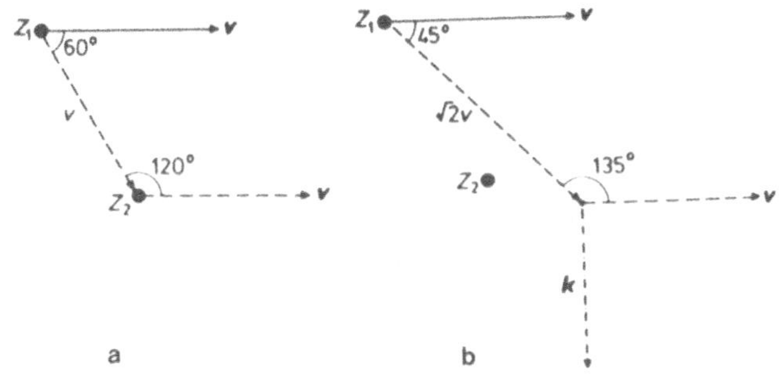

FIGURE 7. Schematic representation of Thomas scattering mechanisms[37] contributing to electron capture (see text) (from Ref. 36).

collisions. The latter results are displayed in Figure 8 together with theoretical calculations.[42,43]

Another double scattering mechanism is illustrated in Figure 7b. Here the electron recoils from the projectile with velocity $\sqrt{2}v$ and at 45°. In a second collision this "primary" electron is scattered off a second electron, whereby both electrons leave the target with velocity v, and one parallel with the projectile and the other perpendicular. In such a collision a free electron should be ejected at 90° in coincidence with a capture event. First experimental results so far did not find any evidence in the ejected electron spectrum in 300 keV H^+–He collisions.[44] A somewhat different conclusion was reached by Horsdal *et al.*,[45] who measured the ratio for production of doubly and singly ionized helium ions in few 100 keV H^+–He charge-changing collisions. They observed a strong increase of the fraction of doubly ionized helium ions near the projectile scattering angle $\theta_s =$ 0.55 mrad, which is the critical angle[40] for the above double-scattering

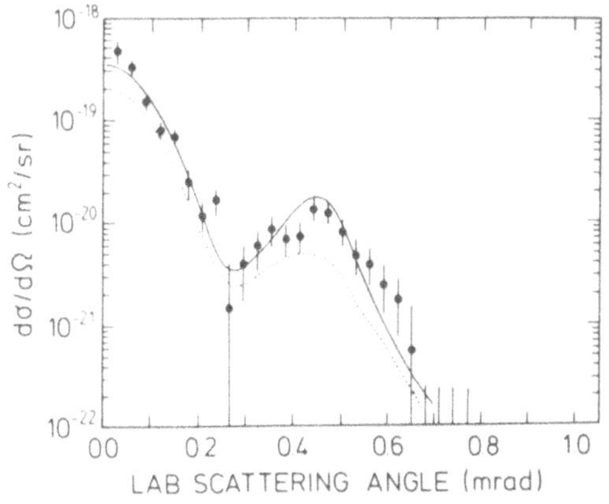

FIGURE 8. Differential capture cross section $d\sigma/d\Omega$ of 5-MeV protons in atomic hydrogen. Experimental data of Vogt *et al.*[42] are compared with calculations based on the strong-potential Born (dotted line) and the strong-potential Born in full-peaking approximation (solid line)[43] (from Ref. 42).

mechanism (Figure 7b). While the position of this peak was in agreement with expectation, the observed peak fraction of He^{++} compared to He^{+} was surprisingly large (about 20% at 500 keV proton energy) and thus much larger than expected at this rather low incident velocity.[45]

It was recently pointed out by Dewengan and Eichler[46] that first-order charge-exchange theories are significantly improved if the Coulomb boundary conditions are properly taken into account. Unlike OBK, which is unable to reproduce the experimental cross sections, their "boundary-corrected" first-order Born calculation yields remarkable agreement with experiment. This holds for both symmetric and asymmetric collisions and for intermediate velocities where the Thomas double-scattering mechanism does not dominate.

4. SLOW COLLISIONS

In slow atomic collisions, i.e., if the relative motion of the two nuclei is much slower than the orbiting velocity of active electrons, the interaction between two atomic particles may be viewed as within a quasimolecule transiently formed during the collision (e.g., Refs. 11–14). Within this model the binding energy of a single electron moving in the fields of two nuclei is calculated as a function of the internuclear separation R. Figure 9 displays a so-called (schematic) correlation diagram for the H^{+}–H collision system.[47] For clarity, the l-degeneracy of states with the same main quantum number n has been removed. In this molecular-orbital (MO) picture excitation of H($2p$) takes place when the two nuclei approach each other and an electron initially in the $2p\sigma$ MO may be transferred at small R via rotational coupling into the $2p\pi$ MO. Here σ and π refer to the projection of the

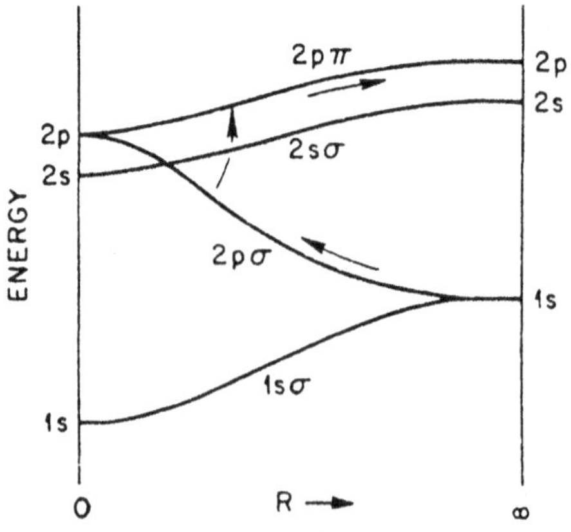

FIGURE 9. Schematic correlation diagram for H_2^+. For clarity, the l-degeneracy has been removed.

angular momentum on the internuclear axis. Thus, H(2p) excitation via the 2$p\pi$ MO should result in H(2p_1) excitation only. First experimental results for the H^+-H collision system confirm this prediction.[48] Measuring the degree of linear polarization of Lyman-α radiation emitted during the decay of the excited H(2p) state, detailed information is obtained about the relative population of H(2p_0) and H(2p_1) magnetic substates (e.g., Ref. 47). The integral alignment A_{20} is defined as[47]

$$A_{20} = (\sigma_1 - \sigma_0)/\sigma(2p) \tag{6}$$

where $\sigma(2p) = \sigma_0 + 2\sigma_1$, and σ_0 and σ_1 are the partial cross sections for H(2p_0) and H(2p_1) production, respectively. A_{20} is, at low incident energies of a few keV, close to the maximum value of +50% predicted by the 2$p\sigma$–2$p\pi$ rotational coupling theory (Figure 10).[48] This result is also in agreement with recent numerical calculations.[49–51]

A somewhat different picture of H(2p) excitation emerges for H^+-He collisions. There the 1$s\sigma$ and 2$p\sigma$ MOs correlate to the H(1s) and He(1s) states, respectively, and are no longer asymptotically degenerate. Charge exchange excitation into H(2p) in H^+-He collisions is therefore the result of a two-step process.[47,52,53] The first step takes place at large internuclear separations of few atomic units and radially couples the 1$s\sigma$ ground state to the 2$p\sigma$ MO. Again, H(2p) excitation is then the result of a 2$p\sigma$–2$p\pi$ rotational coupling at small internuclear distances (Figure 11). However, experimental results (Figure 12) for the H^+-He system have shown that in addition to H(2p_1) excitation at incident energies of a few keV an about equal population of H(2p_0) is observed.[52,53] In Figure 13 the relative population of H(2p_0) excitation is plotted versus impact parameter b.[55] The experimental results here were obtained using the so-called polarized photon-scattered projectile coincidence technique[47]; they clearly

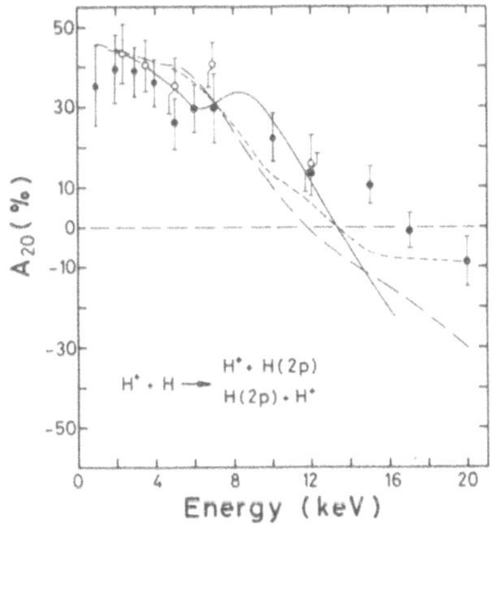

FIGURE 10. Integral alignment A_{20} of H(2p) as a function of incident energy for H^+-H collisions. Experimental results[48] are compared with theoretical calculations of Lüdde and Dreizler[51] (solid line), and Fritsch and Lin (dash-dotted line, Ref. 49; dashed line, Ref. 50).

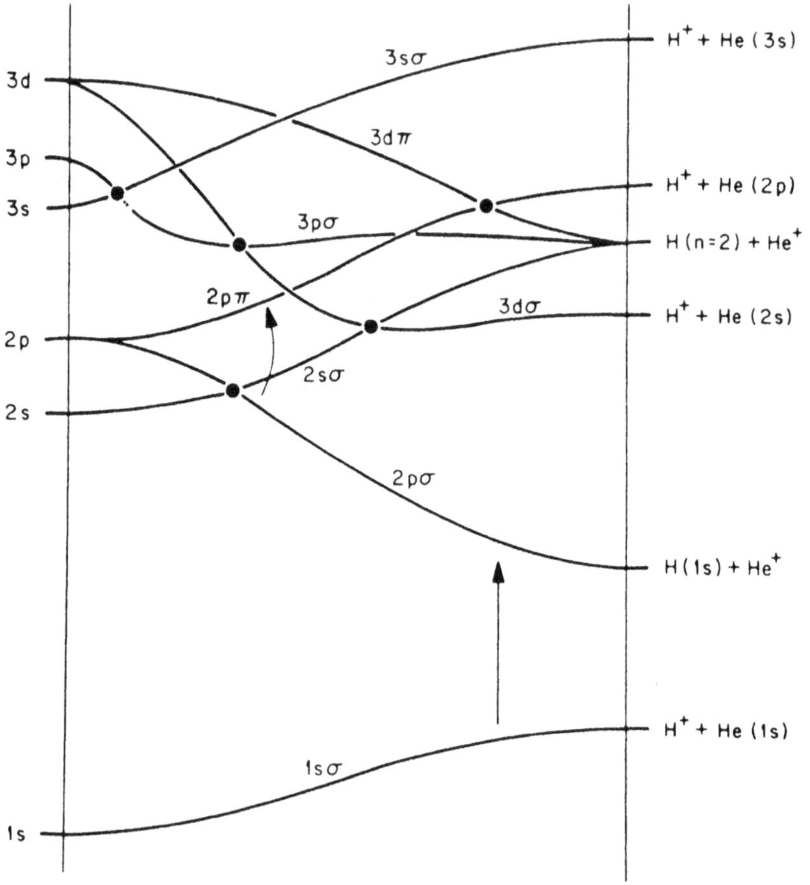

FIGURE 11. Schematic correlation diagram for $(H-He)^+$ (from Ref. 47).

demonstrate that $H(2p_1)$ charge exchange excitation dominates at large impact parameters, whereas $H(2p_0)$ is dominant for small b. The experimental results here are in excellent agreement with recent theoretical calculations by Macek and Wang[56] and with model calculations by Fritsch.[57] Macek and Wang[56] have pointed out that owing to the $2p\sigma$–$2s\sigma$ radial coupling at about $R \approx 0.4$ a.u. neither the diabatic $2s\sigma$ nor the diabatic $2p\sigma$ MO asymptotically converge to adiabatic Σ states for $R \to \infty$. Hence, rotational coupling populates not only the $2p\pi$ but also the $2s\sigma$ state via the double-step rotational coupling Σ–Π–Σ, hence resulting in $H(2p_0)$ excitation. For symmetry reasons this double-step rotational coupling becomes largest at $b = 0$. The single-step rotational coupling has its maximum around $b \approx 1$ a.u., whereas it is zero at $b = 0$.

In quasi-one-electron systems, for example $H(1s) + He \to H(2p) + He$, only a single (outer) electron is active, whereas other (core) electrons remain

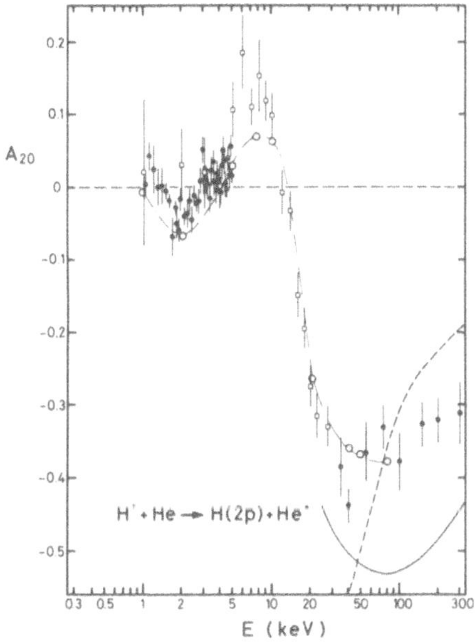

FIGURE 12. Integral alignment A_{20} of H(2p) as a function of incident energy for H^+-He collisions. Experimental results—●, Ref. 52; □, Ref. 53—are compared with theoretical calculations of Kimura and Lin[54] (open circles, dash-dotted line to guide the eye), and with CDW calculations including (dashed line) and excluding (solid line) postcollision interaction effects (from Ref. 52).

passive during the collision. Menner *et al.* (1981) have experimentally and theoretically investigated 2p excitation in Li(2s) + He → Li(2p) + He collisions. Despite its suggested simplicity it was observed that at large impact parameters Li(2p) excitation proceeds via a nonlocalized direct excitation mechanism, whereas at small impact parameters a rotational coupling is effective.

Somewhat simpler is H(2p) excitation in H + He → H(2p) + He collisions. Here, and for not too large projectile energies (below about 5 keV), the dominating excitation mechanism is governed by transitions at small impact parameters between quasimolecular states. Recent calculations by Kimura and Lane[59] demonstrate the importance of a radial Σ-Σ coupling, whereas the rotational Σ-Π coupling is relatively weak. In effect, this results in a negative integral alignment A_{20}, indicating a preferred population of H($2p_0$) compared to H($2p_1$). In Figure 14 experimental and theoretical

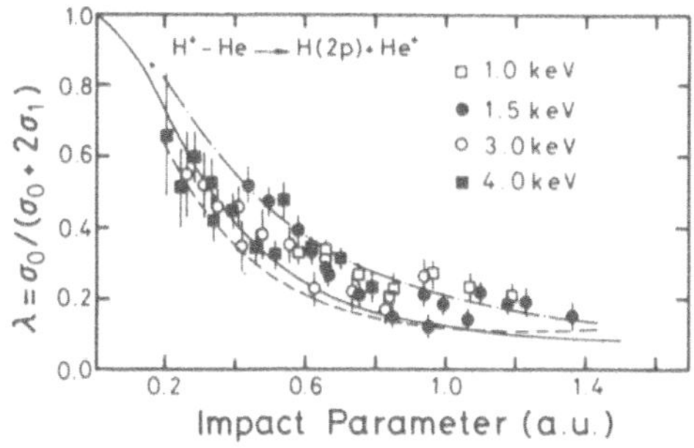

FIGURE 13. Relative population of H($2p_0$) versus impact parameter b in H^+-He collisions at incident energies of 1-4 keV.[55] Solid line is a theoretical calculation of Macek and Wang,[56] the other two lines are from Fritsch.[57]

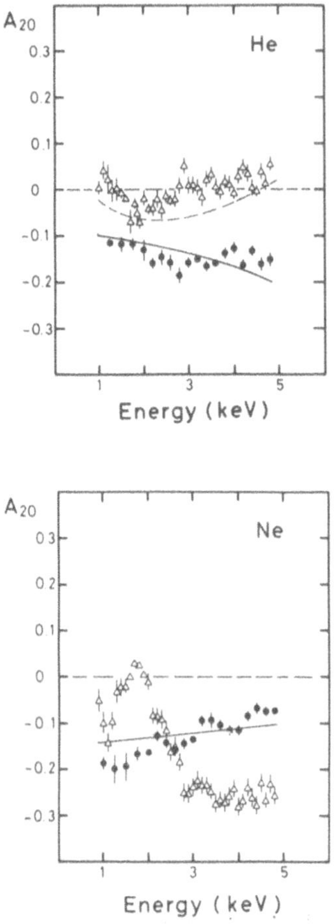

FIGURE 14. Integral alignment A_{20} for H($2p$) excitation in ●, H-He, Ne, and ΔH$^+$-He, Ne collisions. Dashed and solid lines are theoretical calculations by Kimura and Lin[54] and Kimura and Lane,[59] respectively (from Ref. 60).

results for H($2p$) excitation in H + He, Ne and H$^+$ + He, Ne collisions are displayed. As can be seen, good agreement between experiment and theory is obtained for both He and Ne targets.[59]

The orientation, i.e., the transferred angular momentum $\langle L_y \rangle$ perpendicular to the scattering plane was obtained from a coincidence experiment between emitted photons and scattered projectiles for 2-keV and 3-keV deuterium-helium collisions (Figure 15). Experimentally it was observed that, within the experimental error bars, $\langle L_y \rangle$ is independent of the impact parameter but displays a pronounced velocity dependence,[60] in fair agreement with calculations by Kimura and Lane.[59] Using their adiabatic potential energy diagram one can estimate the relative phase that develops if H($2p$) excitation proceeds along either Σ or Π orbitals. The adiabatic phase difference accumulated during the encounter satisfactorily explains the observed magnitude and velocity dependence of $\langle L_y \rangle$.

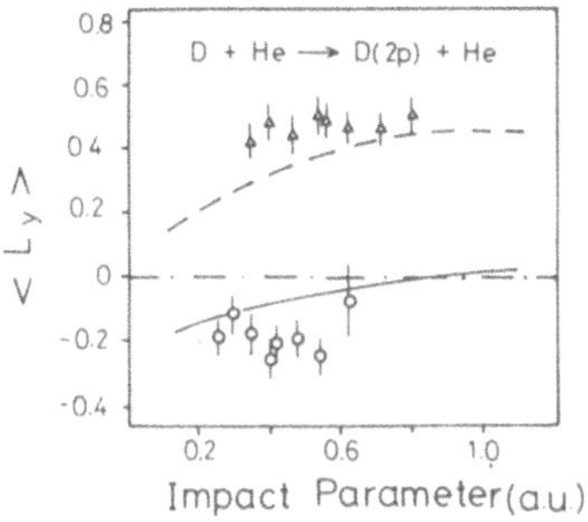

FIGURE 15. Transferred angular momentum $\langle L_y \rangle$ perpendicular to the scattering plane versus impact parameter b. Experimental results for △, 2-keV and ○, 3-keV D-He collisions[60] are compared with theoretical calculations by Kimura and Lane[59] (from Ref. 60).

Inner shell vacancies may be created also by couplings in the transiently formed quasimolecule. For example, $2p\sigma$-$2p\pi$ rotational coupling was found to be responsible for vacancy production in atomic K shells (see, e.g., Refs. 12-14). Similarly, L-shell excitation in, for example, Kr-Kr collisions are induced by rotational $4f\sigma$-$4f\pi$-$4f\delta$-$4f\phi$ couplings.[61] Results for these collision systems are discussed in Chapter 7 of this book.

5. MANY-ELECTRON PROCESSES AND HIGHLY CHARGED ION IMPACT

While in the past most theoretical and experimental work was devoted to one-electron processes, many-electron processes have received considerable attention in the last few years. Part of this increased awareness of many-electron processes comes from collisions of multiply-charged ions with atoms, where it was observed that such processes may have cross sections comparable to or even larger than for one-electron processes. Conceptually, many-electron processes may be considered correlated or uncorrelated. Correlated processes are the result of electron-electron correlation and have been observed in photoionization experiments (see, e.g., Refs. 62 and 63). In contrast, so-called uncorrelated processes use a statistical description for the many-electron process (e.g., Refs. 64 and 65).

Some results for ionization of neon by electron and proton impact are displayed in Figure 16 on an equal velocity scale.[66] The data displayed are for ionization of a single $2s$ and $2p$ electron, and for double ionization of $2s$ and $2p$ electrons. They show that asymptotically the proton data merge with the electron data. In general, cross sections for double ionization are about one order of magnitude smaller than corresponding data for single ionization. Also, the energy dependence is different for single and double ionization.

In Figure 17 experimental results for double ionization of helium atoms by electron, proton, and antiproton impact are displayed.[67] The cross section ratio $R = \sigma^{++}/\sigma^{+}$ for double relative to single ionization is plotted

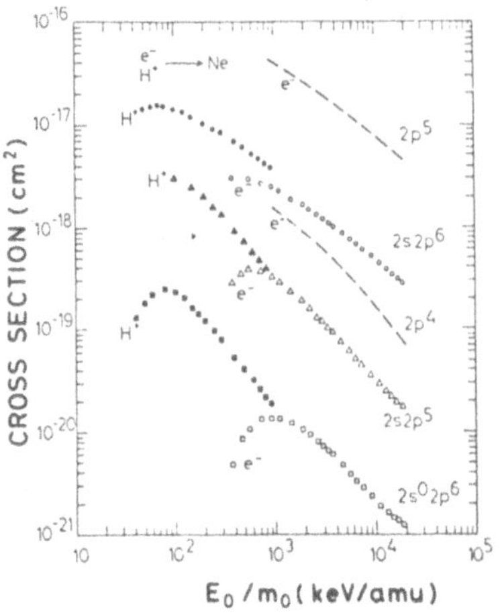

FIGURE 16. Total cross sections for ionization of neon by proton and electron impact. The final ionic states are indicated (see text) (from Ref. 66).

on an equal velocity scale. Since impact ionization by heavy projectiles (protons, antiprotons, bare ions) is similar to ionization by electron impact, cross sections are predicted to become independent of the projectile's mass and the sign of the incident charge for sufficiently large incident velocities.[18] Pronounced differences between, for example, electron and proton impact only occur at low velocities and as a consequence of the different mass-to-charge ratio and of the different charge. While this prediction agrees with the experimental results for one-electron processes like single excitation and single ionization (e.g., Refs. 67 and 68), it does not hold for double ionization of helium at medium and large velocities. Figure 17 shows that for incident velocities of about 10 a.u. (corresponding to an electron energy of about 1.36 keV) electron impact is about 50% more efficient in producing doubly ionized helium than proton impact of the same velocity.

The fact that antiproton impact around 16 a.u. is as efficient as electron impact in producing double ionization (and even more efficient at lower velocities) may have pointed to an interpretation that the observed differences are due to an interference between ionization amplitudes for double ionization of independent electrons and double ionization due to single ionization followed by shake-off.[69] While the first and second amplitudes are proportional to Z_1^2 and Z_1, respectively, the square of both amplitudes added coherently contains a term proportional to Z_1^3 and is thus dependent on the sign of the projectile charge. However, recent calculations by Reading and Ford[70] and Olson[71] have shown that other effects are responsible. As was pointed out, the antiproton due to its negative charge can *push* one electron into the other already for distant collisions, whereas the proton needs to be closer to the nucleus in order to *pull* the first electron into a trajectory that collides with the second electron. In addition, the antiproton

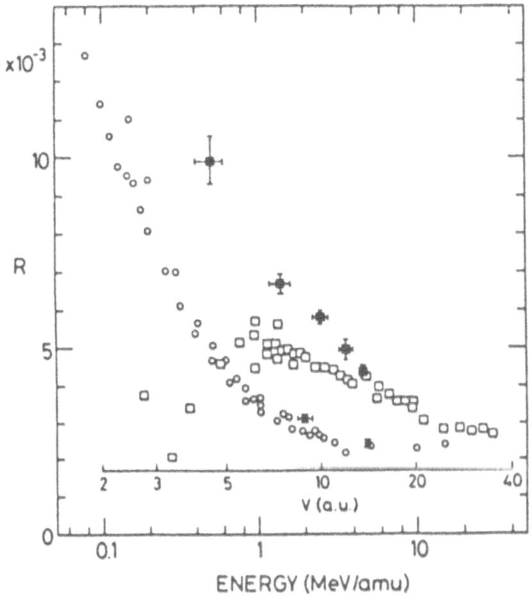

FIGURE 17. The ratio $R = \sigma^{++}/\sigma^{+}$ between double- and single-ionization cross sections for electrons (open squares), protons (open and closed circles), and antiprotons (solid squares) colliding with helium atoms (from Ref. 67).

for small impact parameters screens the helium nucleus, while the proton increases the binding of the two electrons. These calculations also show that for incident energies ≤ 500 keV single ionization by proton impact is more efficient (about 1/3 at 250 keV) than antiproton impact.(71)

It is not clear how the above picture fits into a previous observation that double ionization of helium (and of other rare gas atoms) becomes more likely for "violent" collisions indicated by a large energy transfer. Figure 18 displays such a result for 300-keV H^{+}–He collisions.(72) The experiment was performed by detecting electrons in coincidence with recoil ions whose charge state was determined by time of flight. These data have been discussed within the concept of ionization of uncorrelated electrons; using a statistical description (i.e., a binomial distribution) for multiple ionization, a reasonable agreement of such model calculations with experiment was achieved. Qualitatively, multiple ionization is the result of close collisions (resulting in a large energy transfer) where the probability for ionization is large, whereas single ionization dominates for distant collisions (small energy transfer).(72) It might be interesting to compare these findings with results from photoionization experiments. Experimental data of Wight and Van der Wiel(73) obtained as a function of the energy E_γ of incident

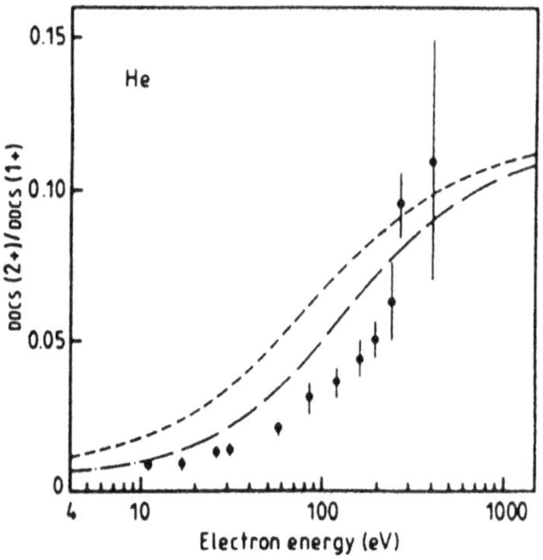

FIGURE 18. The ratio DDCS(2+)/DDCS(1+) for production of doubly and singly charged helium ions by 300-keV proton impact versus the kinetic energy of ejected electrons. Here DDCS is a doubly differential cross section for electron ejection; the ejection angle was 90°. The experimental results are compared with predictions of the independent particle model using the same ionization potential (broken curve) for single and double ionization, or a charge-state-dependent ionization potential (chain curve) (from Ref. 68).

photons indicate that up to $E_\gamma = 200$ eV (corresponding to a kinetic energy of ejected electrons of ≈ 175 eV and $\leqslant 120$ eV for single and double ionization, respectively) photoionization produces a higher fraction of He^{++} than 300-keV proton impact. At larger excess energies the fraction of He^{++} produced by photoionization is believed to saturate,[(62,63)] whereas the proton data show a strong increase.

At low and medium incident velocities ionization of target atoms is to a large extent dominated by charge-changing collisions. This is particularly true for incident highly charged ions. Then, the large potential energy brought into the collision by the highly charged ion rather than its kinetic energy is the driving factor for electronic processes. Experimental results for electron capture in O^{8+}-He collisions at medium velocities are displayed in Figure 19. In the range of incident energies from 10 to 40 MeV the cross section for single capture decreases by more than two orders of magnitude[(74–76)]; the magnitude of the experimental cross section and its energy dependence are here in excellent agreement with theoretical calculations based on the continuum-distorted-wave (CDW) approximation.[(77)] The agreement with calculations based on the classical Bohr–Lindhard model[(78)] or with the empirical scaling rule of Schlachter *et al.*[(79)] is less satisfactory. In the intermediate velocity regime of interest here (5–10 a.u.), both the classical Bohr–Lindhard theory as well as quantum-mechanical theories like OBK and CDW predict a charge-state (q) dependence of the capture cross section $\sigma_c \propto q^3$,[(80)] in fair agreement with experimental results for bare ions.[(74)]

Also shown in Figure 19 are experimental data for capture of two electrons. Typically, in the energy range investigated here and for a helium target, the double capture cross sections are more than two orders of magnitude smaller than corresponding single capture data.[(74)] The measured

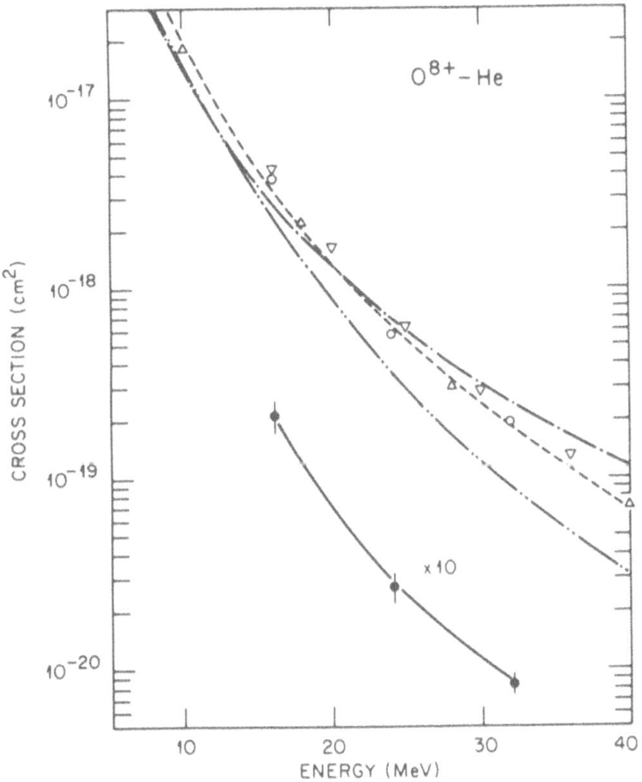

FIGURE 19. Cross sections for single and double electron capture in O^{8+}-He collisions versus projectile energy. Experimental data[74–76] for single capture (○, ▽, △) are compared with theoretical calculations based on the CDW approximation,[77] the Bohr-Lindhard model,[78] and an empirical scaling rule.[79] Also shown are experimental results for double capture (●) (from Ref. 74).

double capture cross sections are also significantly different from what has been estimated on the basis of the independent particle model. Part of this discrepancy could be resolved by considering that a large fraction of capture takes place into excited states with main quantum number $n \geqslant 2$; if both electrons are captured into $n \geqslant 2$ these states may autoionize and thereby escape detection as double capture events.[74]

At low and medium velocities many-electron processes may become comparable to one-electron processes already for distant collisions. Figure 20 shows some results for 0.1-MeV/u Au^{21+}-He[81] and 0.25-1-MeV/u U^{26+}-He collisions.[82] Total cross sections for single electron capture (SC), single (SI) and double ionization (DI), and transfer-ionization (TI) are obtained from a coincidence measurement between projectile-ion and recoil-ion including charge-state analysis of both ions. It is noted from Figure 20

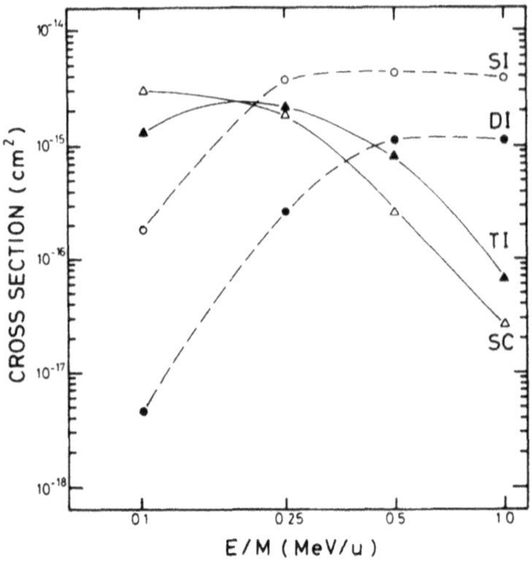

FIGURE 20. Partial cross sections for △, single capture (SC); ○, single ionization (SI); ●, double ionization (DI); and ▲, transfer ionization (TI) in 20-MeV Au^{21+}-He[81] and 60-240-MeV U^{26+}-He[82] collisions (solid and dashed lines to guide the eye only).

that, for sufficiently large incident charge states, transfer ionization (i.e., one electron is captured by the projectile plus a second electron of the target is ionized) is about equal to or larger than the corresponding single capture cross section. This two-electron process is thus, for this collision system, of comparable magnitude to the one-electron capture process. At the same time, capture of two electrons is comparatively small, amounting to about 10% or less of the corresponding single capture cross section only. Andersen *et al.*,[83] for the Au^{q+}-He collision system, investigated electron emission in the forward direction (cusp-peak) in coincidence with projectiles of a given charge state. As it turned out, most cusp electrons were observed with projectiles that had captured one electron. While the cross section for production of cusp electrons was large enough to account for the entire transfer-ionization cross section, the shape of the cusp was that for electron loss (ELC) rather than electron capture (ECC) to the continuum. This finding indicates transfer of two electrons to the projectile, one of which is subsequently lost to the projectile's continuum. While the independent-electron approximation was found to be invalid for this collision, the cusp peak was thought to arise from the creation and decay of a highly excited, highly correlated two-electron state.[83,84]

Evidence for correlated two-electron transfer was recently also provided from a study of electrons ejected in 60-keV O^{6+}-He collisions.[85] By means of high-resolution electron spectroscopy it was found that O^{6+} impact on He creates intense Coster-Kronig lines which were attributed to two-electron transfer processes producing $O^{4+}(1s^2\,2p\,nl)$ configurations with $n \geq 6$. This result was interpreted as being caused by a correlated double-capture process, where resonantly one electron is transferred into $2p$ and another electron into a Rydberg state.[85]

Simultaneous transfer and excitation (TE) in ion-atom collisions has received considerable attention in the last few years (e.g., Ref. 86). In such a collision, one (projectile) electron is excited and a second electron is captured by the projectile. In so-called nonresonant transfer and excitation (NTE), the projectile nucleus interacts with the target electron resulting in capture and the target nucleus interacts with the projectile electron resulting in excitation. Hence, NTE is the result of electron–nucleus interactions. Unlike NTE, in resonant transfer and excitation (RTE) both electrons directly interact and exchange energy and momentum with each other. The RTE process is similar to dielectronic recombination (DR), except that in DR the captured electron is initially free.

Both processes, RTE and NTE, lead to the same final states. Hence it is not possible to distinguish experimentally between RTE and NTE,[87] except for the energy dependence, which is rather different. Figure 21 shows results of Tanis *et al.*[87] for $S^{13+} + He \rightarrow S^{12+*} + He^{+}$ collisions. Here S^{12+*} is a doubly excited ion with a *K*-shell vacancy that eventually decays via emission of characteristic X rays or Auger electrons. The results displayed in Figure 21 were obtained by analyzing the final charge state of the projectile in coincidence with emitted characteristic X rays. Alternatively, one could have used high-resolution X-ray or Auger electron spectroscopy.[88] The cross section for coincident S-*K* X-ray emission exhibits two maxima, with

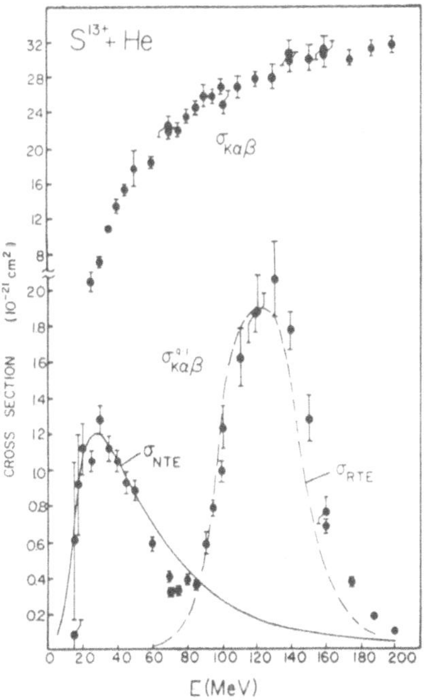

FIGURE 21. Sulfur-*K* X-ray cross sections in S^{13+}-He collisions. $\sigma_{K_{\alpha\beta}}$ is the total (noncoincident) S-K X-ray production, and $\sigma^{q-1}_{K_{\alpha\beta}}$ is for S-*K* X rays in coincidence with single electron transfer. Dashed curve is a RTE calculation of Brandt,[89] multiplied by 0.85. Solid curve is the calculated NTE contribution,[90] normalized to the experimental data (from Ref. 87).

the high- and low-energy maximum being attributed to RTE and NTE, respectively. At the same time, noncoincident S-K vacancy production was found to increase monotonically with projectile energy. The results are here in qualitative agreement with theoretical calculations.(89-91)

In the last few years a large amount of dielectronic recombination (DR) data has accumulated (e.g., Refs. 92-94). Similar to RTE, in dielectronic recombination a *free* electron is captured by an incident ion, with the excess energy and the momentum being exchanged with a core electron. Thus, DR, being a type of inverse Auger process, shows a resonant behavior when the free electron energy matches the excitation energy of a bound electron. A typical DR process for a sodiumlike ion is

$$Mg^+(3s) + e^-(E) \rightleftarrows Mg(3p, n_f) \rightarrow Mg(3s, n_f) + h\nu \qquad (7)$$

The different steps in DR are pictorially represented in Figure 22.(93) An initially free electron with kinetic energy $E = \Delta E - \varepsilon$ interacts with a bound electron. If the energy exchange between the two electrons equals the excitation energy ΔE of a bound electron, the bound electron will be excited and at the same time the initially free electron will find itself attached to the ion and in a Rydberg state with energy ε. The doubly excited atomic state thus formed may either autoionize, in which case we have again the initial situation, or it can radiatively stabilize, in which case dielectronic recombination has occurred.

Experimental results for the sodiumlike ion S^{5+} are displayed in Figure 23.(95) The measurements were performed in a merged-beam experiment; results are given as a function of the kinetic energy of electrons in the rest frame of the ion. The experimental results are here compared with theoretical calculations for DR, showing the influence of an external electric field on the dielectronic recombination rate. As was pointed out (e.g., Ref. 93),

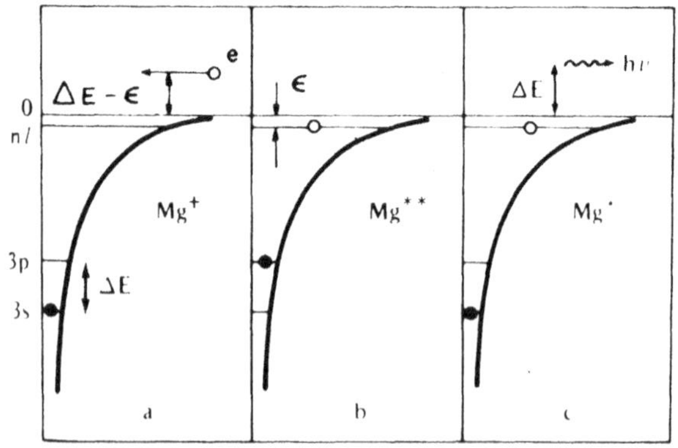

FIGURE 22. Schematics of dielectronic recombination (from Ref. 93).

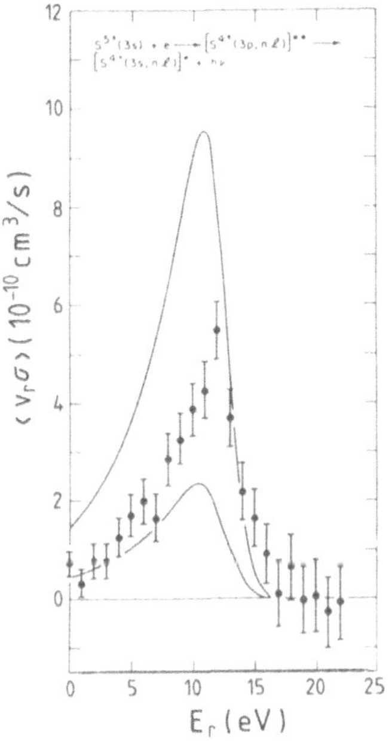

FIGURE 23. Dielectronic recombination rate $\langle v_r\sigma \rangle$ for S^{5+} versus electron energy E_r. Upper and lower curves are theoretical calculations accounting for full mixing and no mixing of angular momentum states, respectively (see text) (from Ref. 95).

mixing of angular momentum states by external electric fields affects radiative and autoionizing rates and hence the dielectronic recombination rate. The upper and lower curves are the theoretical results with full mixing and without mixing, respectively. Best agreement between experiment and theory is here obtained for an electric field of 25 V/cm.[94,96]

The influence of electric fields on DR of Mg^+ was studied in some detail by Müller *et al.*[97,98] In Figure 24 their results for a fixed incident electron energy of 4.43 eV are plotted versus Rydberg quantum number n_f [cquation (7)]. Measurements are presented for three electric fields in the

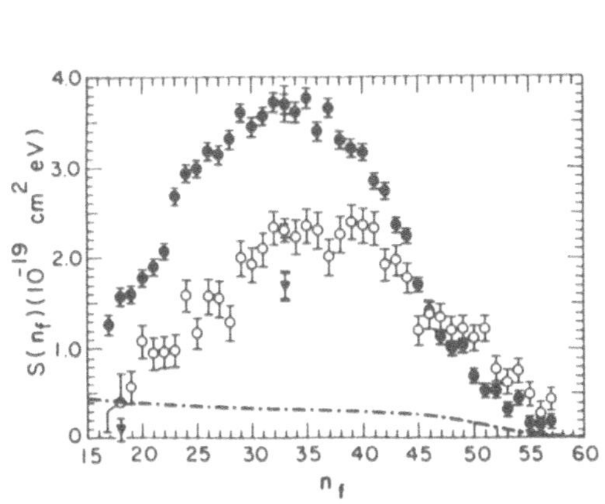

FIGURE 24. Collision strength $S(n_f)$ versus Rydberg quantum number n_f for dielectronic recombination of Mg^+ with electrons of kinetic energy 4.43 eV. Here $S(n_f) = \sigma(n_f)\Delta E_e$ is the DR cross section folded with the energy spread ΔE_e of the electron beam. Experimental results are for external electric fields of 3.6 V/cm (inverted triangles), 7.2 V/cm (open circles), and 23.5 V/cm (closed circles). The dash-dotted curve is a theoretical calculation for no electric field (0 V/cm) (from Ref. 98).

interaction region of 3.6, 7.2, and 23.5 V/cm. A detailed analysis of these data[98] shows that for the total cross sections experiment and theory agree within about 20%, whereas for the n_f-state distributions the comparison is less satisfactory.

6. RADIATIVE COLLISIONS

The origin of a continuous part in the X-ray spectra produced by electron impact is due to the acceleration of incident electrons in the nuclear field of atoms (electron–nucleus, or atomic-field bremsstrahlung) or in the field of other electrons (electron–electron bremsstrahlung). Classically, an electron moving along a parabolic path in a purely Coulombic field feels an acceleration $a = Ze^2/(mr^2)$, with Ze the electric charge of the nucleus, m the electron mass, and r the distance between incident electron and nucleus. This accelerated electron gives rise to the emission of a spectrum of electromagnetic radiation with intensity $dI/d\Omega_\gamma$[99]

$$\frac{dI}{d\Omega_\gamma} = \frac{a^2e^2}{4\pi c^3}\sin^2\theta_\gamma \tag{8}$$

with c the speed of light and θ_γ the angle of emission with respect to the direction of the incident electron. The result for the differential bremsstrahlung cross section $d\sigma/dk$, obtained by integrating over the directions of the outgoing electron, is[100]

$$\frac{d\sigma}{dk} = \frac{16\pi\alpha^5Z^2a_0^2}{3\sqrt{3}k\beta_1^2} \tag{9}$$

where $\beta_{1,2} = v_{1,2}/c$, $\alpha = 1/137.04$, and a_0 is the Bohr radius. It follows from equation (9) that $d\sigma/dk$ is proportional to Z^2 and inversely proportional to the photon energy k. A somewhat different result is obtained from a quantum-mechanical treatment. Sommerfeld and Maue, treating the problem nonrelativistically, have calculated matrix elements f in dipole approximation, where f is given by[101]

$$f = \langle\psi_f|e\mathbf{r}|\psi_i\rangle \tag{10}$$

and where the incident and scattered electron is described by pure Coulomb wave functions ψ_i and ψ_f, respectively. The result of Sommerfeld and Maue may be expressed as[101,99]

$$\frac{d\sigma}{dk} = \frac{16\pi^2Z^2\alpha^5a_0^2X_0}{3k\beta_1^2[\exp(2\pi n_1)-1][1-\exp(-2\pi n_2)]}\frac{d}{dX_0}|F(x_0)|^2 \tag{11}$$

where $n_{1,2} = \alpha Z/\beta_{1,2}$, $X_0 = -4n_1n_2/(n_1 - n_2)^2$, and $F = F(in_1, in_2, 1, X_0)$ is the hypergeometric function. In equation (11) relativistic effects, e.g., retardation of the electromagnetic potential, and screening effects, e.g., the screening of the nuclear field by surrounding electrons, are neglected. The currently most accurate theoretical results for atomic field bremsstrahlung are numerical calculations,[99,102] in which the bremsstrahlung process is treated as a single electron transition in a relativistic self-consistent (screened) potential, and where the electron wave functions are obtained in partial-wave series by numerically integrating the radial Dirac equation.

In Figure 25 experimental results for $d\sigma/dk$ for different gaseous targets are compared with results from equation (11) and with calculations of Kissel *et al.*[102] The experiment was performed by bombarding atomic (He, Ne, Ar, Kr, Xe) or molecular (N_2, UF_6) targets with electrons of incident energies $T = 2.5$, 4, and 10 keV and photon energies of $k = 2$, 3.6, and 9 keV, respectively.[103] Scaled cross sections $Z^{-2}\, d\sigma/dk$ are displayed to suppress the strong Z dependence of $d\sigma/dk$. For low-Z targets the two theoretical calculations give identical results. Screening effects become important for high-Z targets and low incident electron energies. There, the results from equation (11) are larger than the cross sections of Kissel *et al.*,[102] the latter taking screening into account. While the experimental data are relative cross sections and have been normalized to the calculations of

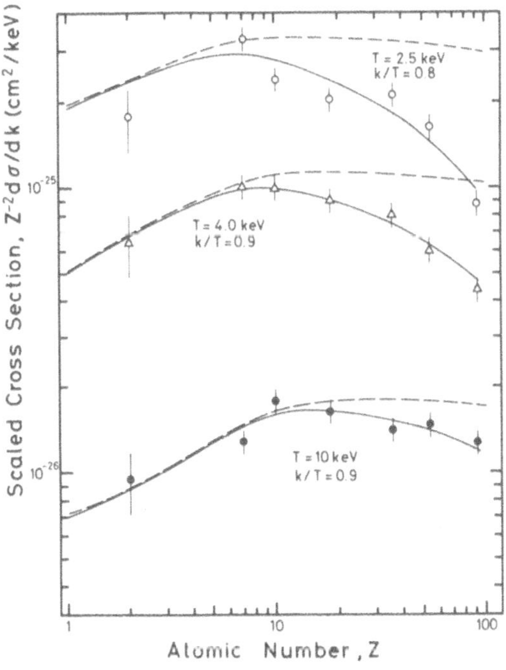

FIGURE 25. Scaled cross section, $Z^{-2}\, d\sigma/dk$, for electron bremsstrahlung versus atomic number Z. The experimental data for incident electron energies $T =$ 2.5, 4, and 10 keV and photon energies of $k = 2.0$. 3.6, and 9 keV, respectively,[103] are compared with results from equation (11) (dashed line), and with calculations of Kissel *et al.*[102]

Kissel *et al.*,[102] it follows from the comparison that the Z dependence of $d\sigma/dk$ is well described by this calculation (see also Ref. 104), but not by equation (11). So far, discrepancies between experiment and Kissel *et al.*'s calculations were reported for uranium targets and photon energies corresponding to $k/T \approx 0.3$, whose origin whether of experimental or theoretical nature is not known and deserves further investigation.[103]

It follows from equation (8) that bremsstrahlung radiation is emitted anisotropically. At low incident velocities $v_1 \ll c$ the angular distribution follows a dipole pattern and may be expressed as

$$I(\theta_\gamma)/I(90°) = 1 - P\cos^2\theta_\gamma \tag{12}$$

where $P = (I_\parallel - I_\perp)/(I_\parallel + I_\perp)$ is the degree of linear polarization, and $I_\parallel$ and $I_\perp$ the light intensities with electric vector parallel and perpendicular to the incident direction, respectively. In equation (12) retardation effects are neglected. However, already at incident energies of a few keV the observed angular distributions differ significantly from the one predicted by equation (12) (see, e.g., Ref. 105). Because of retardation of the electromagnetic potential the observed angular distributions no longer peak at $\theta_\gamma = 90°$ but in a forward direction. Figure 26 shows results for 50-keV electrons incident on (solid) aluminum. The angular distribution (shape function) of photons with energy $k = 20$ keV ($k/T = 0.4$) and $k = 47.5$ keV ($k/T = 0.95$) are plotted versus emission angle.[106] The angular distribution is most pronounced for large photon energies. The angle of maximum emission occurs around 50° for $k = 47.5$ keV, and around 40° for $k = 20$ keV. These observations are in agreement with theoretical calculations by Tseng *et al.*[107] and with calculations in first Born approximation.[108]

Electron-electron bremsstrahlung, i.e., bremsstrahlung created by the interaction of incident electrons with bound electrons, were investigated by Nakel and Pankau.[109] Using the electron-photon coincidence technique in a kinematically overdetermined experiment they were able to separate electron-electron bremsstrahlung from electron-nucleus bremsstrahlung. Figure 27 displays results of Komma and Nakel[110] for an incident electron energy of 300 keV, electron scattering angle $\theta_e = 20°$, and photon angle $\theta_\gamma = -35°$. The triply differential cross section $d^3\sigma/(d\Omega_e\, d\Omega_\gamma\, dE_e)$ for a thin (solid) copper target is plotted versus outgoing electron energy E_e. The experimental results for electron-electron (e-e) and electron-nucleus (e-n) bremsstrahlung are compared with theoretical curves calculated after formulas of Bethe and Heitler,[108] Elwert and Haug,[111] and Haug.[112] The calculation of Haug[110] for e-e bremsstrahlung is in Born approximation and for collisions between free electrons and agrees well with the experimental data. For the e-n process, both Bethe-Heitler (BH) and Elwert-Haug (EH) calculations agree well with experiment.

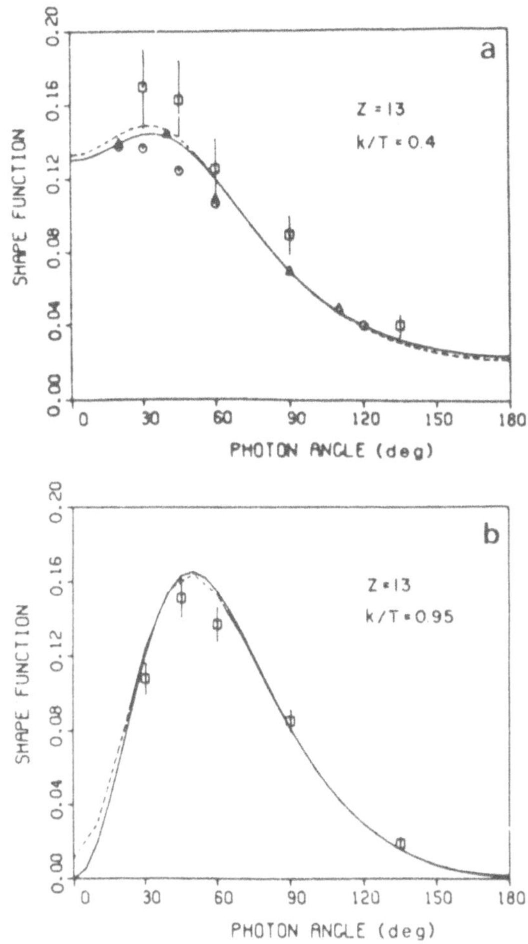

FIGURE 26. Angular distribution of electron bremsstrahlung for 50-keV electrons incident on aluminum, and for (a) $k/T = 0.4$ and (b) $k/T = 0.95$. The solid curve is from Tseng *et al.*,[107] the dashed curve is calculated in first Born approximation[108] (from Ref. 106).

This only holds for low-Z targets, however. For high-Z elements the Elwert-Haug calculation still agrees well with experiment, whereas the Bethe-Heitler results differ considerably from the experimental data.[110]

The coupling of the electron motion to the radiation field may also result in stimulated emission or absorption of photons.[113,114] In strong radiation fields, for example, such as provided by powerful lasers, an incident electron may absorb or emit photons while it is scattered off the atomic ground state. So-called free-free transitions have been observed few years ago for electron-argon collisions.[115] In Figure 28 the energy-loss

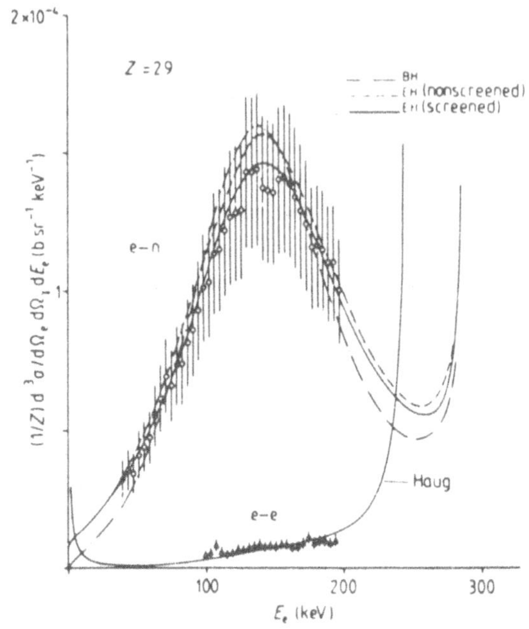

FIGURE 27. Triply differential cross sections for electron-electron (e–e) and electron-nucleus (e–n) bremsstrahlung of copper versus outgoing electron energy E_e. The theoretical curves are calculated after formulas of Bethe-Heitler (BH),[108] Elwert-Haug (EH),[111] and Haug[112] (from Ref. 110).

spectrum of 11-eV electrons scattered from argon atoms is displayed.[116] The pronounced peak without laser (B) is the result of elastic scattering. With laser (A) the intensity of this peak is considerably reduced, and, while the total scattered electron intensity remains unchanged, other peaks in the electron spectrum occur. The position of these peaks corresponds to kinetic energies $E_e = E_0 \pm nE_\gamma$, where n is the number of absorbed or emitted photons and E_0 is the incident and E_γ is the photon energy.

Electron-atom scattering in the presence of a strong electromagnetic field may also result in excitation of the target atom. This "simultaneous" electron-photon excitation of atoms has been the subject of theoretical investigations.[117,118] In Figure 29 theoretical calculations for simultaneous electron-photon excitation of H(2s) as a function of the kinetic energy of the incident electron and for a photon energy of 1.17 eV are displayed.[117] Total cross sections are of the order of πa_0^2, for an incident photon flux corresponding to 10^{13} W/cm^2. The total cross section was calculated in second-order perturbation theory, with the interaction V being given as $V = V_1 + V_2$, where $V_1 = -1/r_1 + 1/r_{12}$ is the Coulomb interaction of the incident electron with target electron and nucleus. $V_2 = -(\mathbf{A}/c)(\mathbf{p}_1 + \mathbf{p}_2)$ is

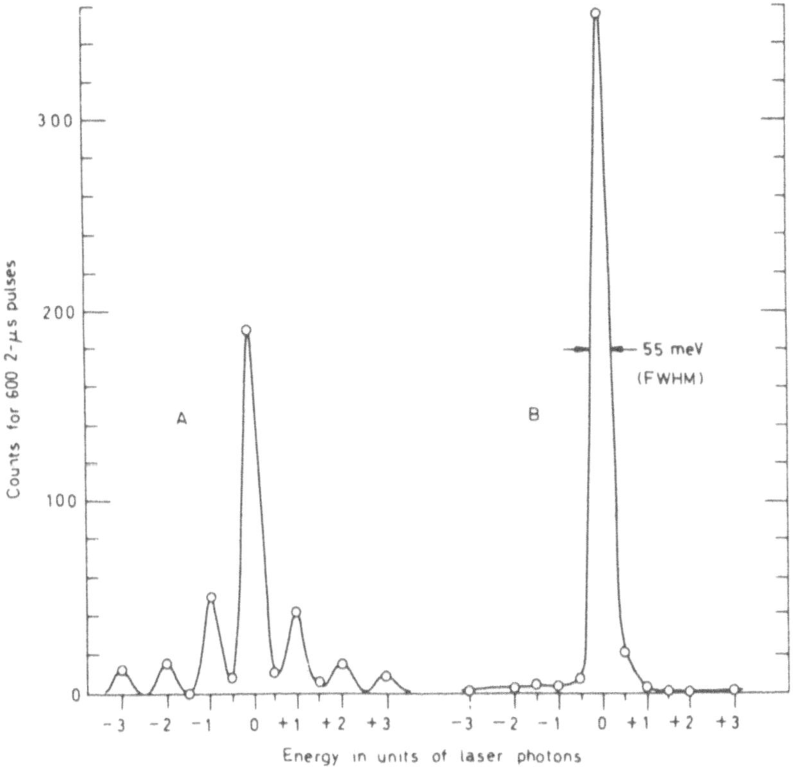

FIGURE 28. Energy-loss spectrum from 11-eV electron-argon scattering with (A) and without (B) laser field (from Ref. 116).

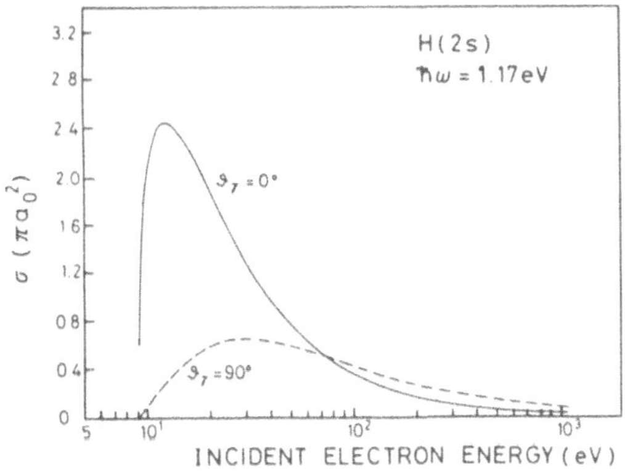

FIGURE 29. Total cross section for simultaneous electron-photon excitation of H(2s) as a function of incident electron energy and for two polarization angles $\theta_\gamma = 0°$ and 90°. The photon energy is 1.17 eV; the photon flux corresponds to 10^{13} W/cm^2 (from Ref. 117).

the interaction of the electromagnetic field with both projectile and target electrons. The two calculations displayed in Figure 29 differ by the linear polarization angle θ_γ of the incident light. The pronounced polarization dependence is easily explained recalled that in dipole approximation the atomic transition amplitude for this process is proportional to $\mathbf{q}\boldsymbol{\varepsilon}$, where $\mathbf{q}$ is the momentum transfer, $\boldsymbol{\varepsilon}$ is the polarization vector, and $\mathbf{z}$ is the direction of the incident electron. At small incident energies we have $\mathbf{q} \parallel \mathbf{z}$ and a large cross section for $\boldsymbol{\varepsilon} \parallel \mathbf{z}$ $(\theta_\gamma = 0°)$. For large incident energies $\mathbf{q} \perp \mathbf{z}$, and the cross section is larger for $\theta_\gamma = 90°$.

First experimental evidence for this process was recently provided by Mason and Newell.[119] They investigated simultaneous electron–photon excitation of helium in the threshold region, where the onsets for direct (i.e., no interaction with the photon field) and the simultaneous excitation process differ by the photon energy. In the subthreshold region for direct excitation a statistically significant signal was observed when the photon field was switched on. This was interpreted as simultaneous electron–photon excitation of the metastable He($2\,^3S$) state.

ACKNOWLEDGMENTS. The author acknowledges numerous discussions with Professor Dr. H. O. Lutz, Dr. S. Datz, and Professor Hans Kleinpoppen. Part of this work was supported by the Deutsche Forschungsgemeinschaft.

REFERENCES

1. H. S. W. MASSEY AND E. H. S. BURHOP, *Electronic and Ionic Impact Phenomena*, Vol. I (Clarendon Press, Oxford, 1969).
2. C. G. BARKLA, *Proc. Cambridge Phil. Soc.* **15**, 257 (1909); W. H. BRAGG AND W. L. BRAGG, *Proc. R. Soc. London* **88A**, 428 (1913).
3. J. FRANCK AND G. HERTZ, *Phys. Z.* **14**, 1115 (1913); **17**, 409, 430 (1916).
4. N. BOHR, *Phil. Mag.* **25**, 10 (1913); **30**, 581 (1915).
5. H. BETHE, *Ann. Phys.* (*Leipzig*) **5**, 325 (1930); *Z. Phys.* **76**, 293 (1932).
6. H. S. W. MASSEY AND C. B. O. MOHR, *Proc. R. Soc. London Ser. A* **140**, 613 (1933).
7. E. H. S. BURHOP, *Proc. Cambridge Phil. Soc.* **36**, 43 (1940).
8. A. M. ARTHURS AND B. L. MOISEIWITSCH, *Proc. R. Soc. London, Ser. A* **247**, 550 (1958).
9. H. S. PERLMAN, *Proc. Phys. Soc.* (*London*) **76**, 623 (1960).
10. J. BANG AND J. M. HANSTEEN, *Mat.-Fys. Medd. K. Dan. Vidensk. Selsk.* 31 (1959).
11. U. FANO AND W. LICHTEN, *Phys. Rev. Lett.* **14**, 627 (1965).
12. J. S. BRIGGS, *Rep. Prog. Phys.* **39**, 217 (1976).
13. H. O. LUTZ, *S. Afr. J. Phys.* **6**, 83 (1983).
14. U. WILLE AND R. HIPPLER, *Phys. Rep.* **132**, 129 (1986); U. WILLE, this book, Chapter 7.
15. D. H. MADISON AND E. MERZBACHER, in *Atomic Inner-Shell Processes*, Vol. I (Academic, New York, 1975), p. 1.
16. R. HIPPLER AND K.-H. SCHARTNER, *J. Phys. B* **8**, 2528 (1975).
17. R. HIPPLER, in *Progress in Atomic Spectroscopy*, Part C, edited by H. J. Beyer and H. Kleinpoppen (Plenum Press, New York, 1984), p. 511.

18. M. INOKUTI, *Rev. Mod. Phys.* **43**, 297 (1971).
19. W. JITSCHIN, this book, Chapter 6.
20. R. HIPPLER AND W. JITSCHIN, *Z. Phys. A* **307**, 287 (1982).
21. M. R. H. RUDGE AND M. J. SEATON, *Proc. R. Soc. London Ser. A* **283**, 262 (1965).
22. W. L. FITE AND R. T. BRACKMANN, *Phys. Rev.* **112**, 1141 (1958); R. L. F. BOYD AND A. BOKSENBERG, *Proc. IVth Int. Conf. Ioniz. Phenom. Gases*, Vol. 1 (North-Holland, Amsterdam, 1960), p. 529; E. W. ROTHE, L. L. MARINO, R. H. NEYNABER, AND S. M. TRUJILLO, *Phys. Rev.* **125**, 582 (1962).
23. R. HIPPLER, H. KLAR, K. SAEED, I. MCGREGOR, A. J. DUNCAN, AND H. KLEINPOPPEN, *J. Phys. B* **16**, L617 (1983).
24. D. L. MOORES, L. B. GOLDEN, AND D. H. SAMPSON, *J. Phys. B* **13**, 385 (1980).
25. G. H. WANNIER, *Phys. Rev.* **90**, 817 (1953); **100**, 1180 (1955).
26. F. H. READ, in *Atomic Physics 7*, edited by D. Kleppner and F. M. Pipin (Plenum Press, New York, 1981), p. 429.
27. H. KLAR, *J. Phys. B* **14**, 3255 (1981).
28. H. GENZ, *Comments At. Mol. Phys.* **15**, 173 (1984).
29. H. KOLBENSTVEDT, *J. Appl. Phys.* **38**, 4785 (1967); **46**, 2271 (1975).
30. J. H. SCOFIELD, *Phys. Rev. A* **18**, 963 (1978).
31. P. ESCHWEY AND P. MANAKOS, *Z. Phys. A* **308**, 199 (1982).
32. C. D. LIN AND L. N. TUNNELL, *J. Phys. B* **12**, L485 (1979).
33. H. TAWARA AND A. RUSSEK, *Rev. Mod. Phys.* **45**, 178 (1973).
34. M. RØDBRO, E. PEDERSEN, C. L. COCKE, AND J. R. MACDONALD, *Phys. Rev. A* **19**, 1936 (1979).
35. J. R. MACDONALD, C. L. COCKE, AND W. W. EIDSON, *Phys. Rev. Lett.* **32**, 648 (1974).
36. K. TAULBJERG, in *Fundamental Processes in Energetic Atomic Collisions*, edited by H. O. Lutz, J. S. Briggs, and H. Kleinpoppen (Plenum Press, New York, 1983), Vol. B 103, p. 349.
37. L. J. DUBÉ, in *Electronic and Atomic Collisions*, edited by D. C. Lorents, W. E. Meyerhof, and F. R. Peterson (North-Holland, Amsterdam, 1986), p. 345.
38. R. M. DRISKO, Ph.D. thesis, Carnegie Institute of Technology (1955).
39. L. H. THOMAS, *Proc. R. Soc. London* **114**, 561 (1927).
40. J. S. BRIGGS, *Nucl. Instrum. Methods A* **240**, 577 (1985).
41. E. HORSDAL-PEDERSEN, C. L. COCKE, AND M. STOCKLI, *Phys. Rev. Lett.* **50**, 1910 (1983).
42. H. VOGT, W. SCHWAB, R. SCHUCH, M. SCHULZ, AND E. JUSTINIANO, *Phys. Rev. Lett.* **57**, 2256 (1986).
43. J. H. MCGUIRE AND N. C. SIL, *Phys. Rev. A* **28**, 3679 (1983); J. H. MCGUIRE, R. E. KLETKE, AND N. C. SIL, *Phys. Rev. A* **32**, 815 (1985).
44. R. HIPPLER, G. SCHIWIETZ, AND J. BOSSLER, *Phys. Rev. A* **35**, 485 (1987).
45. E. HORSDAL, B. JENSEN, AND K. O. NIELSEN, *Phys. Rev. Lett.* **57**, 519 (1986).
46. D. P. DEWANGAN AND J. EICHLER, *J. Phys. B* **19**, 2939 (1986).
47. R. HIPPLER, in *Fundamental Processes in Atomic Collision Physics*, edited by H. Kleinpoppen, J. S. Briggs, and H. O. Lutz (Plenum Press, New York, 1985), p. 181.
48. R. HIPPLER, H. MADEHEIM, W. HARBICH, AND H. O. LUTZ, unpublished.
49. W. FRITSCH AND C. D. LIN, *Phys. Rev. A* **26**, 762 (1982).
50. W. FRITSCH AND C. D. LIN, *Phys. Rev. A* **27**, 3361 (1982).
51. H. J. LÜDDE AND R. M. DREIZLER, *J. Phys. B* **15**, 2703 (1982).
52. R. HIPPLER, W. HARBICH, M. FAUST, H. O. LUTZ, AND L. J. DUBÉ, *J. Phys. B* **19**, 1507 (1986).
53. P. J. O. TEUBNER, W. E. KAUPPILA, W. L. FITE, AND R. J. GIRNIUS, *Phys. Rev. A* **2**, 1763 (1970).

54. M. KIMURA AND C. D. LIN, *Phys. Rev. A* **34**, 176 (1986); and private communication.
55. R. HIPPLER, M. FAUST, R. WOLFE, H. KLEINPOPPEN, AND H. O. LUTZ, *Phys. Rev. A* **36**, 4644 (1987).
56. J. MACEK AND C. WANG, *Phys. Rev. A* **34**, 176 (1986).
57. W. FRITSCH, in Ref. 47.
58. B. MENNER, TH. HALL, L. ZEHNLE, AND V. KEMPTER, *J. Phys. B* **14**, 3693 (1981).
59. M. KIMURA AND N. F. LANE, in preparation; M. KIMURA, private communication.
60. R. HIPPLER, W. HARBICH, M. FAUST, AND H. O. LUTZ, *J. Phys. B* **21**, 103 (1988).
61. R. SHANKER, U. WILLE, R. BILAU, R. HIPPLER, W. R. MCMURRAY, AND H. O. LUTZ, *J. Phys. B* **17**, 1353 (1984).
62. J. A. R. SAMSON, in *Handbuch der Physik*, Vol. XXXI, edited by W. Mehlhorn (Springer-Verlag, Berlin, 1982), p. 123.
63. A. F. STARACE, in *Handbuch der Physik*, Vol. XXXI, edited by W. Mehlhorn (Springer-Verlag, Berlin, 1982), p. 1.
64. J. M. HANSTEEN, *Adv. At. Mol. Phys.* **11**, 299 (1975).
65. R. L. BECKER, A. L. FORD, AND J. F. READING, *Nucl. Instrum. Methods* **214**, 49 (1983).
66. M. ECKHARDT, Ph.D. thesis, University of Giessen (1982).
67. L. H. ANDERSEN, P. HVELPLUND, H. KNUDSEN, S. P. MØLLER, K. ELSENER, K.-G. RENSFELT, AND E. UGGERHØJ, *Phys. Rev. Lett.* **57**, 2147 (1986).
68. R. HIPPLER, K.-H. SCHARTNER, *J. Phys. B* **7**, 618 (1974).
69. J. H. MCGUIRE, *Phys. Rev. Lett.* **49**, 1153 (1982).
70. J. F. READING AND A. L. FORD, *Phys. Rev. Lett.* **58**, 543 (1987).
71. R. E. OLSON, *Phys. Rev. A* **36**, 1519 (1987).
72. R. HIPPLER, J. BOSSLER, AND H. O. LUTZ, *J. Phys. B* **17**, 2453 (1984).
73. G. R. WIGHT AND M. J. VAN DER WIEL, *J. Phys. B* **9**, 1319 (1976).
74. R. HIPPLER, S. DATZ, P. D. MILLER, P. L. PEPMILLER, AND P. F. DITTNER, *Phys. Rev. A* **35**, 585 (1987).
75. T. R. DILLINGHAM, J. R. MACDONALD, AND P. RICHARD, *Phys. Rev. A* **24**, 1237 (1981).
76. J. R. MACDONALD AND F. W. MARTIN, *Phys. Rev. A* **4**, 1965 (1971).
77. L. DUBÉ, *J. Phys. B* **17**, 641 (1984); J. BURGDÖRFER, private communication.
78. H. KNUDSEN, H. K. HAUGEN, AND P. HVELPLUND, *Phys. Rev. A* **23**, 597 (1981).
79. A. S. SCHLACHTER, J. W. STEARNS, W. G. GRAHAM, K. H. BERKNER, R. V. PYLE, AND J. A. TANIS, *Phys. Rev. A* **27**, 3372 (1983).
80. D. S. F. CROTHERS AND N. R. TODD, *J. Phys. B* **13**, 2277 (1980).
81. H. DAMSGAARD, H. K. HAUGEN, P. HVELPLUND, AND H. KNUDSEN, *Phys. Rev. A* **27**, 112 (1983).
82. S. DATZ, R. HIPPLER, L. H. ANDERSEN, P. F. DITTNER, H. KNUDSEN, H. F. KRAUSE, P. D. MILLER, P. L. PEPMILLER, T. ROSSEEL, N. STOLTERFOHT, Y. YAMAZAKI, AND C. R. VANE, *Nucl. Instrum. Methods* **A262**, 62 (1987).
83. L. H. ANDERSEN, M. FROST, P. HVELPLUND, H. KNUDSEN, AND S. DATZ, *Phys. Rev. Lett.* **52**, 518 (1984); L. H. ANDERSEN, M. FROST, P. HVELPLUND, AND H. KNUDSEN, *J. Phys. B* **17**, 4701 (1984).
84. L. H. ANDERSEN, P. HVELPLUND, H. KNUDSEN, AND J. O. PEDERSEN, in *XIV ICPEAC Electronic and Atomic Collisions, Book of Abstracts*, edited by M. J. Coggiola, D. L. Huestis, and R. P. Saxon (Palo Alto, California, 1985), p. 525.
85. N. STOLTERFOHT, C. C. HAVENER, R. A. PHANEUF, J. K. SWENSON, S. M. SHAFROTH, AND F. W. MEYER, *Phys. Rev. Lett.* **57**, 74 (1986).

86. J. A. TANIS, E. M. BERNSTEIN, M. W. CLARK, W. G. GRAHAM, R. H. MCFARLAND, T. J. MORGAN, A. MULLER, M. P. STOCKLI, K. H. BERKNER, P. GOHIL, A. S. SCHLACHTER, J. W. STEARNS, B. M. JOHNSON, K. W. JONES, M. MERON, AND J. NASON, in *XIV ICPEAC Electronic and Atomic Collisions*, Book of Invited Papers, edited by D. C. Lorents, W. E. Meyerhof, and J. R. Peterson (North-Holland, Amsterdam, 1986), p. 394.
87. J. A. TANIS, E. M. BERNSTEIN, M. W. CLARK, W. G. GRAHAM, R. H. MCFARLAND, T. J. MORGAN, B. M. JOHNSON, K. W. JONES, AND M. MERON, *Phys. Rev. A* **31**, 4040 (1985).
88. A. ITOH, T. ZOUROS, D. SCHNEIDER, U. STETTNER, W. ZEITZ, AND N. STOLTERFOHT, *J. Phys. B* **18**, 9091 (1985); J. K. SWENSON, Y. YAMASAKI, P. D. MILLER, H. F. KRAUSE, P. F. DITTNER, P. L. PEPMILLER, S. DATZ, AND N. STOLTERFOHT, *Phys. Rev. Lett.* **57**, 3042 (1986).
89. D. BRANDT, *Phys. Rev. A* **27**, 1314 (1983).
90. T. MCABEE, *Nucl. Instrum. Methods* **214**, 89 (1983); D. BRANDT, *ibid.* **214**, 93 (1983).
91. T. M. REEVES AND J. M. FEAGIN, in Ref. 87; J. M. FEAGIN, J. S. BRIGGS, AND T. M. REEVES, *J. Phys. B* **17**, 1057 (1984).
92. S. DATZ, in *Electronic and Atomic Collisions*, Book of Invited Papers, edited by J. Eichler, I. V. Hertel, and N. Stolterfoht (North-Holland, Amsterdam, 1984), p. 795.
93. G. H. DUNN, in *XIV ICPEAC Electronic and Atomic Collisions*, Book of Invited Papers, edited by D. C. Lorents, W. E. Meyerhof, and J. R. Peterson (North-Holland, Amsterdam, 1986), p. 23.
94. P. F. DITTNER (1987), to be published.
95. P. F. DITTNER, S. DATZ, P. D. MILLER, P. L. PEPMILLER, AND C. M. FOU, *Phys. Rev. A* **33**, 124 (1986).
96. D. C. GRIFFIN, M. S. PINDZOLA, AND C. BOTTCHER, *Phys. Rev. A* **33**, 3124 (1986).
97. A. MÜLLER, D. S. BELIC, B. D. DEPAOLA, N. DJURIC, G. H. DUNN, D. W. MUELLER, AND C. TIMMER, *Phys. Rev. Lett.* **56**, 127 (1986).
98. A. MÜLLER, D. S. BELIC, B. D. DEPAOLA, N. DJURIC, G. H. DUNN, D. W. MUELLER, AND C. TIMMER, *Phys. Rev. A* **36**, 599 (1987).
99. R. H. PRATT, in *Fundamental Processes in Energetic Atomic Collisions*, edited by H. O. Lutz, J. S. Briggs, and H. Kleinpoppen (Plenum Press, New York, 1983), p. 145; R. H. PRATT AND I. J. FENG, in *Atomic Inner-Shell Physics*, edited by B. Crasemann (Plenum Press, New York, 1985), p. 533.
100. H. A. KRAMERS, *Phil. Mag.* **46**, 836 (1923).
101. A. SOMMERFELD AND A. W. MAUE, *Ann. Phys. (Leipzig)* **23**, 489 (1935).
102. L. KISSEL, C. MACCALLUM, AND R. H. PRATT, SAND81-1337, Albuquerque: Sandia National Laboratories (1981).
103. R. HIPPLER, K. SAEED, I. MCGREGOR, AND H. KLEINPOPPEN, *Phys. Rev. Lett.* **46**, 1622 (1981).
104. M. SEMAAN AND C. QUARLES, *Phys. Rev. A* **26**, 3152 (1982).
105. M. AYDINOL, R. HIPPLER, I. MCGREGOR, AND H. KLEINPOPPEN, in *Coherence and Correlation in Atomic Collisions*, edited by H. Kleinpoppen and J. F. Williams (Plenum Press, New York, 1980), p. 205; and *J. Phys. B* **13**, 989 (1980).
106. C. A. QUARLES AND D. B. HEROY, *Phys. Rev. A* **24**, 48 (1981).
107. H. K. TSENG, R. H. PRATT, AND C. M. LEE, *Phys. Rev. A* **19**, 187 (1979).
108. H. A. BETHE AND W. HEITLER, *Proc. R. Soc. London Ser. A* **146**, 83 (1934).
109. W. NAKEL AND E. PANKAU, *Z. Phys. A* **274**, 319 (1975).
110. M. KOMMA AND W. NAKEL, *J. Phys. B* **15**, 1433 (1982).
111. G. ELWERT AND E. HAUG, *Phys. Rev.* **183**, 90 (1969).
112. E. HAUG, *Z. Naturforsch.* **30a**, 1099 (1975); and in Ref. 110.

113. M. H. MITTLEMAN, *Comments At. Mol. Phys.* **11**, 91 (1982).
114. F. H. M. FAISAL, *Comments At. Mol. Phys.* **15**, 119 (1985); *Multiphoton Processes* (Plenum Press, New York, 1987).
115. D. ANDRICK AND L. LANGHANS, *J. Phys. B* **9**, L459 (1976); **11**, 2355 (1978).
116. A. WEINGARTSHOFER, J. K. HOLMES, G. CAUDLE, E. M. CLARKE, AND H. KRÜGER, *Phys. Rev. Lett.* **39**, 269 (1977); A. WEINGARTSHOFER, E. M. CLARKE, J. K. HOLMES, C. JUNG, *Phys. Rev. A* **19**, 2371 (1979); A. WEINGARTSHOFER, J. K. HOLMES, J. SABBAGH, AND S. L. CHIN, *J. Phys. B* **16**, 1805 (1983).
117. S. JETZKE, F. H. M. FAISAL, AND R. HIPPLER, H. O. LUTZ, *Z. Phys. A* **315**, 271 (1984).
118. S. JETZKE, Ph.D. thesis, University of Bielefeld (1986).
119. N. J. MASON AND W. R. NEWELL, *J. Phys. B* **20**, L323 (1987).

CHAPTER 6

Inner-Shell Vacancy Creation

Perturbative Approximations and Beyond

Wolfgang Jitschin

1. INTRODUCTION

Inner-shell processes may be of fundamental simplicity: the electrostatic potential in the vicinity of an atomic nucleus is Coulomb-like and has considerable strength, thus the properties of inner-shell electrons are governed by a hydrogenic character and many external perturbations are comparatively weak, allowing a treatment in a first-order approach. Basic processes have been identified and investigated several decades ago. In recent years much progress has occurred in a more comprehensive and more detailed understanding of excitation and ionization processes. For example, developments of instruments and the availability of new experimental devices resulted in total cross-section values of high accuracy; advances in both experiment and theoretical analysis of the measuring process have led to new types of investigations, such as, for example, studies of differential cross sections and angular shapes of collision-induced vacancies. The more detailed experiments provide a more crucial test of theoretical models and specific assumptions of them. In this way deviations from predictions by perturbative theories have been observed that partly bear the signature of coherence and correlation effects.

Comprehensive reviews on inner-shell physics and related processes are available.[1-4] The current work in this active field is documented in well-established conference series, i.e., the International Conference on the Physics of Electronic and Atomic Collisions (ICPEAC)[5] and the International Conference on X-Ray and Inner-Shell Processes (e.g., X 84).[6]

Wolfgang Jitschin • Physikalisch-Technische Bundesanstalt, Institut Berlin, D-1000 Berlin 10, West Germany.

The present chapter tries to give a self-consistent treatment of the fundamental processes and to discuss some recent highlights of the field. The selection of subjects is a personal choice and reflects partly the author's activities in this field. In the chapter, first experimental aspects are discussed; then a fast trip through the broad field of electron and light-ion impact ionization is made where basic principles and various collisional effects are discussed. Finally, an excursion is made to photoionization, where the ionization process is best known and where subtle effects such as the intrinsic response of an atom to a perturbation or the influence of the atomic environment on inner-shell processes can be studied.

2. EXPERIMENTAL ASPECTS

2.1. Cross-Section Measurements

In a typical ion–atom collision experiment a beam of energetic particles hits a target. The primary beam is well characterized in its properties, i.e., the particle's mass, energy, and charge state. The target may consist of atoms (molecules) in gaseous or solid state. Collision processes between impinging particles and the target atoms result in a change of the particle beam as well as in the ejection of reaction products, i.e., X rays, electrons, or ions. The experimental information on the collision process is obtained by analyzing the transmitted beam and/or analyzing the reaction products (Figure 1).[7]

The strength of a particular process in which a reaction product is emitted into a certain solid angle $d\Omega$ at angles θ, ψ is described by the differential cross section $d\sigma/d\Omega$:

$$\frac{d\sigma(\theta, \psi)}{d\Omega} = \frac{\text{flux emitted into solid angle } d\Omega}{\text{incident flux per unit area}} \tag{1}$$

Integration over all solid angles yields the total cross section σ:

$$\sigma = \int_0^{2\pi} d\psi \int_0^{\pi} d\theta \sin\theta \frac{d\sigma(\theta, \psi)}{d\Omega} \tag{2}$$

The rate R of reaction products encountered by a detector is given by

$$R = \phi \cdot n \cdot \varepsilon \int_{\Delta\Omega} d\Omega \frac{d\sigma}{d\Omega} \tag{3}$$

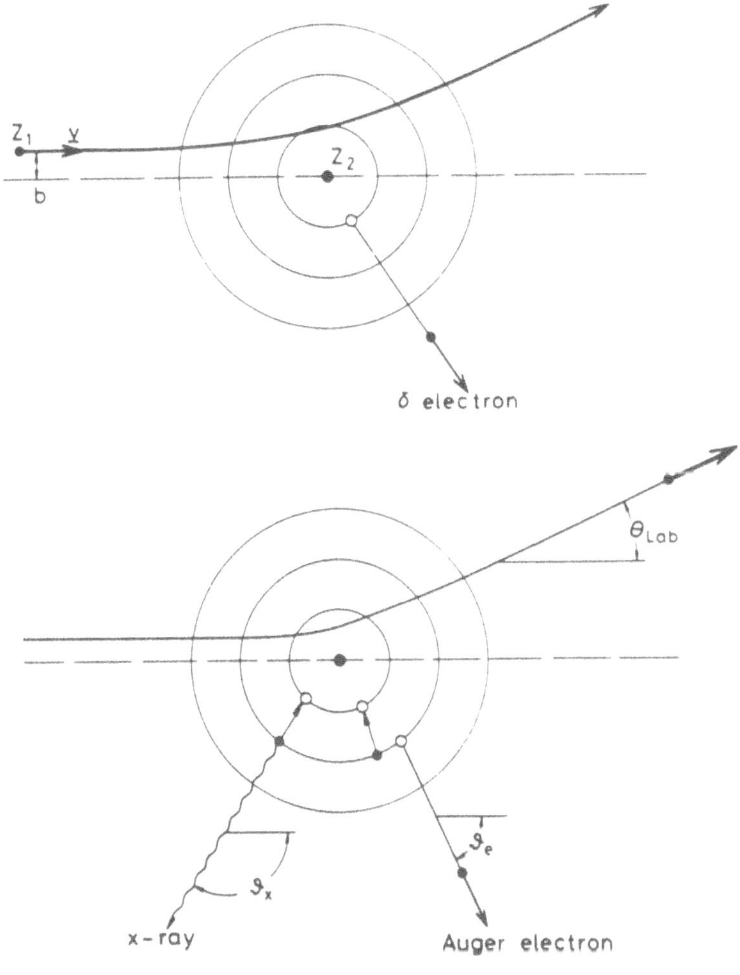

FIGURE 1. Direct Coulomb interaction between the passing projectile ion and an initially bound atomic electron may cause the ejection of this electron, which is then called δ electron (upper half). Subsequently, the created inner-shell vacancy decays by emission of an X ray or an Auger electron (lower half) (from Ref. 7).

where ϕ is the primary flux, n the number of target atoms per area, ε and $\Delta\Omega$ efficiency and acceptance angle of the detector.

In many cases the uncertainty of deriving a cross section is dominated by that of the number of target atoms. To overcome this problem, one may perform a comparative measurement in which a second process is additionally studied whose cross section is known with sufficient accuracy. Thus by normalizing the signal for the first process to the signal for the second process the number of atoms (and the primary flux) cancel.

For ion impact, the nuclear Rutherford scattering is frequently used for normalization. One detects those of the scattered primary particles that have experienced a sizable deflection. Such a deflection occurs mainly at small internuclear distances, i.e., at distances smaller than the radii of inner shells where the actual potential in the atom is not so different from the potential of the unscreened nucleus. In this case the scattering cross section is given by the Rutherford formula, which reads in a convenient form[8]

$$\frac{d\sigma}{d\Omega} = \left(a_0 Z_1 Z_2 \frac{Ry}{2E}\right)^2 \frac{1}{\sin^4(\theta/2)} \tag{4}$$

where Z_1 and Z_2 denote charge of projectile and target nucleus, a_0 is the Bohr radius, Ry the Rydberg energy, E the collision energy, and θ the projectile's deflection angle. Also potentials different from the Coulomb potential can be treated straightforwardly by integrating the equations of nuclear motion.[9]

For electron impact, one can employ another process for normalization: the Bremsstrahlung resulting from the interaction between impinging electron and the atomic potential. This is a process that again occurs in the vicinity of the nucleus, where the potential is well known. The Bremsstrahlung cross section can be accurately calculated[10] and allows an easy way for normalization.[11]

There are other processes that can be used for normalization. For example, one can make use of nuclear reactions.[12] These depend on the element and the isotope and are, therefore, not generally applicable.

2.2. Detectors

Energetic ions can be detected by various processes occurring in their interaction with matter.[13] First, if they hit a surface, they can eject one or more electrons from the surface. These electrons can be amplified by a secondary electron multiplier (multiple dynode multiplier, channeltron, spiraltron, channel plate). This kind of ion detection provides high time resolution (below 1 ns), but the information gained on the energy of the ion is zero or very modest.

Second, ions traveling through a semiconductor or through a gas-filled volume create a large number of free charge carriers on their path. The induced charge can be collected; it is approximately proportional to the energy loss of the ions in the detector, and, in case of complete stopping, to the ion energy. Thus these detectors may provide good energy resolution.

On the other hand, drift times of the charge carriers to the electrodes as well as electronic noise in the analog signal amplifiers cause a poorer time resolution than the one for the above-mentioned electron amplification. Third, ions traveling through a gas or through a transparent solid material create light, which can be detected by a photomultiplier. Such a detector provides good time resolution and modest energy resolution.

The ion detectors can be built in such a way that additionally information on the ions' paths is obtained and are then called position sensitive. Two approaches are feasible to get the spatial resolution: the ion detector may be simply built with multiple electrodes and parallel data acquisition; this scheme ensures a high count-rate capability at the expense of a multiplicity of the basic electronics. Alternatively, one may use a signal readout employing a division of analog signals or a position-dependent time delay; this scheme requires more specialized electronics and has a poorer count-rate capability.

Electrons are created in a collision by various processes, whereby the electron energy is characteristic for the process. For the energy-selective detection of the electrons, whose energies are generally below a few keV, electrostatic spectrometers are employed.(14–16) Various types of spectrometers with different geometries are available as, e.g., the parallel-plate analyzer, the cylindrical mirror analyzer, the hemispherical analyzer, and the 127° cylinder analyzer. The proper choice of the type best suited for a particular application depends on source geometry and required energy resolution. Electrons having passed the energy analyzer are finally detected by an electron multiplier.

X rays created in a collision can be detected by their interaction with matter. This interaction is comparatively weak, so that the electron yield from a surface is tiny.(17) However, they can be effectively detected in the bulk of gaseous, scintillating, or semiconducting materials.(18) A primary X ray is dominantly absorbed in the material via the photoeffect, resulting in an ejected photoelectron and an atom with an (inner-shell) vacancy. The photoelectron ionizes another atom, thereby creating another free electron–ion pair. Also the decay of the vacancy in the atom results in the emission of an Auger electron or X ray with subsequent creation of further electron–ion pairs. In a cascade of electron-induced and photon-induced processes a large number of electron–ion pairs is produced until the primary energy of the photon is converted to the generation of free charge carriers. The number of carriers is roughly proportional to the photon energy. Carriers with opposite charge are separated by an applied electric field and the collected charge is measured.

The maximum number N of charge carriers that can be created is just the ratio of photon energy E and the energy required for the generation of one pair of charge carriers [26 eV for an Ar-filled proportional counter,

3.8 eV for a Si(Li), and 3.0 eV for an intrinsic Ge detector]. The cascade process has statistical character, which results in statistical fluctuations of the number of created pairs. The fluctuations would be $\Delta N = \sqrt{N}$ if the individual processes are uncorrelated. In reality there are correlations and the fluctuations are smaller by the so-called Fano factor F, which amounts to 0.1–0.2. The energy resolution of a detector may be worse as compared to the collection statistics $F\Delta N$, if other fluctuations such as, e.g., electronical noise, are non-negligible.

Available Si(Li) and Ge semiconductor detectors achieve a working energy resolution of ca. 150 eV FWHM at a photon energy of 5.9 keV. Recently, an X-ray detector based on a Giaever tunneling junction consisting of a Sn-oxide barrier between superconducting Sn films has been tested.[19] Here the primary photon creates quasiparticles in the barrier leading to an extra tunneling current. The energy required for the generation of a quasiparticle is as small as 0.6 meV, and one expects a much improved energy resolution of the detector. A value of 65 eV FWHM at 5.9 keV has been achieved so far (Figure 2).

The time resolution of the X-ray detectors is limited at low photon energies by electronic noise,[20] at high photon energies by the drift velocities of the electrons and holes. For example, 60-keV photons passing through a Si(Li) detector of a few millimeters thickness are absorbed with comparable probability at all depths. The charge carriers are created in the vicinity of the point of the primary absorption, and depending on the location of this point, their drift time to the electrodes is different. Figure 3 shows a measured time correlation between an α particle and a 60-keV γ ray emitted in the decay of ^{241}Am.[21] The sharp rise with $\Delta t = 8 \pm 1$ ns is attributed mainly to the drift time spread of the charge carriers in the Si(Li) detector; the electron mobility in Si at 77 K amounts to $2 \cdot 10^4\ \mathrm{cm^2\ V^{-1}\ s^{-1}}$.[18] The smooth exponential decay with $T_{1/2} = 70 \pm 4$ ns results from the lifetime of a nuclear level.

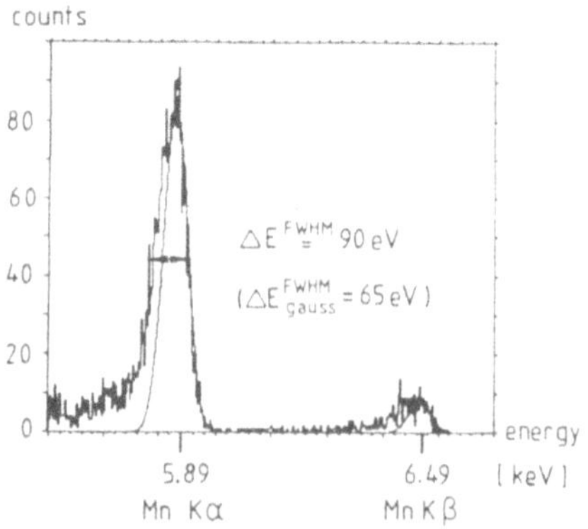

FIGURE 2. Energy spectrum of Mn K X rays as recorded with a Giaever tunneling junction detector. A Gaussian curve is fitted to the high-energy sides of the peaks; the low-energy tail of the peaks is due to quasiparticle diffusion away from the tunneling region (from Ref. 19).

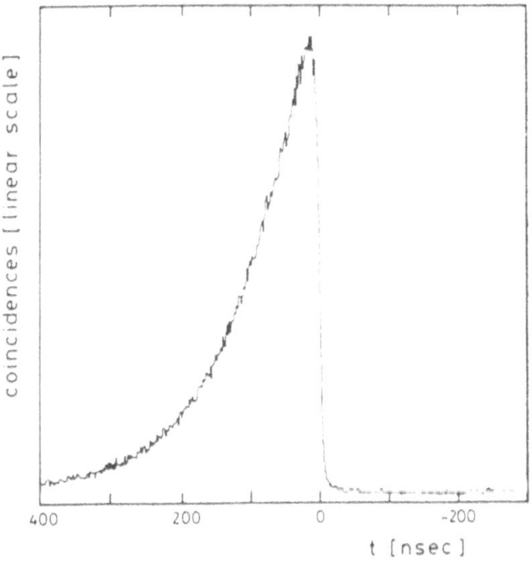

FIGURE 3. Recorded time correlation between the α particle and the 59.5-keV γ quantum emitted in the radioactive decay of ^{241}Am. The steep rise reflects the instrumental time resolution, whereas the exponential tail results from the half life $T_{1/2} = 67$ ns of an excited state of ^{237}Np which is populated in the α decay of ^{241}Am (from Ref. 21).

High-resolution X-ray spectroscopy can be performed by crystal spectrometers that make use of the Bragg X-ray reflection.[22] This reflection exhibits the property of polarization sensitivity, allowing also a polarization analysis.[23] In a crystal spectrometer the incoming X rays hit a crystal where they are reflected and the outgoing X rays are detected. According to Bragg's law, the reflection is specular and occurs only if the incidence angle θ and the X-ray wavelength λ obey the relation

$$\lambda = 2d_n \sin\theta \tag{5}$$

where d_n denotes the crystal lattice spacing. The achievable energy resolution of the spectrometer is obtained from the derivative of equation (5):

$$\Delta E/E = -\Delta\theta \cos\theta \tag{6}$$

A high resolution thus requires a small angular spread $\Delta\theta$, which is a convolution of X-ray collimation and the width of the crystal reflectivity peak. The X-ray collimation is achieved in curved crystal instruments by a slit whereby X rays of only a small source area but of some angular divergence are accepted. Plane crystal spectrometers can accept radiation from a large area, but the angular divergence has to be strongly reduced, e.g., by a collimating Soller slit system. A choice between both spectrometer types with respect to intensity depends on the geometry of the source. Plane crystal instruments can be designed to be mechanically simple and compact (Figure 4); however, the achievable resolution is limited by total reflection at the plates of a Soller collimator.[24]

A general drawback of all crystal spectrometers is their small phase space of X-ray acceptance due to the reflection characteristics of crystals. The reflectivity curve of a crystal has maximum values close to unity; however, the width of the maximum is typically as narrow as a few arcsec, yielding an integral reflecting power in the order of 10^{-5} (Figure 5).[25]

2.3. Evaluation of X-Ray Spectra

Si(Li) and Ge semiconductor detectors register X rays of different energies simultaneously, and with appropriate electronics an X-ray spectrum can be simply recorded. In contrast, most crystal spectrometers have to be scanned in order to record a spectrum. Quantitative work suffers from the absolute uncertainty of the X-ray detection efficiency. An estimate for the efficiency can be obtained by calculating the X-ray absorption in the active detector volume, the losses in the detector window, etc. A more reliable method for determining the efficiency is a calibration by radioactive samples of known activity which are available from national standard institutions.[18] This method is restricted to X-ray energies above a few keV owing to the lack of suitable isotopes. A very general and precise calibration method employs synchrotron radiation whose intensity is precisely known.[26] Complications in spectra evaluation may arise from artificial background (resulting from incomplete charge collection in semiconductor detectors and from scattered or fluorescence X rays in crystal spectrometers) as well as artificial

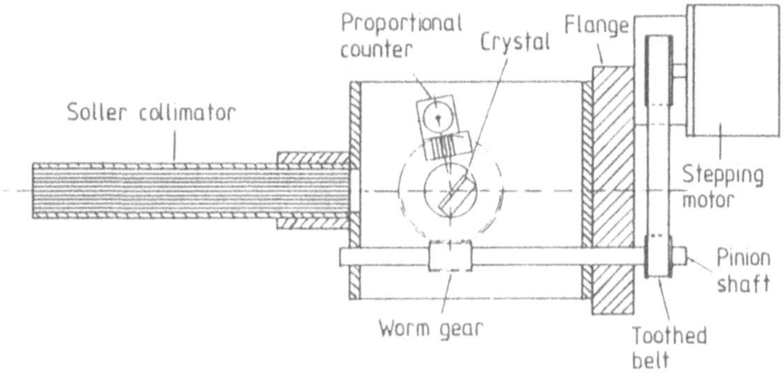

FIGURE 4. Schematic drawing of a compact plane-crystal X-ray spectrometer for ion–atom collision studies. X rays enter from the left; they are energy-selectively reflected by the crystal and finally detected in the proportional counter (from Ref. 24).

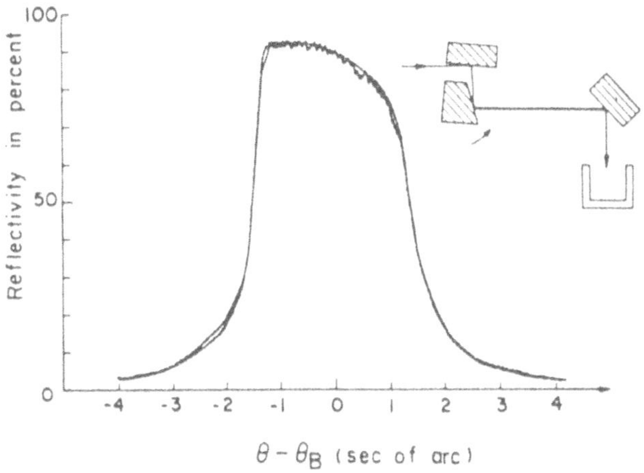

FIGURE 5. Intrinsic diffraction profile for the symmetric 422 diffraction from a silicon crystal at a wavelength of 1.54 Å (8.05 keV). The experimentally measured curve (identifiable by some noise) almost coincides with the theoretically calculated curve; the calculated curve is a convolution with the finite experimental resolution which causes a noticeable broadening at the sides of the peak. The inset depicts the experimental arrangement (from Ref. 25).

lines (resulting from X-ray escape in semiconductor detectors and from higher-order reflection in crystal spectrometers). These instrumental artifacts play only a minor role and can be taken into account reasonably well in most standard applications.[27]

The spectrum of X rays originating from the K shell of an atom, i.e., the K spectrum, consists of only a modest number of individual lines and can easily be well resolved (Figure 6).[24] L spectra consist of significantly more lines and can be resolved for lighter elements ($Z < 50$) only by crystal spectrometers (Figure 7),[24] whereas for heavier elements semiconductor detectors yield reasonable resolution (Figure 8).[28] M spectra consist of a tremendous number of lines requiring again a crystal spectrometer (Figure 9).[29] A widely used nomenclature of the X-ray lines is based on Siegbahn's work[30]; it is of historical origin and somewhat unsystematic. Violent ion–atom collisions can produce not only one, but two or more inner-shell vacancies. X-ray lines originating from a multiply ionized atom (called satellite lines) are energetically shifted as compared to X-ray lines originating from a singly ionized atom (called diagram or parent lines). The shift is usually towards higher energies (Figure 6). Comprehensive calculations for the line shifts in doubly ionized atoms are available.[31-33]

The extraction of ionization cross section σ_I from measured X-ray production cross section σ^X is simple in the case of the K shell, since

$$\sigma^X = \omega_K \cdot \sigma_I \tag{7}$$

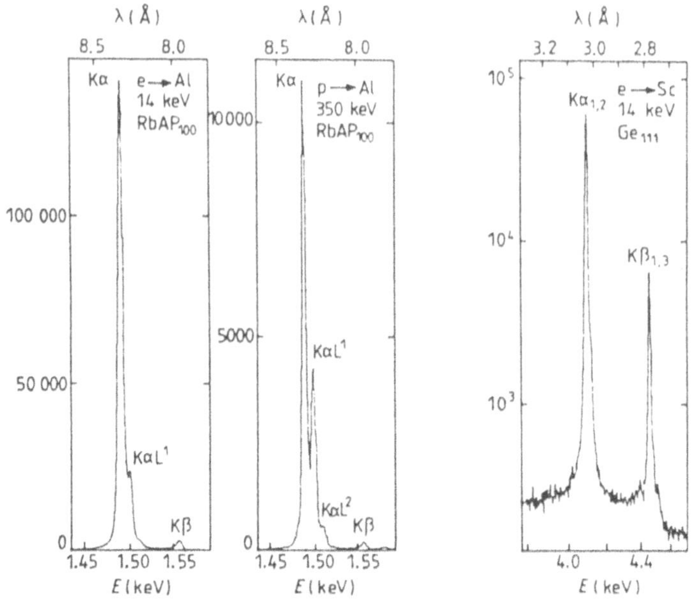

FIGURE 6. *K* X-ray spectra of Al and Sc measured with the crystal spectrometer of Figure 4, note the satellite lines in the Al spectra whose relative intensities depend on the projectile (from Ref. 24).

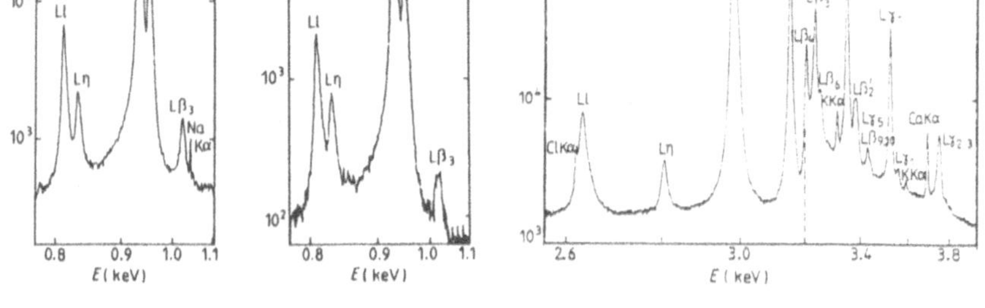

FIGURE 7. *L* X-ray spectra of Cu and Ag recorded with a high-resolution crystal spectrometer (from Ref. 24).

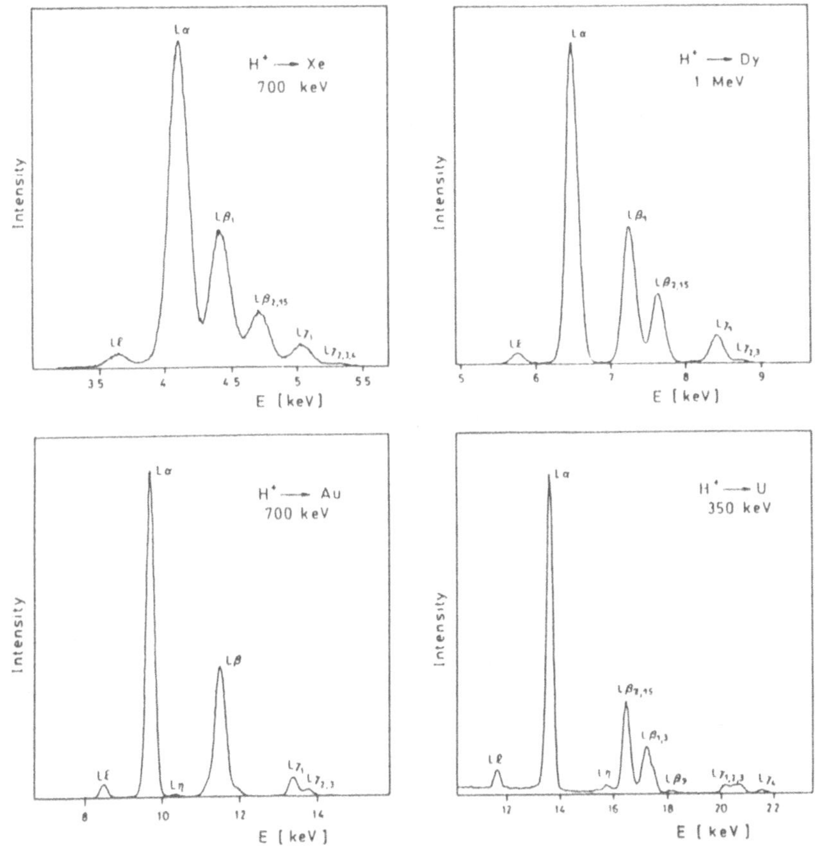

FIGURE 8. *L* X-ray spectra of Xe, Dy, Au, and U as recorded with a Si(Li) semiconductor detector (from Ref. 28).

where ω_K denotes the fluorescence yield. Fluorescence yields have been compiled by various authors.[34,35] *K*-shell yields are quite well known in contrast to yields of higher shells. For these shells, further complications arise by their subshell structure: Vacancies can be transferred from a deeper-lying subshell to a higher one by a kind of cascade process called Coster–Kronig transition. For example, the experimentally observed effective X-ray production cross sections (designated by superscript X) of the individual *L* subshells are given by

$$\begin{aligned} \sigma_{L1}^{X} &= \omega_1 \sigma_{L1} \\ \sigma_{L2}^{X} &= \omega_2(\sigma_{L2} + f_{12}\sigma_{L1}) \\ \sigma_{L3}^{X} &= \omega_3[\sigma_{L3} + f_{23}\sigma_{L2} + (f_{12}f_{23} + f_{13})\sigma_{L1}] \end{aligned} \tag{8}$$

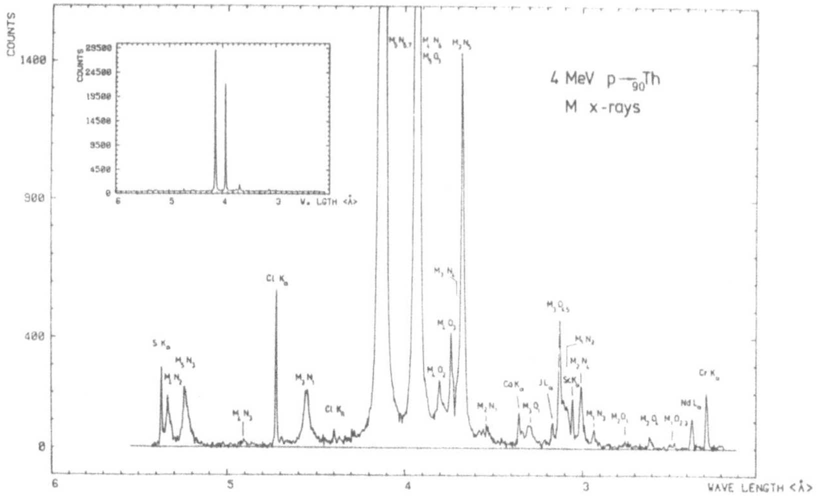

FIGURE 9. *M* X-ray spectrum of Th recorded with a crystal spectrometer; there are only a few strong lines (as shown in the inset) but numerous weak lines (from Ref. 29).

where f_{ij} denotes the vacancy transfer probability from subshell L_i to subshell L_j. The parameters ω_2, ω_3, and f_{23} can be precisely measured by coincidence techniques,[36] whereas the remaining parameters are known with only modest accuracy. This is due to specific problems for the L_1 subshell: experimentally, it is difficult to isolate primary L_1 vacancies and theoretically, calculated decay rates become unreliable for transitions with small energy excess as is the case for some strong L_1 transitions. For example, theory may overestimate f_{12} and f_{13} by a factor of 2.[37] A new powerful and universal method for measuring the various yields is based on vacancy creation by tunable monochromatized synchrotron radiation.[38] By scanning the energy of the radiation across the X-ray absorption edges, the ionization of individual L subshells is switched on sequentially, thus making it possible to isolate primary vacancies in all particular subshells; furthermore, photoionization cross sections are known with comparatively high accuracy (see Section 5.1). The method has been explored for the element Au.[38,39] More comprehensive studies are under way.[40,41]

The large number of partially overlapping L X-ray lines hampers a separation of measured X-ray spectra into the contributions from the different L subshells. In a spectra evaluation one may therefore choose one reasonably well-resolved line for each subshell and derive the total fluorescence originating from the corresponding subshell from this line via the branching ratios. Branching ratios have been calculated, e.g., by Scofield[42] and experimental values have been compiled by Salem.[43] The choice of individual X-ray lines in the spectra evaluation is not unique; also, different sets of branching ratios and atomic yields are employed. This ambiguity leads to substantial differences in subshell ionization cross sections extracted from experimental data, as discussed by Cohen.(44) Thus differences in the data evaluation procedures employed by different groups account for a large fraction of the scatter of published results; much better agreement is obtained if always the same procedure is applied.(45,46) All experimental cross section data for individual L subshells (and, of course, also for higher subshells) bear considerable systematic uncertainties owing to the data evaluation (see Sections 4.3 and 4.4).

3. ELECTRON IMPACT

3.1. *Plane Wave Born Approximation*

The basic mechanism responsible for the vacancy creation in an electron–atom or ion–atom collision is the electric Coulomb interaction: The electromagnetic field of the passing projectile disturbs the atomic electrons which may be thereby excited from a bound state to the continuum. A simple theoretical procedure to treat this process quantitatively is the plane-wave Born approximation (PWBA) which gives remarkably good results.(47) In this approximation the incoming and outgoing projectile is described as a plane wave with wave vectors $\mathbf{k}_i$ and $\mathbf{k}_f$ respectively, where the momentum transfer $\hbar k = \hbar|\mathbf{k}_f - \mathbf{k}_i|$ results from the collision. By the collision a single electron may be knocked out of the atom for which a transferred energy E is needed, which is just the sum of initial binding energy E_B and final energy E_f of the electron. Energy conservation now requires that the projectile's initial impact energy T is reduced to an outgoing energy $T - E$, yielding the following limiting values for the momentum transfer:

$$\hbar k_{\substack{\max \\ \min}} = (2mT)^{1/2} \pm [2m(T - E)]^{1/2} \tag{9}$$

where m is the mass of the projectile. The PWBA gives a simple formula of the cross section for transferring an electron from initial state $|i\rangle$ to final state $|f\rangle$

$$\sigma_{if} = 8\pi\left(\frac{Z_1 e^2}{\hbar v}\right)^2 \int_{k_{\min}}^{k_{\max}} dk \frac{|F_{if}(k)|^2}{k^3} \tag{10}$$

where Z_1 denotes the projectiles charge and v its initial velocity. The form factor $F_{if}(k)$ is a purely atomic parameter

$$F_{if} = \langle f | e^{i\mathbf{kr}} | i \rangle \tag{11}$$

For excitation processes, the energy difference between initial and final state is fixed and the cross section is obtained by a single integration over k [equation (10)]. For the ionization cross section, one has to integrate additionally over all energies of the final state:

$$\sigma_i = \int_0^{T-E_B} dE_f \, \sigma_{if} \tag{12}$$

In order to calculate numerical cross sections, one needs the form factor F. Assuming hydrogenic atomic wave functions, the form factor can be expressed analytically.(48–51) Form factors given in the literature frequently refer to full (sub-) shells, i.e., the single electron form factor times the number of electrons in the corresponding shell.

The described hydrogenic Born approximation is easily computer programmable(52) and even home computers perform the required integrations within a reasonable time. Advanced PWBA calculations employ more realistic electron wave functions; for example, cross sections for the L shell(53) and the M shell(54) have been calculated employing Dirac–Hartree–Slater wave functions.

3.2. Experimental Data

Ionization cross sections for electron impact are zero for impact energies below the ionization threshold, i.e., the minimum energy required for ionization. Above threshold the cross sections increase by a characteristic power of the excess energy,(55,56) exhibit a maximum at about three times the threshold energy, and then fall off continuously with increasing impact energy. The PWBA with hydrogenic form factor as described above is expected to yield fairly accurate cross-section values for the ionization of inner shells (which are quite hydrogenlike) at energies larger than about ten times the threshold energy. At lower projectile velocities various complications may arise. For example, in case of low-velocity electron impact, exchange between incident and bound electron becomes important. The exchange modifies the cross section in a nontrivial way since it opens a new reaction channel which may interfere constructively or destructively with the direct reaction channel described above. A rigorous treatment of the collisional interaction including exchange is impractical, and thus computationally simple approximations, as, e.g., the Ochkur approximation have been explored. Such a treatment gives a reasonable prediction of experimental K-shell ionization cross sections over a broad range of electron impact energies (Figure 10).[57]

For heavy elements, the threshold energy for K-shell ionization is so high that the impact velocity of an ionizing electron is already in the relativistic regime. In the above treatment, we have considered only the disturbance of the atomic electron by the electric field of the projectile, i.e., the longitudinal cross section σ_l. For relativistic velocities, the disturbance by the magnetic field of the projectile, i.e., the transversal cross section σ_t, becomes important. Its cross section increases with increasing impact energy and finally dominates the total cross section $\sigma = \sigma_l + \sigma_t$. The transversal cross section can also be treated in Born approximation.[58] Comparison of calculated total cross sections with experiment in the regime of relativistic collision energies gives reasonable absolute agreement (Figure 11).[59]

In the ultrarelativistic region, i.e., for energies exceeding several hundred MeV, the transversal interaction between projectile and target electron gets effective even at interaction distances larger than the size of an atom. In experimental studies performed with solid targets one therefore expects an influence of the neighboring atoms on the ionization cross section, caused by a screening of the projectile's charge. Available experimental data, however, favor the missing of this so-called density effect; presumably this effect is smaller than theoretical predictions (Figure 12).[58,59]

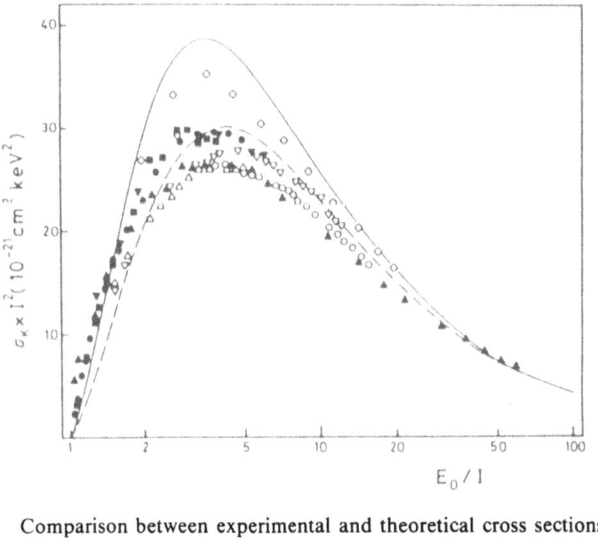

FIGURE 10. Comparison between experimental and theoretical cross sections for K-shell ionization of light atoms (Z = 6–18) by electron impact. ——, PWBA calculation; ---, Born–Ochkur calculation. Symbols represent experimental data for various elements. The axes of the figure are scaled by the binding energy I of the K shell of the corresponding element. E_0 denotes the impact energy (from Ref. 57).

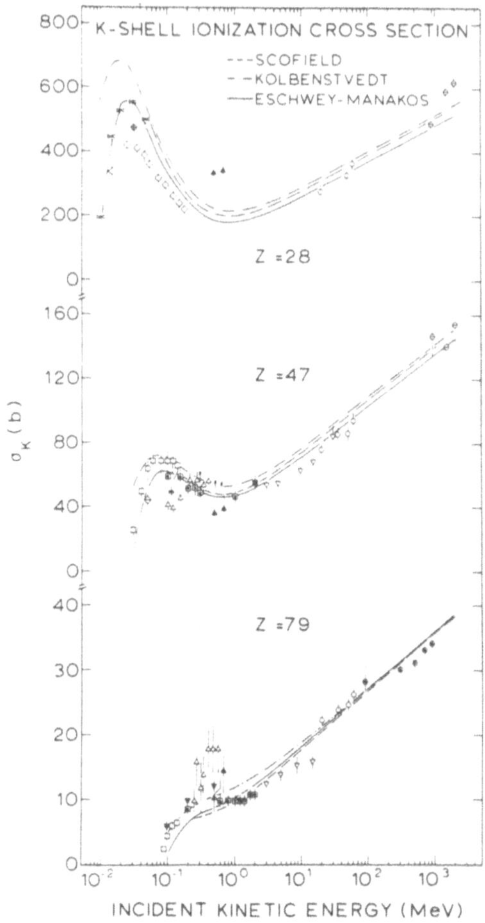

FIGURE 11. Experimental (points) and theoretical (curves) *K*-shell ionization cross sections for electron impact on Ni, Ag, and Au at relativistic energies (from Ref. 59).

4. LIGHT ION IMPACT

4.1. Semiclassical Approximation

The PWBA described in the previous section is a purely quantum-mechanical treatment of the collision process and applicable to electron as well as to ion impact. Its application to ion–atom collisions, however, suffers from a substantial shortcoming: the neglect of the projectile-target nucleus interaction. The trajectory of an ion projectile is hardly affected in a collision by the projectile's interaction with the ionized target electron because of the large ion mass, but mainly by the interaction with the target nucleus, which is not contained in PWBA. This deficiency can be overcome by the

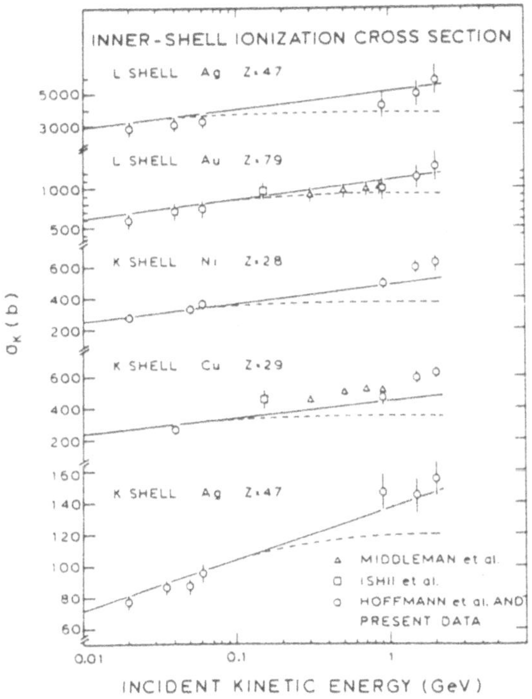

FIGURE 12. Absolute K- and L-shell ionization cross sections for the impact of relativistic electrons on Ni, Cu, Ag, and Au. The curves represent calculations assuming isolated atoms (solid lines) or including the density effect (dashed lines). Points are experimental data for solid targets (from Ref. 59).

distorted-wave extension of the PWBA, which, however, requires tedious calculations. Therefore, a widely employed approach is the semiclassical approximation (SCA).[60] In the SCA, the target electron is described by a quantum-mechanical wave function, whereas the projectile is assumed to be a point charge moving on a classical trajectory in time and space. The disturbance of a target electron by the projectile may induce a transition from the initial state $|i\rangle$ with energy E_i to a final state $|f\rangle$ with energy E_f; the transition amplitude is given by[61]

$$a_{fi} = -\frac{i}{\hbar}\int_{-\infty}^{+\infty} dt\, e^{i(E_f - E_i)t/\hbar}\langle f|\frac{-Z_1 e^2}{|\mathbf{R}(t) - \mathbf{r}|}|i\rangle \tag{13}$$

where $\mathbf{R}(t)$ describes the projectile's trajectory, and the time integration extends over the collision time. The formula is directly applicable for an excitation process to a specified final state $|f\rangle$[62]; for calculating ionization processes one has to integrate over all final states $|f\rangle$ in order to get the trajectory-dependent ionization probability $I(b)$:

$$I(b) = \int_0^{E_{max}} dE_f \, |a_{fi}|^2 \tag{14}$$

The total cross section is obtained by integration over all trajectories

$$\sigma = \int_0^{\infty} db \, 2\pi b I(b) \tag{15}$$

The total cross section decreases towards high velocities since the time period of the perturbation decreases; it decreases also towards small velocities since the perturbations at different times interfere destructively [oscillating exponential factor in equation (13)]. The maximum of the cross section occurs roughly at an ion impact velocity equal to the classical orbiting velocity of the bound atomic electron.

At impact energies above the cross section maximum the main contribution to the total cross section comes from distant collisions for which the projectile trajectory is almost a straight line:

$$\mathbf{R}(t) = \mathbf{v}t + \mathbf{b} \tag{16}$$

$\mathbf{v}$ and $\mathbf{b}$ denote velocity and impact parameter of the projectile. On the other hand, at the lowest impact energies where ionization measurements have been performed, the main contribution comes from close collisions. Here, the small internuclear distance and the small velocity make the projectile's trajectory sensitive to the repulsion by the target nucleus resulting in significant changes of the trajectory and, concomitantly, tremendous effects on the ionization cross sections. These have been termed Coulomb deflection effects, and one may distinguish effects originating mainly from the *deceleration* (reduction of velocity in the field of the target nucleus) or from the *deflection* (change of direction) of the projectile, as discussed in Sections 4.3–4.5.

Assuming a purely Coulomb interaction between projectile and target nucleus, the projectile's trajectory is hyperbolic. This allows an easy analytical description[63]; for example, scattering angle θ and impact parameter b are related by

$$\tan(\theta/2) = d/b \tag{17}$$

where

$$d = Z_1 Z_2 e^2 / 2T \tag{18}$$

is half the distance of closest approach in a head-on collision. The scattering angle θ is of macroscopic size, making experimental studies of the impact-parameter dependence feasible. The implementation of non-Coulombic potentials destroys the simplicity of the formulas but can be done in a straightforward way.

By the interaction between projectile and target nucleus, the target nucleus experiences a recoil. The resulting acceleration of the nucleus relative to its electron cloud may also result in ionization. This effect may be investigated most clearly by neutron impact, where no additional electric projectile–target interaction is present and where the collision is sudden. The cross section for this process is rather small since first the neutron must hit the small nucleus and second this small kick has to lead to ionization. A simple estimate gives the result[64]

$$\sigma = 2\pi a_0^2 \left(\frac{m_e}{m_1 + m_2} Z_1 \frac{v_0}{v_1} \right)^2 \times \int_0^{\infty} \frac{db}{b} \frac{b}{d} \left| \int_{-\infty}^{+\infty} dt\, e^{-iE_i t/\hbar} \frac{|\ddot{\mathbf{R}}|}{v_1} \right|^2 \left| \int_0^{E_{\max}} dE_f \langle f | \frac{Z_2 r}{a_0} | i \rangle \right|^2 \tag{19}$$

Also more sophisticated calculations of ionization by neutron impact have been performed.[65] For ionization by charged particle impact, the two ionization mechanisms of recoil and of Coulomb interaction are both present and interfere. Both can be of the same magnitude; they can be partially separated in experimental studies since they exhibit different impact-parameter dependence (see Section 4.6). Advanced SCA calculations incorporate both mechanisms *ab initio* even for nonhyperbolic trajectories.[66]

In a comparison between PWBA and SCA for the description of ion–atom collisions, the SCA is superior since it makes it possible to include the important projectile–target interaction (i.e., projectile trajectory effects, target nucleus recoil) in a fundamental way. On the other hand, it requires more computational effort, even for simple assumptions.[67] Nevertheless, highly advanced SCA calculations are now feasible employing relativistic Hartree–Fock electron wave functions for the bound as well as for the free electron.[68–71]

4.2. Alignment Theory

A vacancy created in a collision is not necessarily spherically symmetric but may exhibit anisotropy.[72] The angular shape depends on the angular momentum j of the vacancy; e.g., for $j = 1$ it can exhibit dipole and quadrupole characteristics. In general, the anisotropy can be elegantly described by a multipole expansion of the electron density of the vacancy state in terms of spherical harmonics $Y_{kq}(\theta, \psi)$.[73,74] The expansion coefficients are the so-called state multipoles $\langle T(j'j)^{\dagger}_{kq}\rangle$, which are given by

$$\langle T(j'j)^{\dagger}_{kq}\rangle = \sum_{m'm} (-)^{j-m}(j'm', j-m\,|\,kq)\langle j'm'|\rho|jm\rangle \tag{20}$$

The first factor under the sum denotes a Clebsch–Gordan coefficient and the second one the density matrix of the vacancy state. Frequently normalized state multipoles, i.e., alignment tensors $\mathscr{A}_{kq}$ are used:

$$\mathscr{A}_{kq} = \langle T^{\dagger}_{kq}\rangle / \langle T^{\dagger}_{00}\rangle \tag{21}$$

A collisional induced anisotropy may be quite large for an individual collision process, in which, e.g., the projectile trajectory or the final energy of the ejected electron is fixed. Most experiments, however, take an average over individual processes and thereby the observable size of alignment effects is reduced. Furthermore, symmetry conditions may require the vanishing of alignment parameters: in case of axial symmetry all $\mathscr{A}_{kq}$ with $q \neq 0$ (except $\mathscr{A}_{00}$) are zero.

The alignment can be easily derived from the known ionization cross sections of the $|jm\rangle$ magnetic substates. For example, for a $j = 1$ vacancy with axial symmetry one gets

$$\mathscr{A}_{20} = (\sigma_1 - \sigma_0)/(2\sigma_1 + \sigma_0) \tag{22}$$

where σ_1, σ_0 denote the cross section of the $m = 1$ and $m = 0$ magnetic substates.

Calculations of the collisional induced alignment can be performed by various theoretical approaches: PWBA,[75-80] SCA,[81,82] second-order Born approximation,[83] and coupled-channel calculations.[84,85] Experimentally, the alignment can be determined via the vacancy decay by making use of the anisotropy or polarization of the X radiation as well as of the anisotropy of the Auger electrons emitted in the decay. The interrelation between vacancy alignment and emission anisotropy depends on the angular momenta of initial and final states, and can be calculated from angular momentum algebra.[73,23]

4.3. *Total Ionization Cross Sections*

Comprehensive experimental data of absolute ionization cross sections are available. Most work has been done for the K shell,[86,87] and somewhat less for the L shell.[88] The dependence of the cross sections on the impact velocity has a simple shape, and by plotting "scaled" cross sections versus "scaled" velocity the data tend to fall onto a single universal curve.[89] In order to gain some insight into the proper scaling procedure and to improve understanding of the collision process, one can compare the data to theory. The PWBA theory employing hydrogenic wave functions is easily calculable (see Section 3.1). A naive application of PWBA yields disappointing agreement with experiment; therefore Brandt and Lapicki[90–92] have tried to trace the physical origin for discrepancies and to develop corresponding correction procedures to the PWBA.

The nonhydrogenic character of an actual inner-shell wave function is taken into account by replacing the hydrogenic electron binding energy by the real one and by scaling the impact velocity in terms of the classical orbiting velocity of the electron. Thus reasonable agreement at higher impact velocities is achieved. Severe discrepancies between theory and experiment amounting to up to several orders of magnitude are commonly observed at the smallest impact velocities. These are easily attributed to the Coulomb interaction between projectile and target nucleus, causing a deceleration and a deflection of the projectile. Since cross sections at low velocities drop strongly with decreasing velocity, the deceleration results in a substantial reduction of the cross section; in comparison, the deflection has only a minor effect on the cross section. Nevertheless, the total decrease of the cross section by the projectile–target nucleus interaction has unfortunately been named "Coulomb deflection." Experimentally, trajectory effects can be studied by employing different projectiles: Proton and deuteron projectiles produce the same perturbing potential to the atomic electron, but their trajectories are different owing to their different charge-to-mass ratios. For deuteron and alpha particle impact the trajectory is (almost) the same, but the perturbation is different.

For a quantitative description of the Coulomb deflection effect on ionization cross sections a "Coulomb-deflection factor" C as suggested by SCA calculations may be employed. Various forms of C are in use[93]; a simple one is given by

$$C = \exp(-\pi d\omega / v_1) \tag{23}$$

Here d denotes the half-distance of closest approach in a head-on collision [equation (18)]; ω is the classical orbiting frequency of the bound electron and v_1 the collision velocity. Since d/v_1 is a characteristic time of the collision, the argument of the exponential function measures the adiabaticity of the collision.

For heavier ion impact, perturbation theories such as PWBA or SCA predict a scaling of the cross section with Z_1^2, i.e., the squared projectile charge. Observed deviations from such a behavior are due to a higher-order effect, i.e., the disturbance of the atomic eigenstates by the projectile. The disturbance changes both the shape of the wave function and the binding energy of the electron: If the projectile is outside the electron's shell, it tries to retract the electron from its nucleus thereby reducing its binding energy; if the projectile is inside the shell, the electron's binding energy is increased. Clearly a change of the binding energy affects the cross section. This effect has been accounted for by the "perturbed stationary state" correction (PSS).[91] This name is also unfortunately chosen, since the commonly employed PSS only accounts for the energy change, but neglects the deformation of the wave function's shape.

At small collision velocities two further effects become significant; their experimental investigation is hampered by the severe and not precisely known Coulomb deflection effect. The first effect is the relativistic character (R) of inner-shell electrons in heavy atoms. Relativistic wave functions yield higher cross sections than nonrelativistic. This effect is most pronounced at low impact velocities and for wave functions with nonvanishing amplitude at the nucleus, i.e., the K and L_1 shells, as is shown by PWBA calculations.[94-96,53] The second effect is the energy loss (E) of the projectile by the ionization process.[92] A PWBA calculation including all the above-described corrections is called ECPSSR. The ECPSSR model is quite successful in reproducing total cross section data. For example, some 2300 measured values for the ionization of targets with atomic number $10 \leqslant Z_2 \leqslant 92$ induced by proton impact, which increase a million fold with impact velocity, rarely scatter by more than a factor of 2 about the CPSSR prediction; in fact, some 90% of the data agree with the CPSSR within 30%[92] (Figure 13). Systematic problems evident at the smallest impact velocities may not be surprising regarding the number and the enormous size of corrections in this regime (Figure 14).

For heavier projectile ions, ionization can additionally occur by an electron transfer from a bound target state to a bound projectile state. The cross section for such an electron capture process can easily be calculated in the Oppenheimer-Brinkmann-Kramers (OBK) approximation. As for the PWBA, heuristically correction procedures to the OBK have been developed in order to obtain agreement with experiment.[97,98]

4.4. Relative L-Subshell Cross Sections

Relative cross sections, i.e., ratios of cross sections of subshells belonging to the same main shell, provide a sensitive testing ground for various aspects of the ionization mechanism, as is discussed here for the L subshells.[89,99,100] The L_2 and L_3 wave functions are similar, but the corresponding binding energies differ considerably in heavy elements. In contrast, the L_1 and L_2 subshells have similar energies but different shapes. The L_1

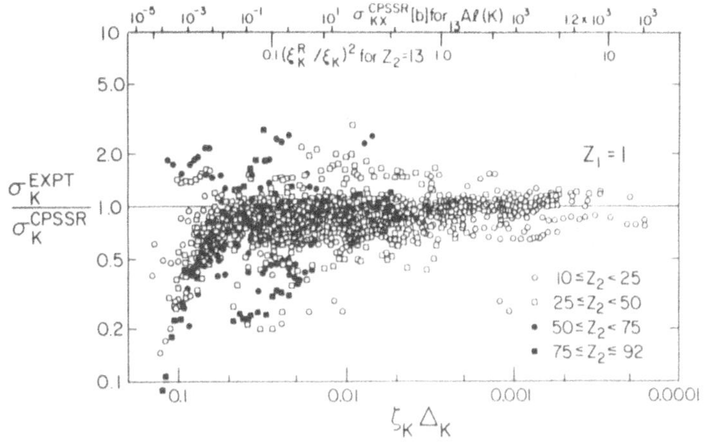

FIGURE 13. Experimental cross sections for the proton-induced K-shell ionization of a large number of target atoms versus scaled projectile impact energy. The different target atoms are classified according to their atomic number Z_2. The experimental data are normalized to the theoretical prediction of the CPSSR model (from Ref. 92).

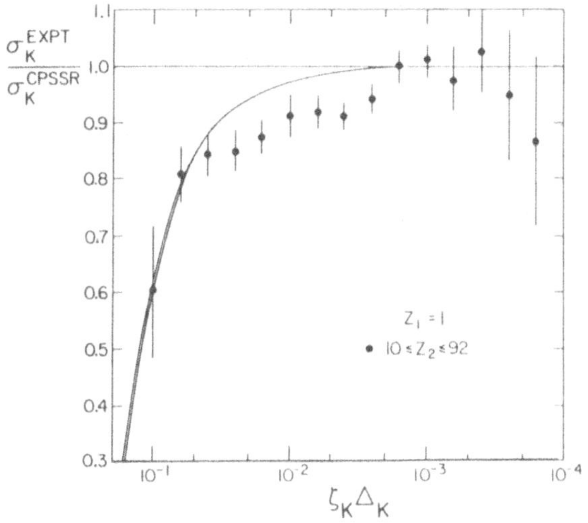

FIGURE 14. Same data as in Figure 13, but the comprehensive experimental data have been statistically averaged for the sake of clarity. A ratio of unity means agreement between experiment and CPSSR theory. The strong deviation from unity toward small impact energies is due to the energy loss effect. The curved lines show the prediction of the ECPSSR model which takes the projectiles energy loss into account (from Ref. 92).

wave function has a particular feature, i.e., a node of the radial wave function.

Measurements of relative subshell cross sections are not affected by uncertainties of the absolute normalization but are still susceptible to various problems. The experimental observed radiative (and nonradiative) decay occurs some time period after the collision. During this time, collisionally induced vacancies in the deeper-lying L_1 and L_2 subshells may have cascaded to the L_3 subshell via Coster–Kronig transitions [equation (8)]. Accordingly, the apparent cross sections have to be corrected for these cascades. There are collision systems where the cross section of the L_1 subshell is the largest of all L subshells and where the L_1 to L_3 Coster–Kronig yield amounts to about 70% (e.g., for Pb[35]); in these systems, most of the observed L_3 vacancies are created indirectly, and a reliable determination of the direct cross sections is hampered by uncertainties of the yields. Further problems of experimental origin arise from the large number of L X-ray emission lines, which are only partially resolved by semiconductor detectors. Besides the more trivial problems of spectrum fitting, the selection of lines for determining the subshell cross sections is arbitrary, and also the branching ratios, i.e., the relative strengths of particular lines, bear uncertainties. Different data evaluation procedures applied to the same experimental data may result in ionization cross sections differing by more than a factor of 2,[44] as discussed in Section 2.3. Fortunately, the "best" evaluation procedures are expected to yield cross sections reliable within some 10%.[38] A consistent evaluation procedure applied to the rough data of various groups gives reasonably consistent cross sections.[45,46] Therefore, a consistent set of relative cross sections for different projectiles and impact velocities provides conclusive information.[100]

We now turn to the discussion of the numerical cross section ratios. For proton impact (Figure 15), the ratio σ_2/σ_3 varies only slowly with impact energy owing to the similarity of the two corresponding wave functions. At high energy, it seems to approach the statistical value 1/2 given by the electron occupancy of the subshells. With decreasing energy, the adiabatic character of the collision becomes more important, and σ_2 drops more rapidly than σ_3 owing to the larger L_2 binding energy, resulting in a decrease of σ_2/σ_3. The σ_1/σ_2 and σ_1/σ_3 ratios display a pronounced energy dependence. There is a minimum at intermediate energies, i.e., at impact velocities roughly equal to the electron orbiting velocity. This can be traced back to the nodal structure of the L_1 radial wave function. With decreasing impact energy, σ_1/σ_2 and σ_1/σ_3 increase and would finally rise to infinity if only the shape of the wave function would determine the behavior. In reality, the adiabatic response of the electrons at small projectile velocities and projectile trajectory effects decrease σ_1 more than σ_2 and σ_3. As a result, σ_1/σ_2 and σ_1/σ_3 exhibit a maximum. The experimental data

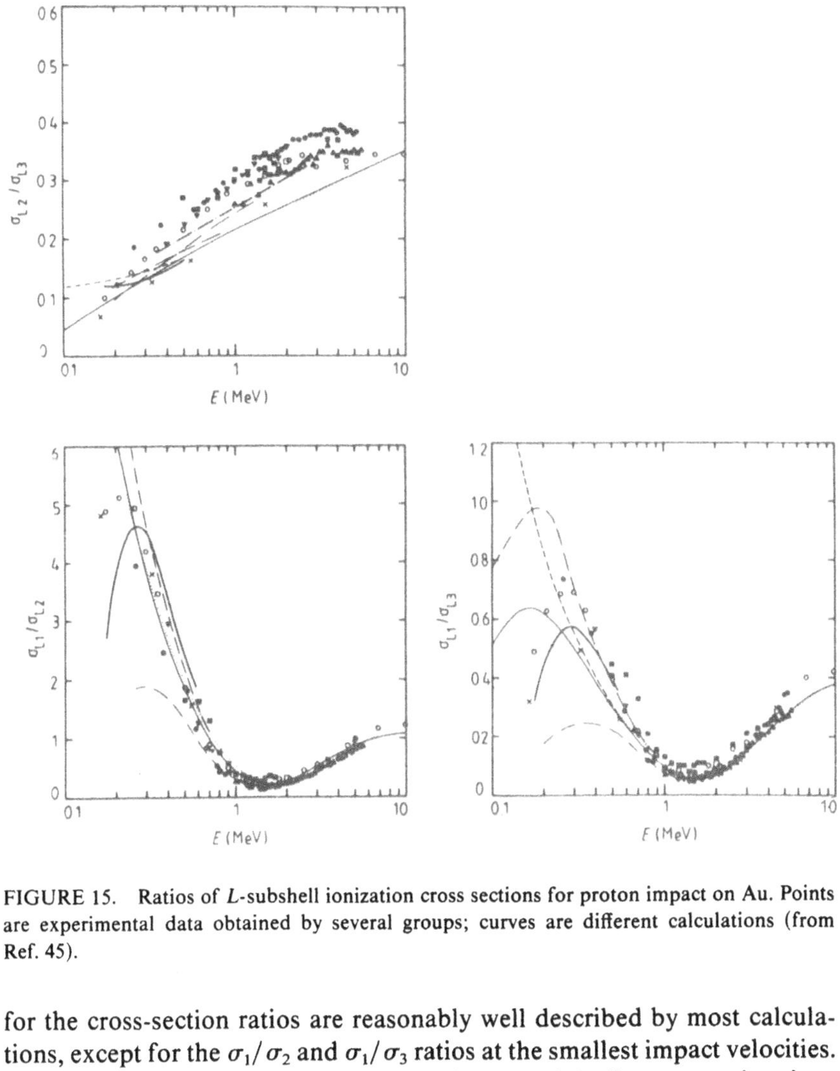

FIGURE 15. Ratios of L-subshell ionization cross sections for proton impact on Au. Points are experimental data obtained by several groups; curves are different calculations (from Ref. 45).

for the cross-section ratios are reasonably well described by most calculations, except for the σ_1/σ_2 and σ_1/σ_3 ratios at the smallest impact velocities. The maximum is not predicted at all from straight-line approximations [equation (16)]. The CPSSR with its simple correction factor for the Coulomb deflection (which amounts to about 10^{-2} at the peak position) gives a remarkably good description. Nevertheless, the position of the maximum is only properly reproduced by those calculations that incorporate a proper *ab initio* treatment of the actual projectile trajectory.

For projectiles with $Z_1 \geqslant 2$, trajectory effects are smaller than for proton impact and almost equal due to the specific charge-to-mass ratios. The maximum of σ_1/σ_3 resulting from the Coulomb deflection is expected to occur at very small impact velocities where cross sections become almost unmeasurably small, and it has not yet been observed. Thus one expects very similar cross-section ratios close to the CPSSR prediction for the different projectiles. This is in contrast to the experimental data (Figure 16); in particular, at small impact velocities a strong unexpected projectile dependence of the ratios σ_2/σ_3 and σ_1/σ_2 is observed.[46] This behavior

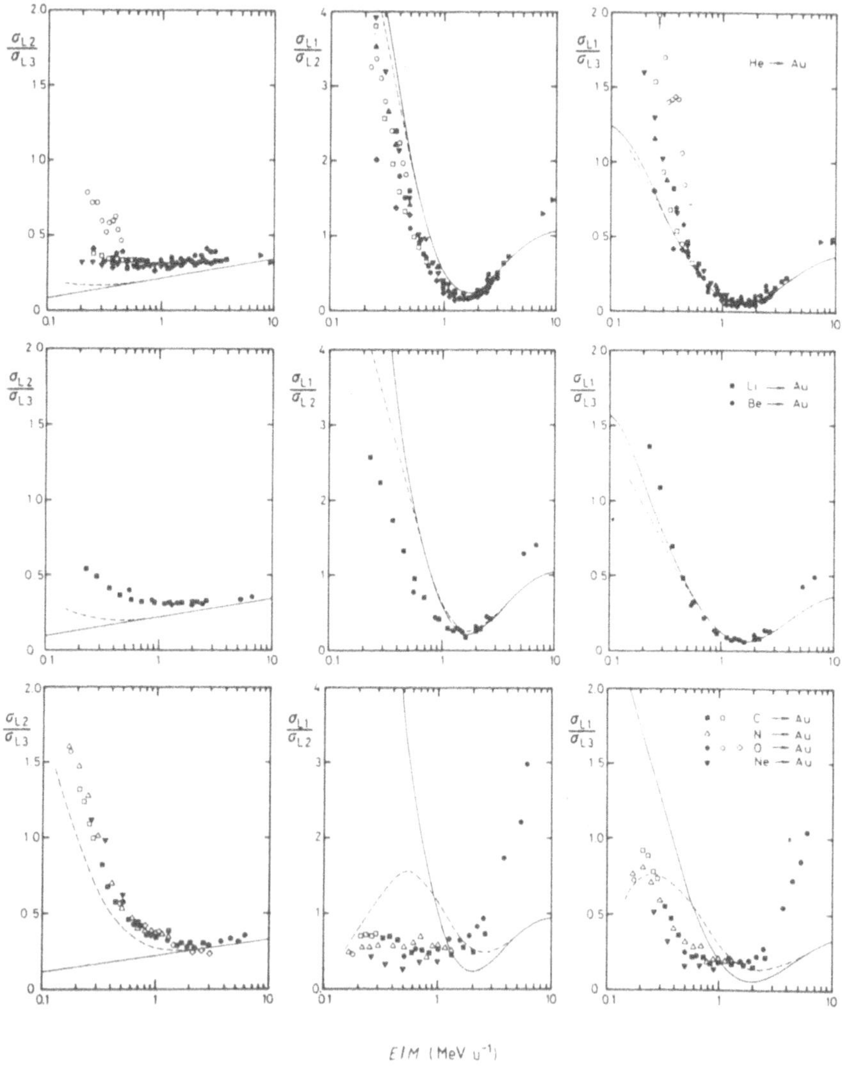

FIGURE 16. Same as Figure 15, but for various projectiles. Top: ^{4}He impact; middle: ^{7}Li and ^{9}Be impact; bottom: C, N, O, and Ne impact. The solid curve gives a CPSSR calculation; the dashed curve shows a calculation in a two-step model (from Ref. 46).

can be traced back to the absolute size of σ_2, which is predicted by theory to be smaller than σ_1 and σ_2, contrary to the experimental findings. The origin of this behavior is a higher-order process[62]: the disturbance of the atomic eigenstates by the field of projectiles with $Z_1 > 1$ is so large that the induced Stark mixing of the states causes a hybridization of the L-substate wave functions, thereby making a classification by subshell labels L_1, L_2, and L_3 during the collision meaningless. The CPSSR does not account for shape changes of wave functions.

In order to estimate the effect of the collisional Stark mixing on ionization cross sections, a two-step model of the collision process has been proposed: In a first step, ionization of atomic eigenstates takes place, and in the second step the vacancies are collisionally transferred to other eigenstates.[62] This model was later refined, and model calculations reveal that the largest intrashell transfer probability exists between the L_1 and L_2 subshells, owing to their comparatively close binding energies.[101] A comprehensive theoretical treatment of the distorted wave-function problem can be performed by more elaborate theories, such as the generalized distortion approximation,[102] the second-order Born approximation,[83] or coupled-state calculations.[84,85,103]

4.5. Alignment and Dealignment

The alignment is a sensitive parameter for angular effects in a collision. Similarly to the relative cross section, it is not affected by normalizing problems. Experiments have been performed by measuring the anisotropy of X-ray emission with a semiconductor detector or a crystal spectrometer as well as by measuring the X-ray polarization with a crystal spectrometer.[23] Anisotropy β and polarization P of dipole radiation are simply related by

$$\beta = 2P/(P-3) \tag{24}$$

Other radiative decay modes, as e.g., the magnetic quadrupole transition, exhibit different relationships; however, due to their small intensities they can be neglected in most cases.[104]

The proton-induced L_3-subshell alignment of heavy elements has been investigated by several groups[105-108,28] (Figure 17). At medium and high impact velocities nice agreement between the comprehensive experimental data and CPSSR calculations is found, whereas at small velocities significant deviations are observed. These are mainly due to the change of the projectiles direction by the interaction with the target nucleus, i.e., the Coulomb deflection: at the smallest velocities at which measurements have been taken, a projectile ionizing the L shell is deflected by more than 90° on the average. The collisional-induced anisotropic shape of the vacancy (alignment) is predicted by straight-line theories as the CPSSR to be aligned parallel to

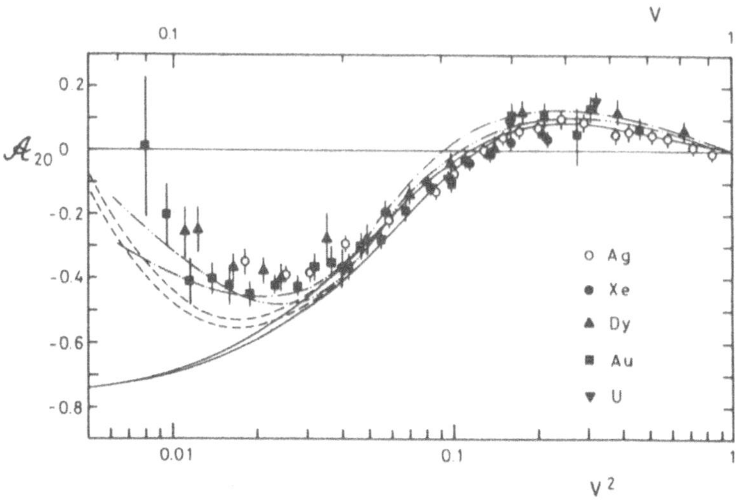

FIGURE 17. L_3 subshell alignment for proton impact on several heavy atoms ($47 \leq Z_2 \leq 92$) as a function of the relative impact velocity. Theory: ——, PWBA; ---, PWBA with correction for Coulomb deflection, - · · - and - · · · -, RSCA (from Ref. 28).

the impact direction. However, the actual severe deflection of the projectiles direction may result in a tilt by an angle θ with respect to the impact direction, where θ depends on the impact parameter b. In a noncoincident experiment where an average over all azimuthal angles is taken, the averaging over different tilt angles effectively reduces the observed alignment $\mathcal{A}'_{20}$ [28]

$$\frac{\mathcal{A}'_{20}}{\mathcal{A}_{20}} = \frac{\int_0^\infty db\, 2\pi b[I_1(b) - I_0(b)]P_2(\cos\theta)}{\int_0^\infty db\, 2\pi b[I_1(b) - I_0(b)]} \tag{25}$$

where $I_1(b)$ and $I_0(b)$ denote the impact-parameter-dependent ionization probabilities of the $m = 1$ and $m = 0$ magnetic substates and P_2 the second Legendre polynomial. Different choices for the angle θ as well as for the ionization probabilities have been used,[106,109,108] all yielding a partially quantitative reproduction of the experimental data.

Advanced SCA calculations incorporating a proper treatment of the projectile trajectory have no difficulties in reproducing the experimental alignment measurements at all collision velocities.[81]

The wave function of the $M_{2,3}$ subshell has a radial node and thus the M_3 alignment exhibits more structure as function of impact velocity: an additional maximum occurs at small impact velocities, as predicted by theory[80] and confirmed experimentally (Figure 18).[29,110]

Studies of the proton-induced L_3 alignment in the neighboring elements $Z = 27$–30 gave surprising results. For example, the measured X-ray

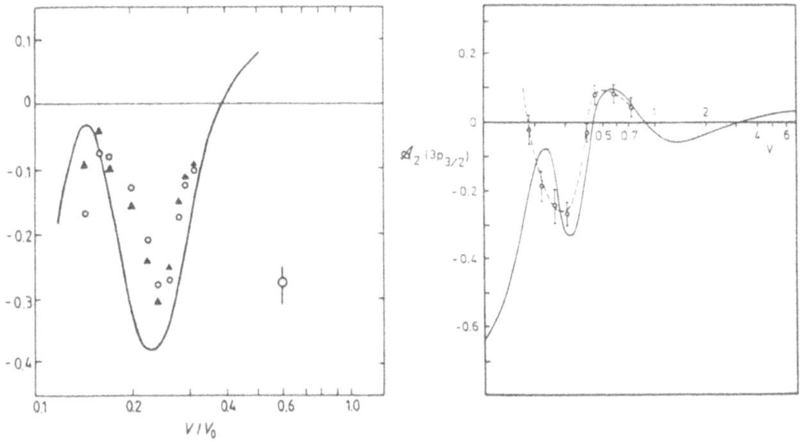

FIGURE 18. Proton-induced M_3 subshell alignment as a function of the relative collision velocity. Left: Au target; experimental results without (○) and with (▲) correction for target thickness. The curve shows the results of Born approximation calculations in the Hartree–Slater model (from Ref. 110). Right: Th target. The dashed curve connects the experimental points. The solid curve shows calculations using hydrogenlike wave functions (from Ref. 29).

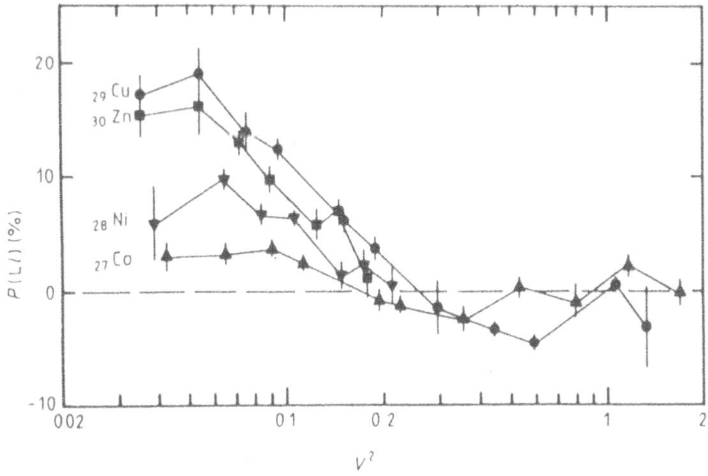

FIGURE 19. Polarization of the L_l line for proton impact on Co, Ni, Cu, and Zn. V denotes the impact velocity in terms of the classical $2p_{3/2}$ electron orbiting velocity. Curves are guides for the eye only (from Ref. 111).

polarization is much smaller for $_{28}$Ni than for $_{29}$Cu (Figure 19).[111] The reason for this result is an effective postcollisional dealignment process: For $_{28}$Ni, there are unoccupied states in the M shell. Angular momentum coupling of the collisional-induced aligned L_3 vacancy with an initially anisotropic M vacancy within its lifetime transfers part of the initial L_3 alignment to the M shell, thereby reducing the L_3 alignment. A model treatment of this dealignment mechanism shows that vacancies in the M shell effectively destroy the L_3 alignment, whereas vacancies in higher shells do not, because of their comparatively weak interaction with the L_3 vacancy.[39,112,113] The decisive criterion for the effectiveness of dealignment is the ratio of vacancy lifetime (inverse levelwidth)[35,114] and angular momentum coupling time (inverse multiplett splitting)[115] (Figure 20).

Experimental alignment data for heavier ion impact exhibit a strong increase towards small impact velocities[116-118] (Figure 21). The onset of the increase occurs at higher impact velocities than expected for trajectory effects; it occurs at approximately the same impact velocities and has similar projectile dependence as the increase of the σ_2/σ_3 cross-section ratio (see Section 4.4). Apparently the reason for this behavior is the Stark mixing of the atomic eigenstates by the projectile's field. Model calculations in second-order perturbation theory and close-coupling approximation are able to describe this increase qualitatively (Figure 22).[83,84]

For heavier ion impact ($Z_1 \geqslant 16$) ionization mechanisms other than direct ionization become important, as electron capture by the projectile or electron promotion via molecular states. These mechanisms should be able to produce large alignment effects. However, only minor L_3 alignments have here been observed so far.[118,119] Presumably an efficient postcollisional dealignment occurs by interaction with M vacancies produced in the same collision. The probability of producing M vacancies simultaneously to an L vacancy easily approaches values of unity (Figure 23).[120]

4.6. *Coincidence Measurements*

The total ionization cross section is an average over all projectile trajectories and an integral over all final states of the ejected δ electron, i.e., over all energies and directions. More detailed and decisive information on the collision process can be experimentally obtained by a selective study of a well-defined collision event (Figure 1). The projectile trajectory can be identified by the detection of the scattered projectile; similarly, energy and direction of the δ electron can be fixed. The detection of a characteristic X-ray or Auger line originating from a particular (sub-)shell makes it possible to ferret out the ionized shell. The simultaneous, i.e., coincident occurrence of detector signals from the projectile, the δ electron, and the characteristic particle allows one to select a single, well-defined collision

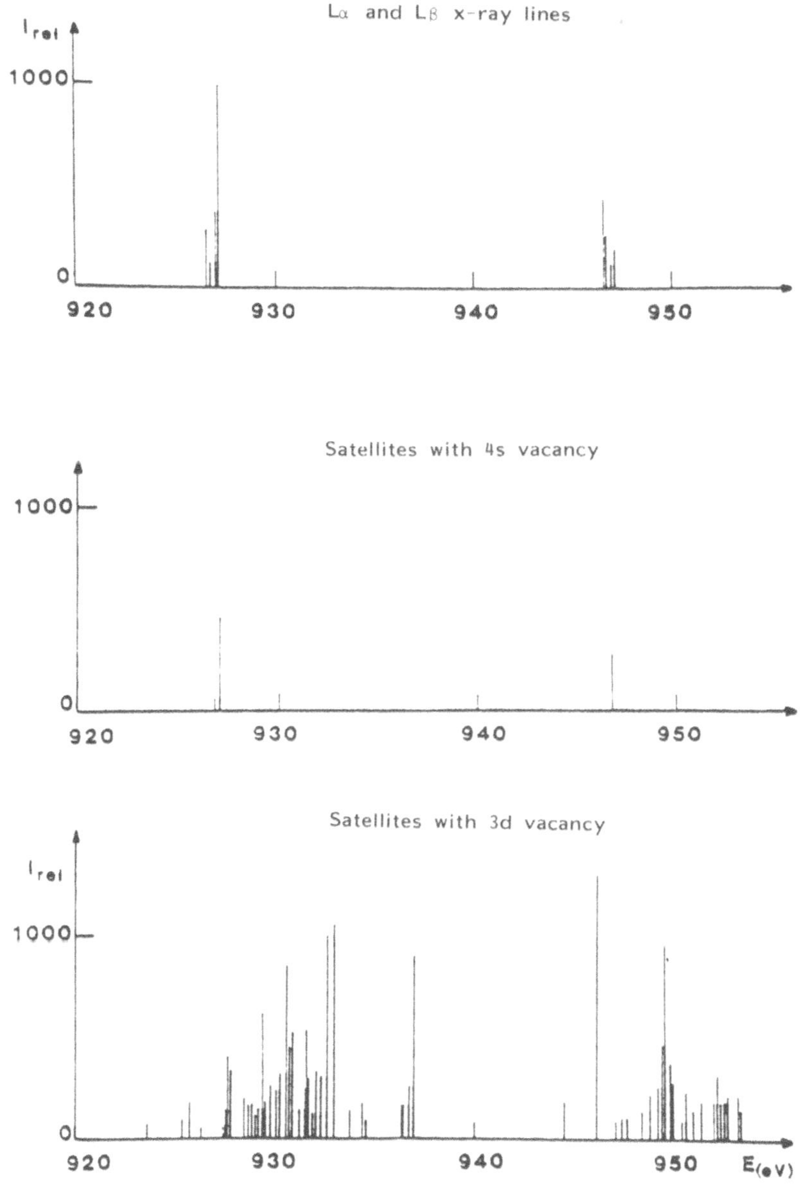

FIGURE 20. Calculated $L\alpha$ and $L\beta$ X-ray lines of free Cu ions emitted in the decay of an L_3 vacancy reflecting the multiplet splitting of initial and final states. Top: initially only one vacancy in the L_3 $(2p_{3/2})$ subshell; middle: with additional N_1 $(4s)$ spectator vacancy; bottom: with additional $M_{4,5}$ $(3d)$ spectator vacancy (from Ref. 115). For comparison, the natural width of a L_3 vacancy amounts to about 0.5 eV (Ref. 35).

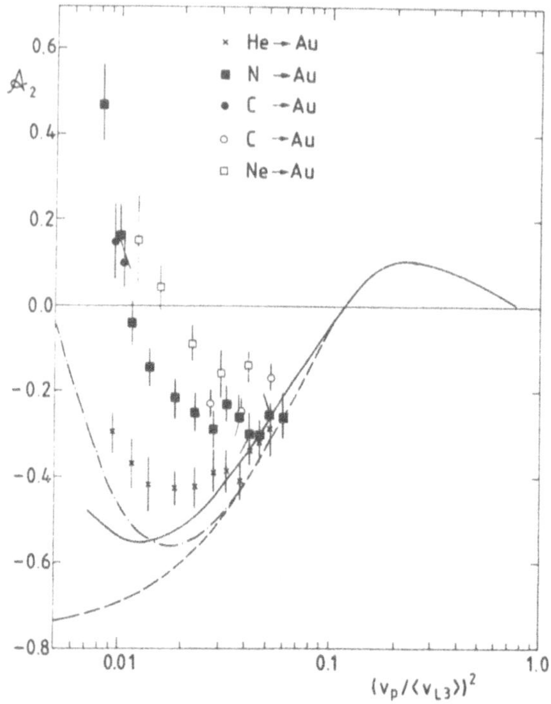

FIGURE 21. Alignment of the L_3 subshell of Au induced by impact of various heavy ion at low impact velocities. Curves: - - -, PWBA; - · -, corrected for Coulomb deflection; ——, SCA (from Ref. 117).

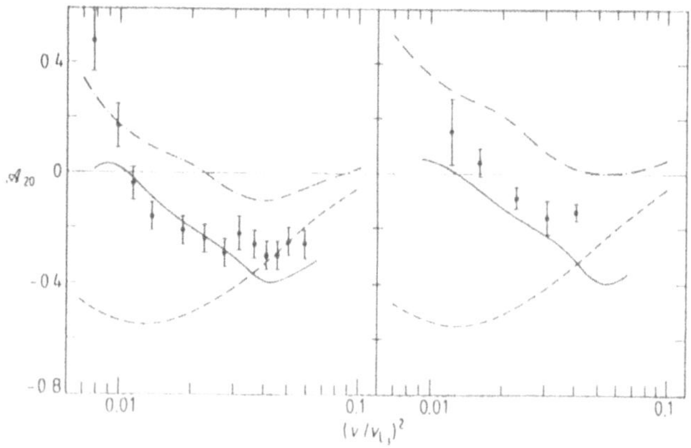

FIGURE 22. L_3-subshell alignment of Au induced by N (left) and Ne (right) ions, as a function of the relative collision velocity. Points are experimental data. Theoretical curves: - - -, first-order Born approximation; - · -, second-order Born approximation; ——, coupled-states theory (from Ref. 84).

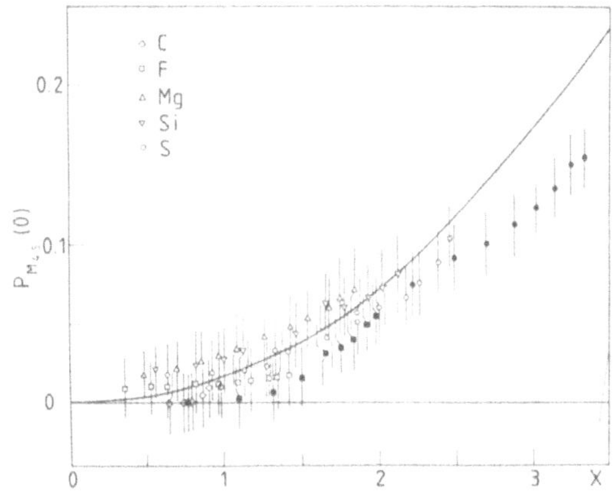

FIGURE 23. Experimental ionization probability (per electron) of the $M_{4,5}$ subshell in central collisions of C, F, Mg, Si, and S ions with Au (solid symbols) and Th (open symbols) as function of the scaled impact velocity, compared to the prediction by a geometrical model (from Ref. 120).

event. For inner shells and light ion impact, the ionization probabilities are small (typically 10^{-3} or less), and double coincidence measurements, i.e., the simultaneous detection of two particles, are difficult and lengthy; triple coincidences are at the limits of practicability. In the following we discuss two types of coincidence experiments: first, studies of the ejected δ electron, and second, ionization studies with fixed projectile trajectory.

The energy of the δ electron emitted in a collision is not fixed, but may lie between zero and the full collision energy. The cross section is largest for a δ-electron energy of zero and drops rapidly with increasing energy. This theoretically predicted behavior can be experimentally investigated by measuring the energy spectrum of the δ electrons or, alternatively, by measuring the energy-loss spectrum of the projectile. The projectile's inelastic energy loss corresponds to the sum of electron binding energy and kinetic energy of the electron; it is rather small compared to the projectile energy. Measurements of the energy loss spectrum thus require ion detectors with high energy resolution. This requirement can be fulfilled partially with cooled semiconductor detectors (see below) and completely with sophisticated magnetic spectrometers.[121] Using such a spectrometer, the K-shell ionization of Pb by 15-MeV protons was studied with an energy resolution of 6 keV. The K-shell binding energy of Pb amounts to 88 keV, and thus the energy loss spectrum is well resolved (Figure 24).[122,123] The measured spectrum is governed by the peaks of elastic scattering from the target material (Pb) and its support; the peak of inelastic K-shell ionization with the δ-electron spectrum appears only as a minor effect. It can be enhanced by requiring a coincidence condition between the scattered projectile and a K X ray, whereby all elastic processes and ionization of higher shells are suppressed. The resulting nice experimental energy spectrum of the δ electrons is in good agreement with results of refined calculations.[124]

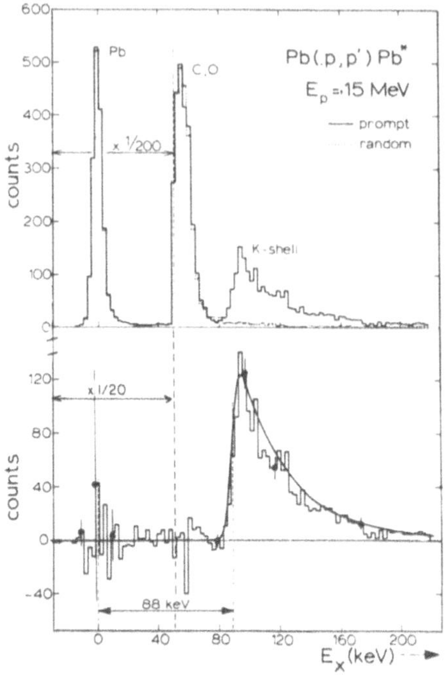

FIGURE 24. Energy spectra for 15-MeV protons scattered by 12 degrees from Pb. (a) The solid line histogram shows the prompt coincidences between scattered protons and K X-rays, the dotted line histogram the random coincidences. The peak denoted with C, O is due to random coincidences between K X-rays and protons scattered from the C, O target backing. (b) Histogram of "true" coincidences; the solid line represents a calculation in the SCA model (from Ref. 123).

The energy spectrum as well as the angular distribution of δ electrons can be directly measured by a rotatable electron spectrometer. Corresponding experimental data[125] contain the contributions from all electron shells. Since the ionization cross sections of intermediate and outer shells are considerably larger than those of inner shells, δ electrons from these shells constitute the main contribution. The δ electrons originating from an inner shell can be separated by a coincidence requirement with a characteristic inner-shell X ray or Auger electron. Corresponding measurements of the proton-induced δ electrons from the K shell of Ar show reasonable agreement with direct ionization theories (Figure 25).[126] In contrast, measurements of the L-shell ionization of the same collision system reveal deviations (Figure 26)[127] that are due to the influence of other ionization mechanisms as a capture of the atomic electron to a continuum state of the projectile.[128]

In the second type of coincidence experiments, the trajectory of the projectile is identified by the detection of the projectile scattered by a specific angle; from the angle the impact parameter b is derived [equation (17)]. As for total cross sections, measurements of L-subshell cross section ratios are sensitive to subtle effects of the collision process. The impact-parameter dependence of the ionization cross sections is different for the L_1 and the

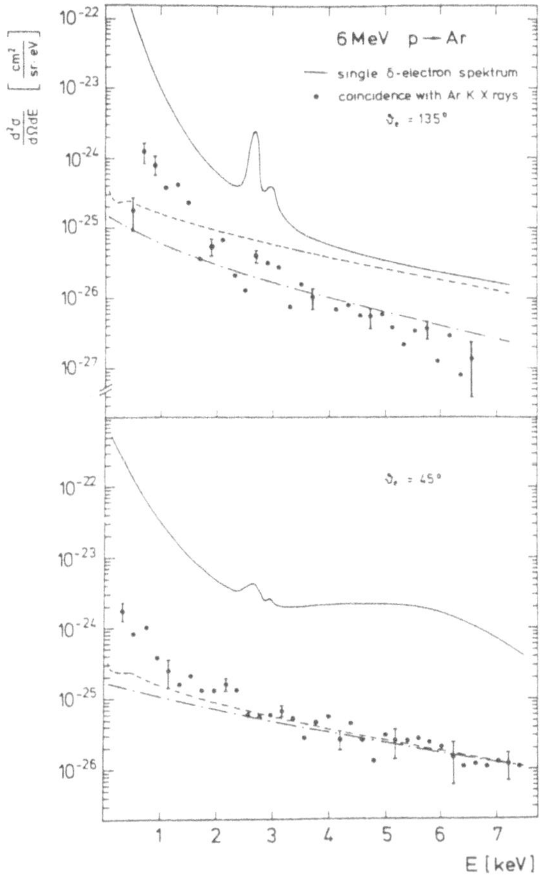

FIGURE 25. Double differential electron emission spectrum of Ar induced by 6-MeV proton impact for two emission angles ϑ_e. The solid curve is the measured single spectrum consisting of the δ electrons and the K-Auger lines at about 2.7 keV. Points are measured coincidences. Dashed curve is a PWBA calculation, dashed-dotted curve a SCA calculation (from Ref. 126).

L_2, L_3 subshells owing to the different radial shapes of the corresponding wave functions. This expected behavior is confirmed experimentally in case of light-ion impact on heavy elements (Figure 27).[129] For a heavier projectile ($_{10}$Ne), however, the measured impact-parameter dependencies for the L_1 and L_2, L_3 subshells are practically equal; this behavior is due to a sharing of vacancies between the different subshells caused by the collisional Stark interaction with the projectile.

An azimuthal resolved detection of the scattered projectile allows us to break the axial symmetry of a collision experiment and thus to observe

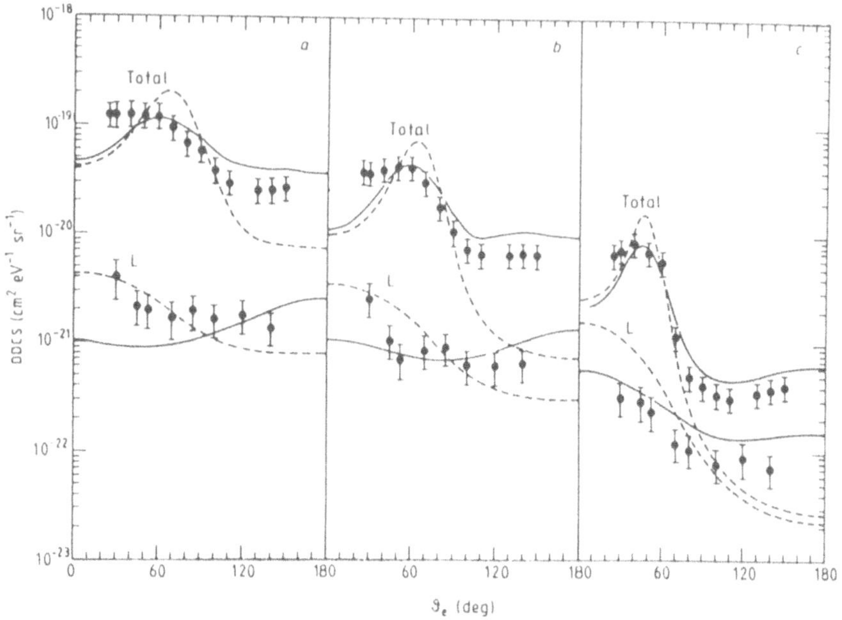

FIGURE 26. Double-differential cross sections for δ-electron emission induced by 350-keV proton impact on Ar as a function of the emission angle. Electron energy: (a) 46 eV, (b) 100 eV, (c) 308 eV. Curves are calculations: ——, PWBA using Hartree-Fock/Hartree-Slater wave functions; - - -, PWBA using screened hydrogenlike wave functions (from Ref. 127).

alignment parameters other than $\mathscr{A}_{20}$, which merely describes the longitudinal-transversal asymmetry of the collisional induced vacancy. In the case of the L_3 subshell, the alignment parameter $\mathscr{A}_{22}$ describes the azimuthal dependence of the vacancy and $\mathscr{A}_{21}$ is a measure for a tilt of the vacancy shape with respect to the impact axis. In recent experiments, the collision systems α + Dy[130] and p + Sm[131] have been studied (Figure 28). In both cases, quite similar results have been found; the maximum of $\mathscr{A}_{20}$ and the minimum of $\mathscr{A}_{22}$ occur at the same impact parameter for the experimental values[130,131] as well as for the theoretical values[82,66]; however, both are shifted against each other. The origin of this shift is not clear. Presumably it is not a trajectory effect, since the deflection of the projectile is tiny and included in the calculations; it is also no effect due to the Stark distortion of the atomic eigenstates as confirmed by coupled-channel calculations.[85]

At small impact parameters, the projectile transfers a sizable amount of momentum to the target nucleus, resulting in a recoil of the nucleus. This recoil effect can be directly investigated by employing projectiles with different mass, such as, e.g., protons and deuterons. Such measurements of the K-shell ionization of Ni and Ag at small impact parameters, i.e., large scattering angles, reveal clear differences for the two projectiles

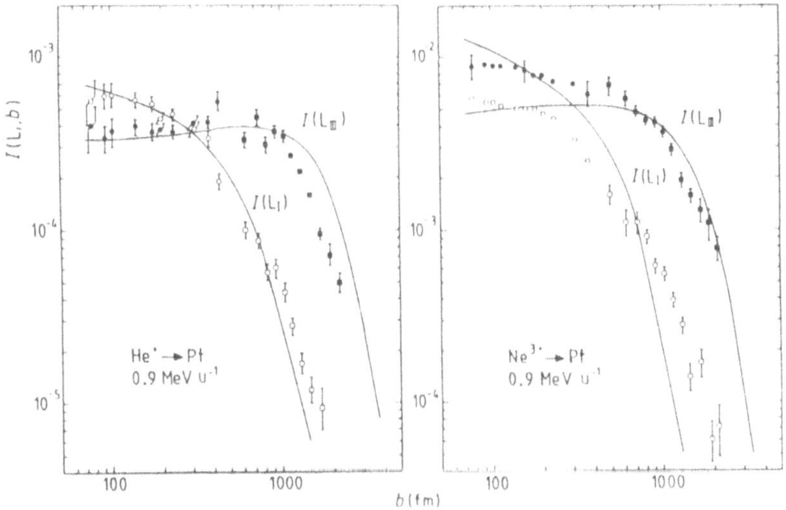

FIGURE 27. L_3- and L_1-subshell ionization probabilities for (left) He impact and (right) Ne impact on Pt. Points are experimental data; curves represent SCA calculations (from Ref. 129).

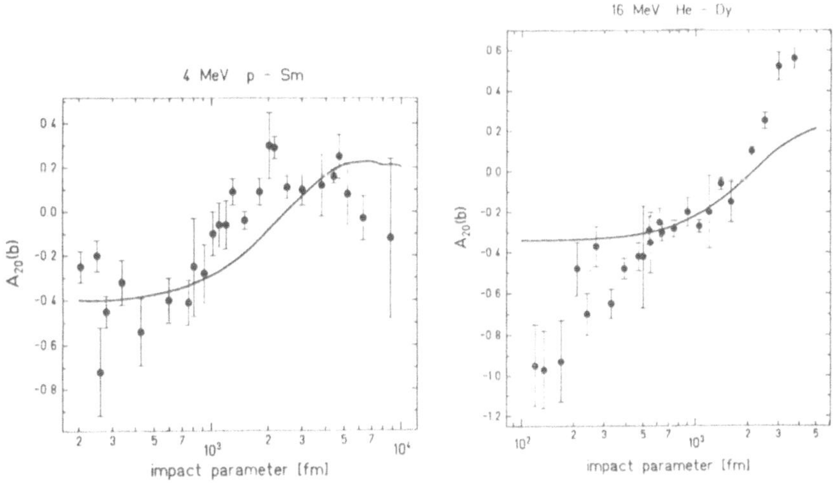

FIGURE 28. Calculated and measured alignment parameter $\mathscr{A}_{20}(b)$ at different impact parameters b for the L_3 subshell. Left: 4-MeV proton impact on Sm; right: 16-MeV He impact on Dy. Calculations have been performed within a coupled-state formalism, using relativistic atomic wave functions as basis states (from Ref. 85).

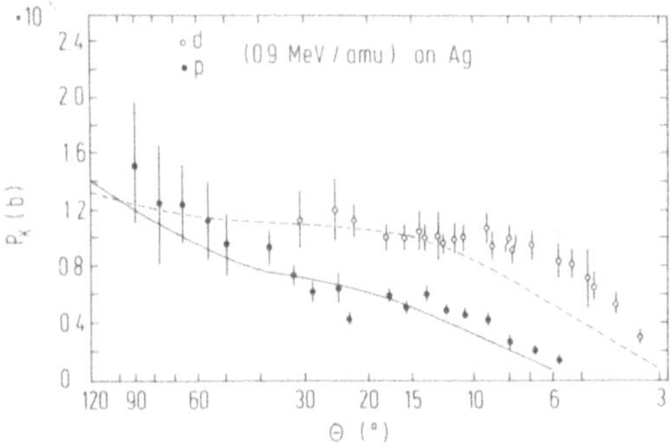

FIGURE 29. K-shell ionization probability of Ag for protons and deuterons at the same impact velocity as function of the projectiles scattering angle. Curves are SCA calculations including the recoil effect (from Ref. 133).

(Figure 29).[132-134] These are predicted by SCA calculations incorporating a proper treatment of the projectile trajectory.[135,136]

Interesting new effects arise by a simultaneous occurrence of nuclear processes and atomic ionization in the same collision. For projectile energies in the vicinity of a nuclear resonance, the interaction between projectile and target nucleus becomes non-Coulombic resulting in a modified transit time of the projectiles passing the atomic nucleus. A treatment of this effect in the framework of the semiclassical approach is straightforward,[137] and now comprehensive calculations are available.[138] The theoretical aspects are discussed in this volume in Chapter 7 by U. Wille. The first experimental observation of a nuclear reaction effect on atomic ionization was reported for the K-shell ionization of ^{58}Ni.[139] Later on, further experiments employing other collision systems have been performed, which suffer from large uncertainties in measuring a relatively small (ca. 20–40%) effect.[140-142] In these experiments, the ionization process is identified by an X-ray detector with large solid angle and with high time resolution (Figure 30). Significant improvement of the measuring statistics has recently been achieved in an experiment in the proton-induced K-shell ionization of ^{138}Ba in the vicinity of the $f_{7/2}$ isobaric analog resonance at about 10 MeV.[143] By the use of a cooled semiconductor ion detector with 20 keV (FWHM) energy resolution it was possible to gate selectively only on those projectiles that had produced a K vacancy (minimum energy loss required is 37 keV). Thus the strong signal from (almost) elastically scattered projectiles and its concommitant disturbing statistical noise were strongly reduced (Figure 31). The experimentally obtained ionization probabilities could be nicely described by a parameter fit to a theoretical model that correlates atomic ionization and nuclear scattering. Thus the atomic data deliver information on nuclear processes, such as the compound elastic and the direct elastic scattering as well as the time delay by nuclear resonance scattering. It has therefore been

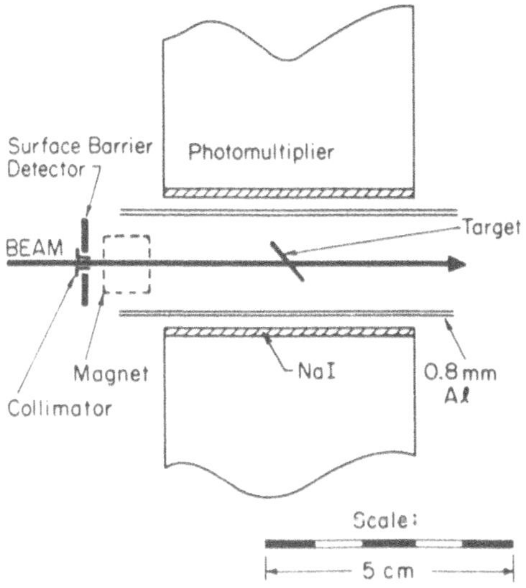

FIGURE 30. Experimental arrangement for measuring coincidences between backscattered projectiles and induced X rays. A set of magnets is mounted in front of the annular surface barrier detector to divert δ electrons emanating from the target away from the particle detector (from Ref. 143).

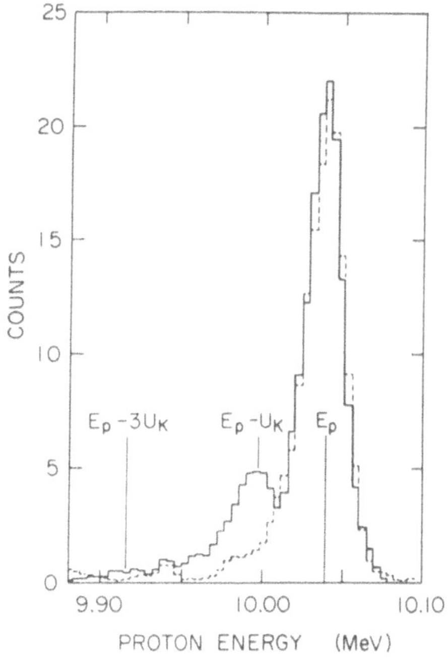

FIGURE 31. Sorted energy spectra of the backscattered protons. Full histogram has been obtained by gating on the peak in the time spectrum of the time-to-amplitude converter (TAC) and thus contains true and random coincidences; the dashed histogram has been obtained by gating off the TAC peak and thus contains only random coincidences (from Ref. 143).

stated that inner-shell ionization studies provide "atomic clocks for nuclear times."[144]

5. PHOTOIONIZATION

5.1. Atomic Photoionization

The ionization of an inner shell by charged particle impact is a complex process, as shown in the previous sections. Cross sections reflect various effects due to the target-induced change of the projectile's trajectory and higher-order effects of the collisional perturbation of the atomic states; furthermore, the energy of the final state, i.e., the ejected δ electron is not fixed. In contrast to the charged particle ionization, the photoionization is of more conceptual simplicity, and photoionization studies provide a unique chance to investigate the intrinsic response of an atom to an external perturbation. Theoretically, the photon–atom interaction is very weak at common intensities warranting the validity of a first-order perturbative approach. Experimentally, photoionization cross sections can be determined directly from photoabsorption studies, and one does not need to worry about postcollisional vacancy redistribution by Coster–Kronig cascades or uncertain fluorescence and Auger yields in evaluating measured emission spectra. The high accuracy of cross sections and the high energy resolution of the final state achievable in photoionization experiments has made it feasible to investigate even tiny (e.g., 1% and below) effects on the cross section. The availability of monochromatized, tunable synchrotron radiation has opened access to new experimental areas. Most of the synchrotron work is concentrated on outer and intermediate shells, but also fascinating studies on inner shells have been performed.[145]

The photoionization cross section for a particular initial state $|i\rangle$ has a sharp onset at the energy threshold (called edge), and decreases above the edge smoothly with increasing photon energy. Quantitatively it is given in the single-electron, single-photon approach by[146,147]

$$\sigma_{if} = \frac{4\pi^2\alpha^5}{\mathbf{k}} |\langle f|\boldsymbol{\alpha}\hat{\boldsymbol{\varepsilon}}\, e^{i\mathbf{kr}}|i\rangle|^2 \delta(E_f - E_i + \hbar\omega) \tag{26}$$

$|f\rangle$ denotes the final state and E_f its energy. $\mathbf{k}$, $\hbar\omega$, and $\hat{\boldsymbol{\varepsilon}}$ stand for momentum, energy, and polarization of the photon. α is the fine-structure constant and $\boldsymbol{\alpha}$ is a 2×2 matrix with Pauli spin matrices as off-diagonal elements.

Simple calculations according to equation (26) using nonrelativistic wave functions[148] predict the absolute values of inner-shell cross sections already with errors not exceeding about 30%. State-of-the-art calculations employ Dirac–Hartree–Slater electron wave functions and include all contributing multipoles of the radiation field as well as retardation effects. Comprehensive tabulations of the individual subshell cross sections of neutral atoms with $Z = 1$–101 for photon energies ranging from 1 to 1500 keV are available.(146) These calculations are presumably the best that can be done within a potential model for the treatment of the electrons, i.e., within a single-electron approach. They show remarkably good agreement with experiment for the total photoelectric cross section; depending on the photon energy, deviations are partly below a few percent and hardly identifiable considering the uncertainties of the experimental data.(149,150) Accordingly, all deviations from this prediction are effects of small size.

The interaction between the individual electrons in an atom is responsible for collective effects in photoionization. For inner shells, effects due to relaxation and dielectric polarization may be important.(151) Static relaxation is the change of the electron eigenstates by the creation of an inner-shell vacancy; for example, after vacancy creation the outer-shell electrons experience a less screened nuclear charge and may contract or even rearrange themselves. Dielectric polarization originates from the response of all electrons to the external field resulting in a (anti)screening of this field. This dynamical polarization depends on the local position in the atom and the photon frequency; it manifests itself in the photoionization cross section usually by a decrease of the threshold value and by a shift of the maximum (if present) towards higher energies. Relaxation and polarization effects have been treated theoretically by various approximations: random-phase approximation with exchange (RPAE),(152,153) many-body perturbation theory (MBPT),(154) R-matrix theory (RMT),(155,156) and the time-dependent local density approximation (TDLDA).(157,158)

In a recent experiment, the photoionization of heavy elements ($72 \leq Z \leq 82$) was studied in the energy regime of the L edges.(159) Comparison of the experimental data with prediction of a single-electron theory reveal minor but significant deviations (Figure 32): the actual total photoelectric cross section is reduced in the vicinity of the L_3 and L_2 edges, and its energy dependence between the L_2 and L_1 edges is unexpectedly small. These dispersionlike deviations are mainly due to the dielectric polarization and can be reproduced partly quantitatively by TDLDA calculations (Figure 33).(160)

The photoionization cross section at an absorption edge shows a steep onset at the edge; measurements with high resolution reveal a fine structure at the edge. The onset is broadened by the finite level width Γ, due to the lifetime of the created vacancy.(35) Integration over the Lorentzian level profile gives for the energy dependence of the ionization cross section in

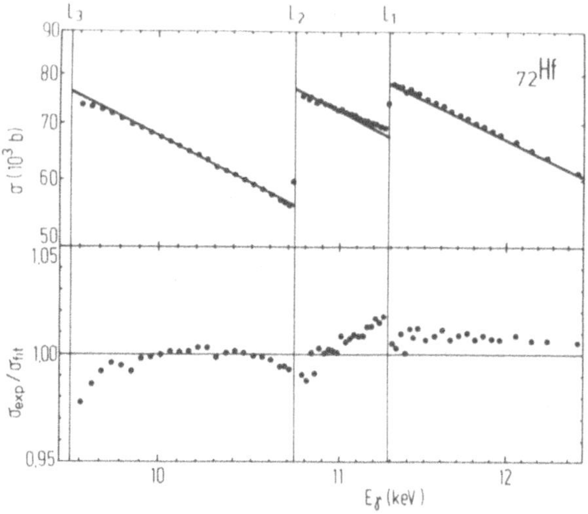

FIGURE 32. X-ray absorption spectrum of Hf in the vicinity of the L edges. Points are experimental data; the curve is a relativistic Hartree-Slater calculation; experimental data have been normalized to theory in the energy regime between the L_3 and L_2 edges. To elucidate the deviations between experiment and theory, the lower part of the figure shows the ratio of experiment to theory (from Ref. 159).

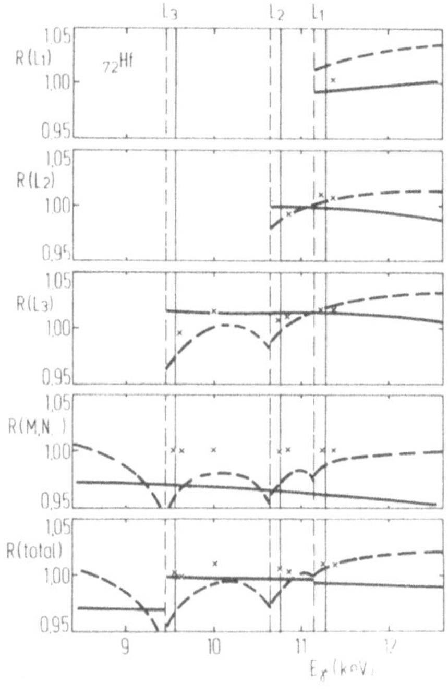

FIGURE 33. Comparison of theoretically calculated photoionization cross sections; for ease of comparison, the various calculations have been normalized to an analytical cross section. ×, relativistic Hartree-Slater; ——, independent particle approximation; - - -, time-dependent local-density approximation (from Ref. 159).

the threshold regime

$$\sigma(\hbar\omega) \propto (1/2) + (1/\pi)\arctan[2(\hbar\omega - E_i - E)/\Gamma] \tag{27}$$

where E_i denotes the inner-shell binding energy and E the kinetic energy of the photoelectron. Inner-shell vacancies can also be created by excitation to an unoccupied outer shell, which is already possible for photon energies below the ionization threshold; the cross section for excitation to a particular outer shell f with binding energy E_f exhibits a Lorentzian profile:

$$\sigma_f(\hbar\omega) \propto \frac{\Gamma/2\pi}{(\hbar\omega - E_i + E_f)^2 + (\Gamma/2)^2} \tag{28}$$

The total photoelectric cross section is just the sum over all contributions (Figure 34).[161] For heavier elements, the detailed features of the cross section spectrum are blurred by the level width, which becomes larger than the energy separation of the excited levels. The structures of L edges even

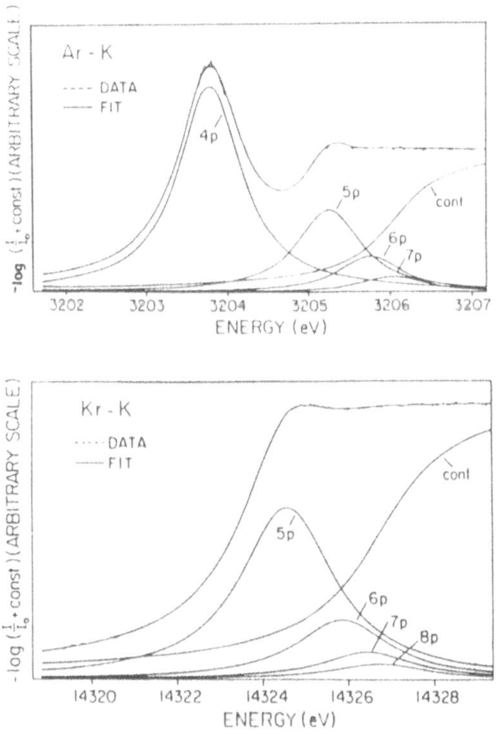

FIGURE 34. Measured Ar and Kr absorption spectra in the vicinity of the K edge. Measured and fitted curves of the total absorption are hardly distinguishable. Also shown is the fitted partition of the total absorption into the various contributing X-ray absorption processes as excitation (Lorentzian shape identified by the upper state) and ionization (identified by "cont.") (from Ref. 161).

of neighboring elements such as, e.g., $_{54}$Xe and $_{56}$Ba are quite different, since they reflect the unoccupied outer shell states.[161,162]

Somewhat above threshold, the measured absorption cross sections exhibit structures as shown for the K-edge regime of Ar (Figure 35),[163,164] and the L_1 edge of Xe (Figure 36).[165] These structures occur at the energy thresholds for the simultaneous excitation of two atomic electrons. Apparently a two-electron transition is induced by the absorption of a single photon; this process can be described approximately by the so-called shake model: The photon absorption transfers one (inner-shell) electron to an

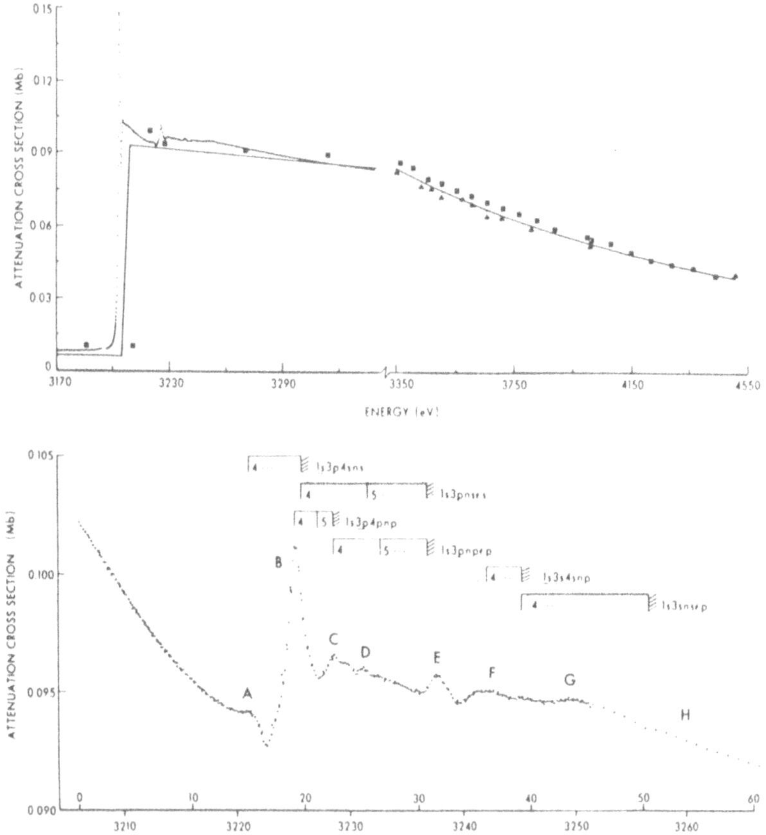

FIGURE 35. Top: Argon X-ray absorption spectrum scanned in a synchrotron-based experiment (dots) and discrete data (squares). Solid line is a calculation with hydrogenic wave functions. Bottom: Expansion of the suprathreshold structure. The upper energy scale starting from zero gives the energy above the onset of the one-electron continuum at 3206.0 eV. Energy positions of several configurations obtained by Hartree-Fock calculations are indicated; underlined electron states in the designation of configurations denote vacancies (from Ref. 164).

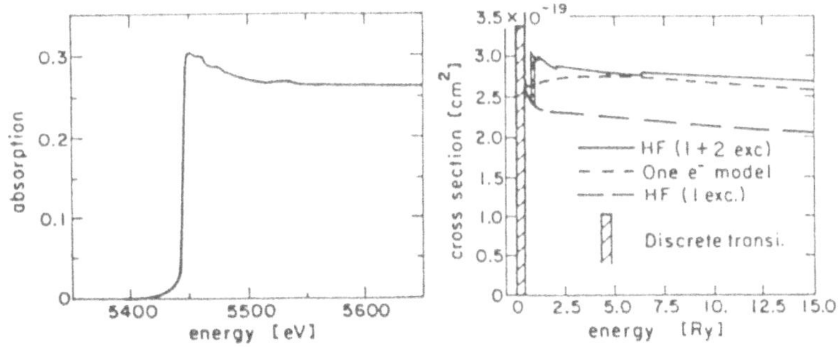

FIGURE 36. X-ray absorption of Xe in the vicinity of the L_1 edge. Left: Experimental data for Xe vapor; the vertical scale is arbitrary. Right: Calculated absorption profile using the sudden approximation for two-electron excitations. The cross-hatched rectangles represent transitions in which all excited electrons are in the bound state; no lifetime broadening is included (from Ref. 165).

unoccupied excited or continuum state, thereby changing the effective potential seen by the outer shell electrons. These may react by adjustments of their wave functions and binding energies towards the final "relaxed" singly ionized state, or alternatively by the transfer of an outer shell to an excited state ("shake up") or to the continuum ("shake off").[166] The shake process is a dynamical process, and the shake probability depends on the velocity of the receding photoelectron and thus on the photon energy.[167]

The shake processes seen in absorption measurements can be investigated in more detail by photoelectron measurements. The energy of an absorbed primary photon is shared by the atom and the photoelectron and thus the energy of the photoelectron determines the atom's (or ion's) energy and final state. A shake satellite photoelectron line has a smaller energy than the corresponding diagram line. Photoelectron spectroscopy thus allows one to measure the onset and threshold behavior of shake processes (Figure 37).[167,168]

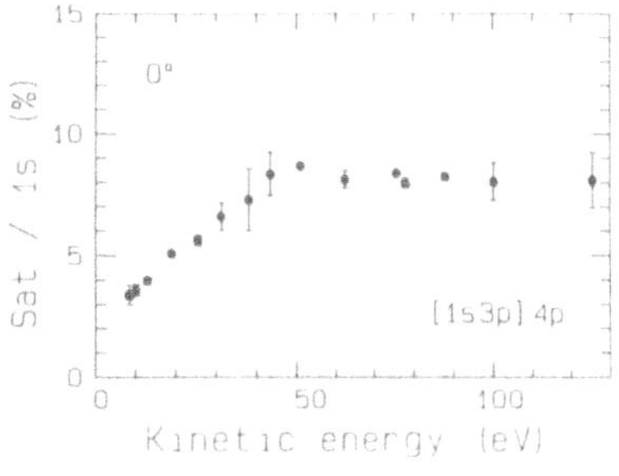

FIGURE 37. Intensity of photoelectrons from the Ar K shell. The intensity of the photoelectron satellite line for an excited final $1s^{-1}3p^{-1}4p$ Ar^+ state relative to the main line for a final $1s^{-1}$ Ar^+ state is plotted. Above threshold, the ratio shows a clear rise with increasing kinetic energy of the photoelectron (from Ref. 168).

5.2. *Environmental Effects*

High-resolution X-ray absorption or emission spectra recorded for free atoms, atoms in molecules, atoms at surfaces, and bulk atoms exhibit characteristic differences (Figure 38).[169,170] The chemical environment influences binding energies and thus the energies of photoelectrons. It is now a well-established technique to employ photoelectron spectroscopy for chemical analysis called ESCA (electron spectroscopy for chemical analysis).[171-173] Because of the small escape depth of photoelectrons,[174] this technique is surface-sensitive and probes a surface layer of only a few monolayer thickness.

Properties of bulk atoms can be investigated by X-ray absorption measurements. The near-edge structure of an absorption edge allows a precise determination of (relative) binding energies, and thus of the valence state. Interesting cases are the transition and rare-earth elements with partly unfilled $3d$ and $4f$ states, respectively. For example, the observed energy shift of the L_3 absorption edge of the rare-earth elements $_{58}$Ce, $_{62}$Sm, $_{64}$Gd, and $_{68}$Er either in the gaseous or in the solid phase is as small as 1–2 eV for Ce and Gd, but as large as about 8 eV for Sm and Er (Figure 39).[175] The origin for the large differences in the energy shifts is the different chemical behavior in solidification: Ce and Gd have the same $4f^n$ configuration in the atom and in the solid (Ce: $4f^1$; Gd: $4f^7$) whereas Sm and Er undergo a configuration change (Sm: $4f^6 \rightarrow 4f^5$; Er: $4f^{12} \rightarrow 4f^{11}$). The

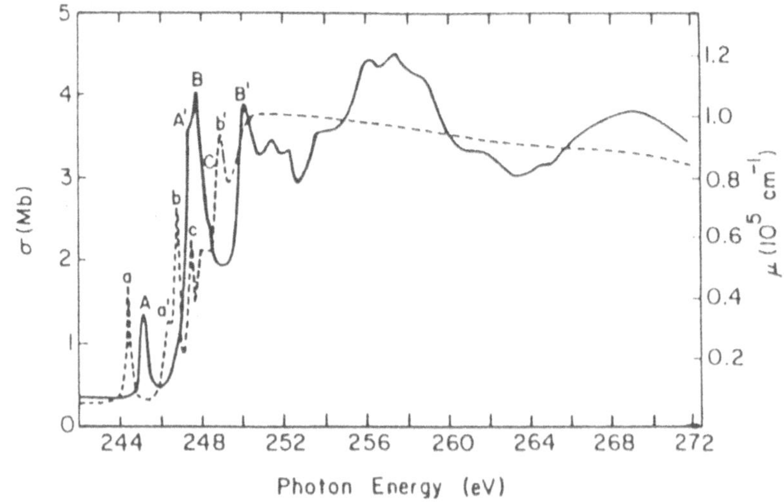

FIGURE 38. X-ray absorption spectrum of solid (solid line) and gaseous (dashed line) Ar in the vicinity of the $L_{2,3}$ edge. Corresponding peaks in the two spectra are labeled by upper and lower case letters, respectively (from Ref. 169).

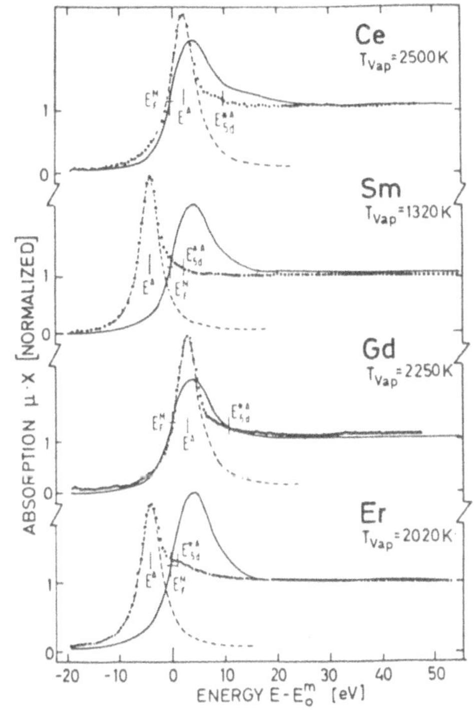

FIGURE 39. L_3 absorption spectra of atomic (dotted lines) and metallic (solid lines) Ce, Sm, Gd, and Er. The positions of the metallic Fermi levels E_F^M, the atomic $5d$ excitation energies E^A and the atomic ionization energies E_{5d}^{*A} are indicated. The dashed lines are fits for the Lorentzian white lines (from Ref. 175).

valence of an element in the solid phase may depend on the environmental conditions: For example, Yb undergoes a structural transition that changes its valence state at an applied pressure of about 100 kbar, as can be clearly seen from its X-ray absorption (Figure 40).[(176)]

X-ray absorption spectra recorded for atoms in molecules or condensed matter frequently exhibit an oscillatory structure, starting at threshold and extending with decreasing amplitude to higher energies.[(177-179)] This extended X-ray absorption fine structure (EXAFS) occurs only when atoms are surrounded by other atoms. The origin for EXAFS is an interference effect: The photoelectron emitted from an atom by the ionization process interacts with the surrounding atoms and may be partially reflected back (Figure 41). Thus its wave function is a sum of an outgoing and a scattered portion. Constructive or destructive interferences between the two waves result in an increase or decrease of the photoelectron wave function in the overlap region with the initial state, thereby increasing or decreasing the ionization cross section. Since the phase between the two waves depends on the photoelectron energy, the cross section exhibits an oscillatory behavior (Figure 42).[(179)] The occurrence of EXAFS requires constant interatomic

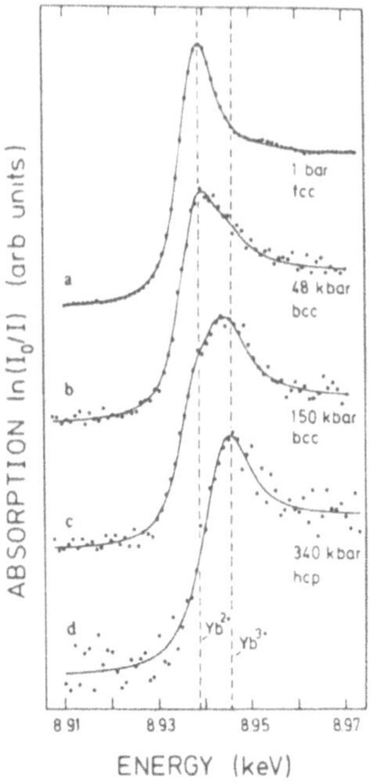

FIGURE 40. X-ray absorption edge of solid Yb in the vicinity of the L_3 edge at room temperature but different ambient pressures. Points are experimental data; the curves are fits. The dashed vertical lines indicate the positions of the L_3-edge peak for divalent and trivalent metallic Yb systems, respectively. Depending on the pressure, the crystal structure and the concommitant valence state changes (from Ref. 176).

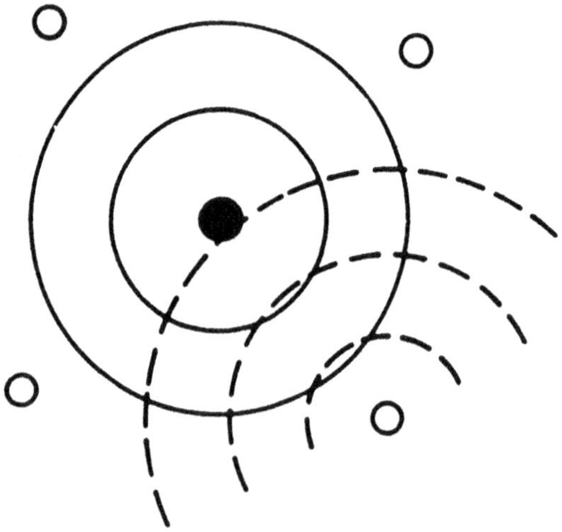

FIGURE 41. Mechanism responsible for EXAFS: photoelectron waves (solid lines) originating from the absorbing atom (solid circle) propagate to neighboring atoms (open circles). The backscattered waves (dashed lines) modify the wave function at the absorbing atom (from Ref. 179).

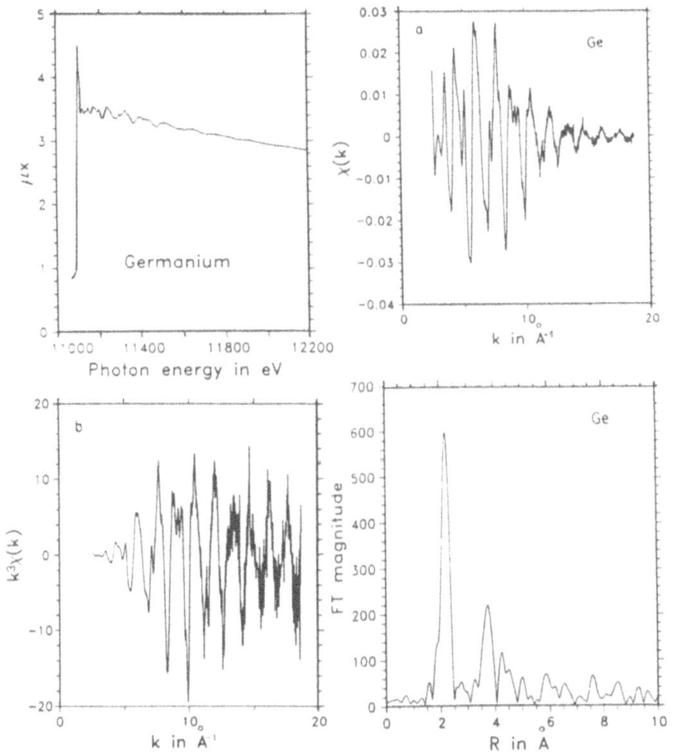

FIGURE 42. X-ray absorption in the vicinity of the k edge of crystalline Ge at a temperature of 100 K. μx: measured X-ray attenuation spectrum; the sharp rise near 11 keV is the edge and the modulation above the edge the EXAFS; $\chi(k)$, where k denotes the wave vector of the photoelectron: derived EXAFS spectrum; $k^2\chi(k)$: EXAFS spectrum multiplied by the third power of k; FT: Magnitude of the Fourier Transform of the $k^3\chi(k)$ curve (from Ref. 179).

distances; at temperatures above the Debye temperature the vibrations of the atoms strongly diminish the oscillatory structure.[180] EXAFS probes the short-range order, i.e., the distance between the central X-ray absorbing atom and its nearest neighbors, since the amplitude of the outgoing electron wave function decreases rapidly with increasing distance and since the free mean path of an electron in a solid is only a few Å. The accuracy of determining distances between neighboring atoms by EXAFS may be as high as 0.01 Å [178] and thus EXAFS measurements are a powerful tool for solid-state structure studies.

In order to apply the EXAFS technique to surface studies, the surface sensitivity has to be enhanced by orders of magnitude. This can be achieved by measuring the induced electron yield instead of the X-ray absorption since the electron escape depth amounts to a few monolayers only. The accuracy of surface EXAFS (SEXAFS) for determining bond lengths and the short-range order at surfaces is unbalanced by other methods.[181] It allows one to determine the site geometry of adsorbates. For example, SEXAFS studies in the vicinity of the L_3 edge of I and Te have shown, that I adatoms occupy the same sites on the surfaces of Si and Ge crystals, but Te adatoms behave differently from I and differently for each substrate (Figure 43).[182]

As we have seen for the photoelectron, an electron moving through a solid interacts with the matter through which it moves. As a result, elastic scattering and reflection may occur. Furthermore, inelastic processes may take place, such as, e.g., the excitation of one (and more) bulk plasmons,

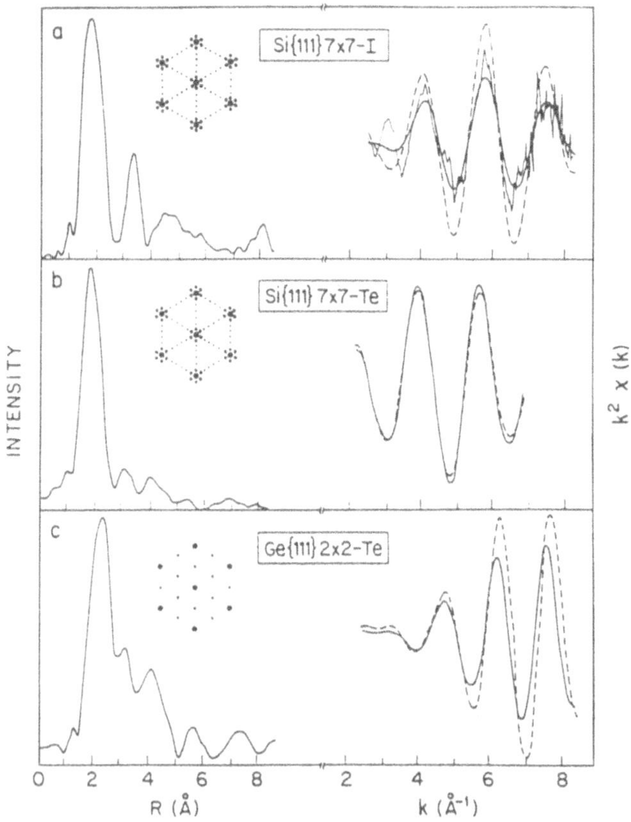

FIGURE 43. I and Te adsorbed on the {111} surface of Si and Ge crystals. Fourier-transformed SEXAFS data with corresponding observed LEED patterns are shown on the left. Filtered SEXAFS spectra of the first nearest neighbors are shown on the right; they depend on the polarization vector of the X radiation with respect to the surface normal (distinguished by the dashed and solid curves). Clearly the differences for the different adsorbate-surface systems (a), (b), and (c) can be seen (from Ref. 182).

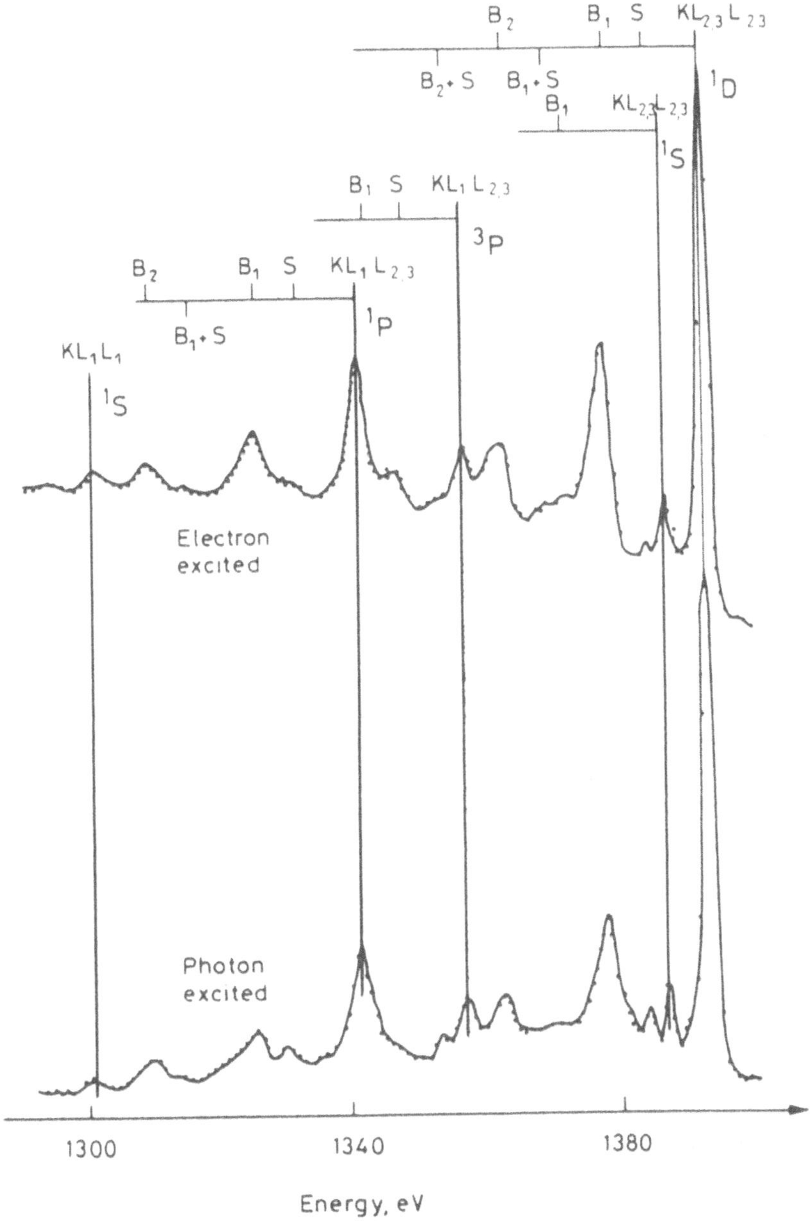

FIGURE 44. Spectrum of the *KLL* Auger electrons of solid Al excited by electrons (top) and photons (bottom). The positions of the parent Auger peaks are designated by the vertical lines. The plasmon loss fine structure is also indicated. The surface plasmons are designated by *S*, the successive bulk plasmons by B_1 and B_2, and coupling between bulk and surface plasmons by $B_1 + S$ and $B_2 + S$ (from Ref. 183).

surface plasmons, and a coupling between them. These processes can be clearly identified in Auger electron spectra via their characteristic energy loss (Figure 44).[183] They also take place for the photoelectron emitted in photoionization[172] and the δ electron emitted in ion-impact ionization. By these processes the final state is affected and, as a consequence, the ionization cross section is modified. This modification is a small but sizeable effect; its influence on the ionization process may be hard to investigate experimentally since it exhibits no fingerprint signature that allows an easy identification.

6. CONCLUSIONS

Inner-shell ionization by electron impact, by light ion impact, and by photons has been investigated both experimentally and theoretically in numerous studies. Measured absolute cross section can be reproduced by perturbation approaches if proper assumptions are made. The uncertainties amount to 10% to 100% for electron and ion impact and 1% to 10% for photoionization. Remaining deviations are often hard to trace: experimentally, the measured data bear statistical uncertainties and may suffer from systematic errors; theoretically, various assumptions may lead to similar results owing to a cancellation of terms. More conclusive information can be obtained from more detailed studies, such as, e.g., of alignment effects, of ratios of subshell cross sections, and of the impact parameter dependence. The high accuracy achieved for photoionization makes it possible to investigate the influence of subtle small effects on the inner-shell ionization, such as electron correlations, multi-electron transitions, and the influence of the atomic environment. Ongoing research will result in a reduction of data errors and improved calculations; the increased accuracy may lead to the discovery of new effects. Nevertheless, present-day knowledge is sufficient for many applications: inner-shell ionization studies provide a powerful tool for quantitative analysis of elemental, chemical, and structural composition of bulk and surface materials.

ACKNOWLEDGMENTS. The author is indebted to Professor H. O. Lutz, Professor H. Kleinpoppen, and Professor B. Crasemann, who attracted his interest to inner-shell processes. Special thanks go to A. Kaschuba, T. Harbach, B. Wisotzki, K. Finck, and U. Werner, who were engaged in the experiments. The manuscript was prepared with the skillful assistance of M. Wölfel.

Note added in proof. A comprehensive review on *Ion-Induced Characteristic X-Ray Emission in Solids* has recently been published by N. M. Kabachnik, H. Kerkow, V. P. Petukhov, E. A. Romanovskii, and R. Wedell [*Phys. Status Solidi B* **139**, 17 (1987)].

REFERENCES

1. B. CRASEMANN, Ed., *Atomic Inner-Shell Processes*, Vols. I and II (Academic, New York, 1975).
2. I. A. SELLIN, Ed., *Structure and Collisions of Ions and Atoms* (Springer, Berlin, 1978).
3. P. RICHARD, Ed., *Atomic Physics, Accelerators* (Academic, New York, 1980).
4. B. CRASEMANN, Ed., *Atomic Inner-Shell Physics* (Plenum Press, New York, 1985).
5. D. C. LORENTS, W. E. MEYERHOF, AND J. R. PETERSON, Eds., *Electronic and Atomic Collisions* (North-Holland, Amsterdam, 1986).
6. A. MEISEL AND J. FINSTER, Eds., *X-Ray and Inner-Shell Processes in Atoms, Molecules and Solids* (Karl-Marx-Universität, Leipzig, 1984).
7. H. O. LUTZ, *S. Afr. J. Phys.* **6**, 83 (1983).
8. R. E. JOHNSON, *Introduction to Atomic and Molecular Collisions* (Plenum, New York, 1982).
9. B. FASTRUP, in *Methods of Experimental Physics*, Vol. 17, edited by P. Richard (Academic, New York, 1980), p. 149.
10. R. H. PRATT, H. K. TSENG, C. M. LEE, L. KISSEL, C. MACCALLUM, AND M. RILEY, *Atom. Data Nucl. Data Tables* **20**, 175 (1977).
11. R. HIPPLER, I. MCGREGOR, M. AYDINOL, AND H. KLEINPOPPEN, *Phys. Rev.* **23**, 1370 (1981).
12. M. PFUETZNER, J. SZERYPO, Z. ZELAZNY, F. M. EL-ASHRY, M. GOCŁOWSKI, L. GŁOWACKA, D. TRAUTMANN, AND M. JASKÓŁA, *J. Phys. B* **20**, 3453 (1987); J. SZERYPO, M. PFUETZNER, W. KRETSCHMER, R. SCHMITT, W. SCHUSTER, A. BIÉNKOWSKY, L. GŁOWACKA, D. TRAUTMANN, AND M. JASKÓŁA, *J. Phys. B* **20**, 5475 (1987).
13. B. MARTIN AND H. STELZER, in *Experimental Methods in Heavy Ion Physics*, edited by K. Bethge (Springer, Berlin, 1978), p. 150.
14. D. L. MATTHEWS, in *Methods of Experimental Physics*, Vol. 17, edited by P. Richard (Academic, New York, 1980), p. 433.
15. N. STOLTERFOHT, in *Fundamental Processes in Energetic Atomic Collisions*, edited by H. O. Lutz, J. S. Briggs, and H. Kleinpoppen (Plenum Press, New York, 1983), p. 295.
16. J. C. RIVIÈRE, in *Practical Surface Analysis by Auger and X-Ray Photoelectron Spectroscopy*, edited by D. Briggs and M. Seah (John Wiley, Chichester, 1983), p. 17.
17. R. H. DAY, P. LEE, E. B. SALOMON, AND D. J. NAGEL, X-Ray Diodes for Laser Fusion Plasma Diagnostics, Los Alamos Scientific Laboratory Report LA-UR-79-1360 (1979).
18. G. BERTOLINI AND G. RESTELLI, in *Atomic Inner-Shell Processes II*, edited by B. Crasemann (Academic, New York, 1975), p. 123.
19. D. TWERENBOLD, Dissertation ETH Zürich No. 8038 (1986).
20. H. SCHMIDT-BÖCKING, I. TSERRUYA, H. ZEKL, AND K. BETHGE, *Nucl. Instrum. Methods* **120**, 329 (1974).
21. K. FINCK AND W. JITSCHIN, *Nucl. Instrum. Methods A* **238**, 479 (1985).
22. Y. CAUCHOIS AND C. BONNELLE, in *Atomic Inner-Shell Processes II*, edited by B. Crasemann (Academic, New York, 1975), p. 84.
23. B. CLEFF, *Acta Phys. Pol. A* **61**, 285 (1982).
24. W. JITSCHIN, B. WISOTZKI, U. WERNER, AND H. O. LUTZ, *J. Phys. E* **17**, 137 (1984).
25. K. KOHRA AND M. ANDO, *Nucl. Instrum. Methods* **177**, 117 (1980).
26. F. RIEHLE AND B. WENDE, *Opt. Lett.* **10**, 365 (1985).
27. J. L. CAMPBELL AND H. H. JORCH, *Nucl. Instrum. Methods* **159**, 163 (1979).
28. W. JITSCHIN, A. KASCHUBA, H. KLEINPOPPEN, AND H. O. LUTZ, *Z. Phys. A* **304**, 69 (1982).
29. J. WIGGER, S. OST, M. BRÜSSERMANN, G. RICHTER, AND B. CLEFF, *Phys. Lett.* **84A**, 110 (1981).

30. J. A. BEARDEN, *Rev. Mod. Phys.* **39**, 78 (1967).
31. D. BURCH, L. WILETS, AND W. E. MEYERHOF, *Phys. Rev. A* **9**, 1007 (1974).
32. F. PARENTE, M. H. CHEN, B. CRASEMANN, AND H. MARK, *At. Data Nucl. Data Tables* **26**, 383 (1981).
33. W. UCHAI, C. W. NESTOR, JR., S. RAMAN, AND C. R. VANE, *At. Data Nucl. Data Tables* **34**, 201 (1986).
34. W. BAMBYNEK, B. CRASEMANN, R. W. FINK, H.-U. FREUND, H. MARK, C. D. SWIFT, R. E. PRICE, AND P. V. RAO, *Rev. Mod. Phys.* **44**, 716 (1972).
35. M. O. KRAUSE, *J. Phys. Chem. Ref. Data* **8**, 307 (1979).
36. J. L. CAMPBELL, P. L. MCGHEE, R. R. GINGERICH, R. W. OLLERHEAD, AND J. A. MAXWELL, *Phys. Rev. A* **30**, 161 (1984).
37. M. H. CHEN, B. CRASEMANN, AND H. MARK, *Phys. Rev. A* **24**, 177 (1981).
38. W. JITSCHIN, G. MATERLIK, U. WERNER, AND P. FUNKE, *J. Phys. B* **18**, 1139 (1985).
39. W. JITSCHIN, U. WERNER, K. FINCK, AND H. O. LUTZ, in *High-Energy Ion-Atom Collisions II*, edited by D. Berényi and G. Hock (Elsevier Scientific, Amsterdam, 1985), p. 79.
40. W. JITSCHIN AND U. WERNER, *J. Vac. Sci. Technol. A* (1987).
41. U. WERNER AND W. JITSCHIN, *J. Phys.* (*Paris*) **5**, 1203 (1988).
42. J. H. SCOFIELD, *Atom. Data Nucl. Data Tables* **14**, 121 (1974).
43. S. I. SALEM, S. L. PANOSSIAN, AND R. A. KRAUSE, *Atom. Data Nucl. Data Tables* **14**, 91 (1974).
44. D. D. COHEN, *J. Phys. B* **17**, 3913 (1984).
45. W. JITSCHIN, A. KASCHUBA, R. HIPPLER, AND H. O. LUTZ, *J. Phys. B* **15**, 763 (1982).
46. W. JITSCHIN, R. HIPPLER, K. FINCK, R. SCHUCH, AND H. O. LUTZ, *J. Phys. B* **16**, 4405 (1983).
47. M. INOKUTI, Y. ITIKAWA, AND J. E. TURNER, *Rev. Mod. Phys.* **43**, 297 (1971).
48. B.-H. CHOI, E. MERZBACHER, AND G. S. KHANDELWAL, *At. Data* **5**, 291 (1973).
49. O. BENKA AND A. KROPF, *At. Data Nucl. Data Tables* **22**, 219 (1978).
50. D. E. JOHNSON, G. BASBAS, AND F. D. MCDANIEL, *At. Data Nucl. Data Tables* **24**, 1 (1979).
51. T. MUKOYAMA AND L. SARKADI, *Bull. Inst. Chem. Res.* (*Kyoto*) **58**, 95 (1980).
52. T. MUKOYAMA AND L. SARKADI, *Bull. Inst. Chem. Res.* (*Kyoto*) **58**, 60 (1980).
53. M. H. CHEN, B. CRASEMANN, AND H. MARK, *Phys. Rev. A* **26**, 1243 (1982).
54. M. H. CHEN, B. CRASEMANN, AND H. MARK, *Phys. Rev. A* **27**, 2358 (1983).
55. W. HINK, L. KEES, H.-P. SCHMITT, AND A. WOLF, in *Inner-Shell and X-Ray Physics of Atoms and Solids*, edited by D. Fabian, H. Kleinpoppen, and L. M. Watson (Plenum Press, New York, 1981), p. 327.
56. R. HIPPLER, H. KLAR, K. SAEED, I. MCGREGOR, A. J. DUNCAN, AND H. KLEINPOPPEN, *J. Phys. B* **16**, L617 (1983).
57. R. HIPPLER AND W. JITSCHIN, *Z. Phys. A* **307**, 287 (1982).
58. P. ESCHWEY AND P. MANAKOS, *Z. Phys. A* **308**, 199 (1982).
59. H. GENZ, in *X-Ray and Atomic Inner-Shell Physics—1982*, edited by B. Crasemann (American Institute of Physics, New York, 1982), p. 85; H. GENZ, *Comments At. Mol. Phys.* **14**, 173 (1984).
60. J. BANG AND J. DEBOER, Eds., *Semiclassical Descriptions of Atomic and Nuclear Collisions* (North-Holland, Amsterdam, 1985).
61. L. KOCBACH, J. M. HANSTEEN, AND R. GUNDERSEN, *Nucl. Instrum. Methods* **169**, 281 (1980).
62. L. SARKADI AND T. MUKOYAMA, *J. Phys. B* **14**, L255 (1981).
63. J. M. HANSTEEN, *Adv. At. Mol. Phys.* **11**, 299 (1975).
64. W. JITSCHIN AND H. O. LUTZ, in *European Conference on Atomic Physics, Book of Abstracts II*, edited by J. Kowalski, G. zu Putlitz, and H. G. Weber (Heidelberg, 1981), p. 782.

65. G. BAUR, F. RÖSEL, AND D. TRAUTMANN, *J. Phys. B* **16**, L419 (1983).
66. A. JAKOB, F. RÖSEL, D. TRAUTMANN, AND G. BAUR, *Z. Phys. A* **309**, 13 (1982).
67. J. M. HANSTEEN, O. M. JOHNSEN, AND L. KOCBACH, *At. Data Nucl. Data Tables* **15**, 305 (1975).
68. M. PAULI AND D. TRAUTMANN, *J. Phys. B* **11**, 667 (1978).
69. M. PAULI, F. RÖSEL, AND D. TRAUTMANN, *Phys. Lett.* **67A**, 28 (1978).
70. D. TRAUTMANN AND F. RÖSEL, *Nucl. Instrum. Methods* **169**, 259 (1980).
71. A. JAKOB, D. TRAUTMANN, F. RÖSEL, AND G. BAUR, *Nucl. Instrum. Methods B* **4**, 218 (1984).
72. W. MEHLHORN, *Phys. Lett.* **26A**, 166 (1968).
73. E. G. BEREZHKO AND N. M. KABACHNIK, *J. Phys. B* **10**, 2467 (1977).
74. K. BLUM, *Density Matrix Theory and Applications* (Plenum Press, New York, 1981).
75. S. C. MCFARLANE, *J. Phys. B* **7**, 1756 (1974).
76. A. SCHÖLER AND F. BELL, *Z. Phys. A* **286**, 163 (1978).
77. E. G. BEREZHKO, N. M. KABACHNIK, AND V. V. SIZOV, *J. Phys. B* **11**, L421 (1978).
78. M. KAMIYA, Y. KINEFUCHI, H. ENDO, A. KUWAKO, K. ISHII, AND S. MORITA, *Phys. Rev. A* **20**, 1820 (1979).
79. J. PÁLINKÁS, L. SARKADI, AND B. SCHLENK, *J. Phys. B* **13**, 3829 (1980).
80. V. V. SIZOV AND N. M. KABACHNIK, *J. Phys. B* **13**, 1601 (1980).
81. F. RÖSEL, D. TRAUTMANN, AND G. BAUR, *Z. Phys. A* **304**, 75 (1982).
82. L. KOCBACH, J. M. HANSTEEN, AND R. GUNDERSEN, *Phys. Scr.* **27**, 54 (1983).
83. L. SARKADI, *J. Phys. B* **19**, 2519 (1986).
84. L. SARKADI, *J. Phys. B* **19**, L755 (1986).
85. G. MEHLER, J. REINHARDT, B. MÜLLER, W. GREINER, AND G. SOFF, *Z. Phys. D* **5**, 143 (1987).
86. H. PAUL, *Nucl. Instrum. Methods B* **4**, 211 (1984).
87. H. PAUL AND J. MUHR, *Phys. Rep.* **135**, 47 (1986).
88. R. S. SOKHI AND D. CRUMPTON, *At. Data Nucl. Data Tables* **30**, 49 (1984).
89. T. J. GRAY, in *Methods of Experimental Physics*, Vol. 17, edited by P. Richard (Academic, New York, 1980), p. 193.
90. W. BRANDT AND G. LAPICKI, *Phys. Rev. A* **10**, 474 (1974).
91. W. BRANDT AND G. LAPICKI, *Phys. Rev. A* **20**, 465 (1979).
92. W. BRANDT AND G. LAPICKI, *Phys. Rev. A* **23**, 1717 (1981).
93. G. LAPICKI AND W. LOSONSKY, *Phys. Rev. A* **20**, 481 (1979).
94. B.-H. CHOI, *Phys. Rev. A* **4**, 1002 (1971).
95. T. MUKOYAMA AND L. SARKADI, *Bull. Inst. Chem. Res. (Kyoto)* **57**, 33 (1979).
96. T. MUKOYAMA AND L. SARKADI, *Phys. Rev. A* **25**, 1411 (1982).
97. G. LAPICKI AND W. LOSONSKY, *Phys. Rev. A* **15**, 896 (1977).
98. F. D. MCDANIEL, *Nucl. Instrum. Methods* **214**, 57 (1983).
99. L. SARKADI, *Nucl. Instrum. Methods* **214**, 43 (1983).
100. W. JITSCHIN, *Nucl. Instrum. Methods B* **4**, 292 (1984).
101. K. FINCK, W. JITSCHIN, AND H. O. LUTZ, *J. Phys. B* **16**, L409 (1983).
102. L. KOCBACH, *Nucl. Instrum. Methods B* **4**, 248 (1984).
103. P. A. AMUNDSEN AND D. N. JAKUBASSA-AMUNDSEN, unpublished.
104. T. PAPP, *Proceedings of the 3rd Workshop on High-Energy Ion–Atom Collision Processes*, Debrecen (1987), unpublished.
105. W. JITSCHIN, H. KLEINPOPPEN, R. HIPPLER, AND H. O. LUTZ, *J. Phys. B* **12**, 4077 (1979).
106. J. PÁLINKÁS, L. SARKADI, AND B. SCHLENK, *J. Phys. B* **13**, 3829 (1980); J. PÁLINKÁS, B. SCHLENK, AND A. VALEK, *J. Phys. B* **14**, 1157 (1981).
107. G. RICHTER, M. BRÜSSERMANN, S. OST, J. WIGGER, B. CLEFF, AND R. SANTO, *Phys. Lett.* **82A**, 412 (1981).

108. C. V. BARROS LEITE, N. V. DE CASTRO-FARIA, R. J. HOROWICZ, E. C. MONTENEGRO, AND A. G. DE PINHO, *Phys. Rev. A* **25**, 1880 (1982).
109. W. JITSCHIN, H. O. LUTZ, AND H. KLEINPOPPEN, in *Inner-Shell and X-Ray Physics of Atoms and Solids*, edited by D. Fabian, H. Kleinpoppen, and L. M. Watson (Plenum Press, New York, 1981), p. 89.
110. A. M. BORISOV, H. KERKOW, V. P. PETUKHOV, AND E. A. ROMANOVSKY, *J. Phys. B* **16**, L495 (1983).
111. U. WERNER, W. JITSCHIN, AND H. O. LUTZ, *J. Phys. B* **18**, 3111 (1985).
112. W. JITSCHIN, in *Fundamental Processes in Energetic Atomic Collisions*, edited by H. O. Lutz, J. S. Briggs, and H. Kleinpoppen (Plenum Press, New York, 1983), p. 575.
113. W. JITSCHIN, *J. Phys. B* **17**, 4179 (1984).
114. O. KESKI-RAHKONEN AND M. O. KRAUSE, *At. Data Nucl. Data Tables* **14**, 139 (1974).
115. H. GUENNOU, These pour le grade de docteur d'etat es sciences physiques, Université de Paris-Sud, Orsay No. 2705 (1983).
116. J. PÁLINKÁS, L. SARKADI, B. SCHLENK, I. TÖRÖK, AND GY. KÁLMÁN, *J. Phys. B* **15**, L451 (1982).
117. J. PÁLINKÁS, L. SARKADI, B. SCHLENK, I. TÖRÖK, GY. KÁLMÁN, C. BAUER, K. BRANKOFF, D. GRAMBOLE, C. HEISER, W. RUDOLPH, AND H. J. THOMAS, X-82 Conference, Eugene, Oregon (1982).
118. W. JITSCHIN, R. HIPPLER, R. SHANKER, H. KLEINPOPPEN, R. SCHUCH, AND H. O. LUTZ, *J. Phys. B* **16**, 1417 (1983).
119. Z. STACHURA, F. BOSCH, F. J. HAMBSCH, BING LIU, D. MAOR, P. H. MOKLER, W. A. SCHÖNFELDT, H. WAHL, B. CLEFF, M. BRÜSSERMANN, AND J. WIGGER, *J. Phys. B* **17**, 835 (1984).
120. A. BERINDE, C. CIORTEA, A. ENULESCU, D. FLUERASU, G. HOCK, I. PITICU, L. SARKADI, B. SULIK, AND V. ZORAN, *J. Phys. B* **20**, L481 (1987).
121. A. G. DRENTJE, H. A. ENGE, AND S. B. KOWALSKI, *Nucl. Instrum. Methods* **122**, 485 (1974).
122. J. H. VAN DIJK, H. W. WILSCHUT, A. G. DRENTJE, AND A. VAN DER WOUDE, *Z. Phys. A* **314**, 1 (1983).
123. J. H. VAN DIJK, Proefschrift, Rijksuniversiteit te Groningen (1984).
124. D. TRAUTMANN AND G. BAUR (1981), cited in Ref. 123.
125. M. E. RUDD, L. H. TOBUREN, AND N. STOLTERFOHT, *At. Data Tables* **23**, 406 (1976).
126. T. WEITER AND R. SCHUCH, *Z. Phys. A* **305**, 91 (1982).
127. L. SARKADI, J. BOSSLER, R. HIPPLER, AND H. O. LUTZ, *J. Phys. B* **16**, 71 (1983).
128. L. SARKADI, J. BOSSLER, R. HIPPLER, AND H. O. LUTZ, in *Forward Electron Ejection in Ion Collisions*, edited by K. O. Groeneveld, W. Meckbach, and I. A. Sellin (Springer, Berlin, 1984), p. 91.
129. K. DEXHEIMER, J. ULLRICH, K. E. STIEBING, W. SCHADT, S. KELBCH, C. KELBCH, R. SCHUCH, S. ZEHENDNER, AND H. SCHMIDT-BÖCKING, *J. Phys. B* **19**, 3083 (1986).
130. J. KONRAD, R. SCHUCH, R. HOFFMANN, AND H. SCHMIDT-BÖCKING, *Phys. Rev. Lett.* **52**, 188 (1984).
131. S. ZEHENDNER, G. B. BAPTISTA, E. JUSTINIANO, J. KONRAD, H. SCHMIDT-BÖCKING, AND R. SCHUCH, in *6. Arbeitsbericht der Arbeitsgruppe Energiereiche Atomare Stösse*, p. 38 (1985).
132. N. LÖCHTER, K. BETHGE, G. GRUBER, A. GRUPPE, W. SCHADT, H. SCHMIDT-BÖCKING, R. SCHUCH, AND K. E. STIEBING, in *2. Arbeitsbericht der Arbeitsgruppe Energiereiche Atomare Stösse*, p. 15 (1981).
133. G. GRUBER, K. BETHGE, S. KELBCH, N. LÖCHTER, W. SCHADT, K. E. STIEBING, H. SCHMIDT-BÖCKING, R. SCHUCH, F. RÖSEL, AND D. TRAUTMANN, in Wissenschaftlicher Ergebnisbericht 1981, Hahn-Meitner-Institut für Kernforschung, Bereich Kern- und Strahlenphysik (1981), p. 121.

134. H. SCHMIDT-BÖCKING, K. E. STIEBING, W. SCHADT, N. LÖCHTER, G. GRUBER, S. KELBCH, K. BETHGE, R. SCHUCH, AND I. TSERRUYA, *Nucl. Instrum. Methods* **192**, 71 (1982).
135. M. KLEBER AND K. UNTERSEER, *Z. Phys. A* **292**, 311 (1979).
136. F. RÖSEL, D. TRAUTMANN, AND G. BAUR, *Nucl. Instrum. Methods* **192**, 43 (1982).
137. G. CIOCCHETTI AND A. MOLINARI, *Nuovo Cimento* **40B**, 69 (1965).
138. P. A. AMUNDSEN AND K. AASHAMAR, *J. Phys. B* **19**, 1657 (1986).
139. J. S. BLAIR, P. DYER, K. A. SNOVER, AND T. A. TRAINOR, *Phys. Rev. Lett.* **41**, 1712 (1978).
140. J. F. CHEMIN, W. E. MEYERHOF, R. ANHOLT, J. D. MOLITORIS, AND CH. STOLLER, *Phys. Rev. A* **26**, 1239 (1982).
141. M. DOST, R. LOREK, S. RÖHL, J. SEIDEL, AND W. KÖNIG, *Phys. Rev. A* **32**, 2077 (1985).
142. R. ANHOLT, in *Atomic Inner-Shell Physics*, edited by B. Crasemann (Plenum Press, New York, 1985), p. 581.
143. D. W. SPOONER, CH. STOLLER, J.-F. CHEMIN, W. E. MEYERHOF, J.-N. SCHEURER, AND X.-Y. XU, *Phys. Rev. Lett.* **58**, 341 (1987).
144. W. E. MEYERHOF, in *Electronic and Atomic Collisions*, edited by J. Eichler, I. V. Hertel, and N. Stolterfoht (North-Holland, Amsterdam, 1984), p. 31.
145. W. JITSCHIN, in *Progress in Atomic Spectroscopy, Part D*, edited by H. J. Beyer and H. Kleinpoppen (Plenum, New York, 1987) p. 295.
146. J. H. SCOFIELD, Lawrence Livermore Radiation Laboratory Report UCRL-51326 (1973).
147. J. W. COOPER, in *Atomic Inner-Shell Processes I*, edited by B. Crasemann (Academic, New York, 1975), p. 159.
148. H. A. BETHE AND E. E. SALPETER, *Quantum Mechanics of One- and Two-Electron Atoms* (Plenum Press, New York, 1977).
149. J. H. HUBBELL, W. H. MCMASTER, N. KERR DEL GRANDE, AND J. H. MALLETT, in *International Tables for X-Ray Crystallography IV*, edited by J. A. Ibers and W. C. Hamilton (Kynoch Press, Birmingham, England, 1974), p. 47.
150. J. H. HUBBELL AND WM. J. VEIGELE, National Bureau of Standards, Technical Note 901, Washington, D.C. (1976).
151. M. YA. AMUSIA, in *X-84 Conference Proceedings*, edited by A. Meisel and J. Finster (Karl-Marx-Universität, Leipzig, 1984), p. 33.
152. M. YA. AMUSIA, *Comments At. Mol. Phys.* **8**, 61 (1979).
153. G. WENDIN, in *Photoionization and Other Probes of Many Electron Interactions*, edited by F. J. Wuilleumier (Plenum Press, New York, 1976), p. 61.
154. H. P. KELLY, in *Atomic Physics 8*, edited by I. Lindgren, A. Rosén, and S. Svanberg (Plenum Press, New York, 1983), p. 305.
155. P. G. BURKE, in *Electronic and Atomic Collisions*, edited by G. Watel (North-Holland, Amsterdam, 1978), p. 201.
156. W. D. ROBB, in *Electronic and Atomic Collisions*, edited by G. Watel (North-Holland, Amsterdam, 1978), p. 231.
157. A. ZANGWILL, in *Atomic Physics 8*, edited by I. Lindgren, A. Rosén, and S. Svanberg (Plenum Press, New York, 1983), p. 339.
158. Z. LEVINE AND P. SOVEN, *Phys. Rev. Lett.* **50**, 2074 (1983).
159. W. JITSCHIN, U. WERNER, G. MATERLIK, AND G. D. DOOLEN, *Phys. Rev. A* **35**, 5038 (1987).
160. D. A. LIBERMAN AND A. ZANGWILL, *Comp. Phys. Comm.* **32**, 75 (1984).
161. M. BREINIG, M. H. CHEN, G. E. ICE, F. PARENTE, B. CRASEMANN, AND G. S. BROWN, *Phys. Rev. A* **22**, 520 (1980).
162. O. KESKI-RAHKONEN, G. MATERLIK, B. SONNTAG, AND J. TULKKI, *J. Phys. B* **17**, L121 (1984).

163. R. FRAHM, R. HAENSEL, W. MALZFELDT, W. NIEMANN, AND P. RABE, in HASYLAB Jahresbericht 1982 (1983), p. 121.
164. R. D. DESLATTES, R. E. LAVILLA, P. L. COWAN, AND A. HENINS, *Phys. Rev. A* **27**, 923 (1983).
165. KE ZHANG, E. A. STERN, AND J. J. REHR, in *EXAFS and Near Edge Structure III*, edited by K. O. Hodgson, B. Hedman, and J. E. Penner-Hahn (Springer, Berlin, 1984), p. 18.
166. T. ÅBERG, *Phys. Rev.* **156**, 35 (1967).
167. P. H. KOBRIN, S. SOUTHWORTH, C. M. TRUESDALE, D. W. LINDLE, U. BECKER, AND D. A. SHIRLEY, *Phys. Rev. A* **29**, 194 (1984).
168. P. A. HEIMANN, T. A. FERRETT, D. A. SHIRLEY, U. BECKER, AND H. G. KERKHOFF, Stanford Synchrotron Radiation Laboratory, Activity Report for 1985, edited by K. Cantwell (1985), p. IX-9.
169. R. HAENSEL, G. KEITEL, N. KOSUCH, U. NIELSEN, AND P. SCHREIBER, *J. Phys. (Paris) Colloq.* **32**, C4, 236 (1971).
170. F. C. BROWN, in *Synchrotron Radiation Research*, edited by H. Winick and S. Doniach (Plenum Press, New York, 1980), p. 61.
171. K. SIEGBAHN, in *X-84 International Conference on X-Ray and Inner-Shell Processes in Atoms, Molecules and Solids*, edited by A. Meisel and J. Finster (Karl-Marx-Universität, Leipzig, 1984), p. 1.
172. C. D. WAGNER, W. M. RIGGS, L. E. DAVIS, G. E. MUILENBERG, AND J. F. MOULDER, *Handbook of X-Ray Photoelectron Spectroscopy* (Perkin-Elmer, Eden Prairie, Minnesota, 1979).
173. M. P. SEAH, in *Practical Surface Analysis by Auger and X-Ray Photoelectron Spectroscopy*, edited by D. Briggs and M. P. Seah (John Wiley, Chichester, 1983), p. 181.
174. M. P. SEAH AND W. A. DENCH, *Surf. Interface Anal.* **1**, 2 (1979).
175. G. MATERLIK, B. SONNTAG, AND M. TAUSCH, *Phys. Rev. Lett.* **51**, 1300 (1983).
176. K. SYASSEN, G. WORTMANN, J. FELDHAUS, K. H. FRANK, AND G. KAINDL, *Phys. Rev. B* **26**, 4745 (1982).
177. D. R. SANDSTROM AND F. W. LYTLE, *Ann. Rev. Phys. Chem.* **30**, 215 (1979).
178. F. W. LYTLE, G. H. VIA, AND J. H. SINFELT, in *Synchrotron Radiation Research*, edited by H. Winick and S. Doniach (Plenum Press, New York, 1980), p. 401.
179. P. A. LEE, P. H. CITRIN, P. EISENBERGER, AND B. M. KINCAID, *Rev. Mod. Phys.* **53**, 769 (1981).
180. W. BÖHMER AND P. RABE, *J. Phys. C* **12**, 2465 (1979).
181. D. NORMAN, in *X-Ray and Atomic Inner-Shell Physics—1982*, edited by B. Crasemann (American Institute of Physics, New York, 1982), p. 745.
182. P. H. CITRIN, P. EISENBERGER, AND J. E. ROWE, *Phys. Rev. Lett.* **48**, 802 (1982).
183. G. DUFOUR, J.-M. MARIOT, P.-E. NILSSON-JATKO, AND R. C. KARNATAK, *Phys. Scr.* **13**, 370 (1976).

CHAPTER 7

Excitation in Ion–Atom Collisions

Uwe Wille

1. INTRODUCTION

The subject of electronic excitation in ion–atom collisions has a long-standing history and still continues to be one of the most active fields of research in atomic physics. Any attempt to give in the present chapter a complete survey of this subject would be a hopeless task. Accordingly, the selection of the material discussed here has been guided solely by the author's personal interests and by the desire to address, to a certain extent, topics that bear a direct relation to the title of this volume, that is, coherence in atomic collision physics. Strictly speaking, the wave-mechanical nature of atomic physics entails any collision process to be at least partially coherent. In most situations, however, this coherence is not directly reflected in the observed features of the collision. Our aim therefore is to emphasize those cases in which coherence properties of the collision are explicitly revealed, e.g., through interference effects due to the (coherent) superposition of contributions to the excitation amplitude from different excitation mechanisms. Following the author's inclination, a major part of the specific examples considered deals with inner-shell processes in heavy-ion–atom collisions.

We proceed by giving, in the next section, a brief overview of the mechanisms of electronic excitation in ion–atom collisions. Subsequently, in the third section, we consider the impact parameter dependence of inner-shell vacancy production. Studies of this dependence have become in the past few years the principal tool for unraveling the excitation mechanisms of inner-shell processes. In particular cases, the impact parameter dependence may reflect coherence properties of the collision process.

Uwe Wille • Bereich Kern- und Strahlenphysik, Hahn–Meitner-Institut, D-1000 Berlin 39, West Germany.

Examples pertaining to this situation will be discussed in Section 4. In the final section of this article, we leave behind the field of inner-shell processes and survey some recent work on the effect of external (electric or magnetic) fields in ion-atom collisions. Coherence properties of collisionally excited states will be discussed by considering the evolution of Rydberg states in fast, highly charged ions exposed to an external electric field. Ion-atom collisions proceeding in the presence of a very strong magnetic field will also be touched upon.

2. EXCITATION MECHANISMS IN ION-ATOM COLLISIONS

The collision of two heavy ions or atoms in general constitutes a many-body scattering problem of formidable complexity. The dynamics of the collision is likely to be governed in most cases by a large number of strongly coupled, open channels for excitation into bound states, for ionization and for electron transfer. Simple excitation mechanisms can prevail only under favorable conditions. Such conditions can be realized by appropriately selecting the initial and final channels observed and the values of the external parameters that control a collision process. Table I summarizes these parameters and simultaneously establishes the notation used in this article.

Conventionally,[1,2] one achieves a rough classification of the excitation mechanisms by considering the relevant parameters to be Z_1/Z_2, the "asymmetry" of the collision system, and v/v_{el}, the "adiabaticity" parameter. In this way, one is led to distinguish two limiting cases: (1) $Z_1/Z_2 \ll 1$, $v/v_{el} \gg 1$, i.e., fast collisions induced by "light" projectiles: (2) $Z_1/Z_2 \simeq 1$, $v/v_{el} \ll 1$, i.e., slow collisions between ions and atoms with nearly equal nuclear charges. Case (1) corresponds to the situation where, because of the small strength of the perturbing interaction and short interaction time, the influence of the projectile on the target electrons is weak. Excitation and

TABLE I

Parameters Controlling an Ion-Atom Collision Process

Parameter	Symbol
Nuclear charge number of projectile (target)	Z_1 (Z_2)
Mass of projectile (target)	M_1 (M_2)
Projectile ion charge state	q_1
Collision velocity	v
Impact parameter	b
Typical electron velocity in projectile or target	v_{el}
Time delay due to nuclear reactions	T
Strength of external electric (magnetic) field	F (B)

ionization of the target are in this case* appropriately described in terms of direct (first-order) transitions induced by the Coulomb field of the projectile. In case (2), the strong interaction between projectile and target results in a large mutual distortion and polarization of the collision partners, which is viewed to lead ultimately to the formation of a transient (diatomic) quasimolecular complex.[3] The coupling between electronic and nuclear motion in this complex gives rise to electronic transitions which constitute the basic excitation mechanisms in the quasimolecular picture of slow ion–atom collisions.

The observed features of an arbitrary collision process in general result from a subtle interplay of direct Coulomb excitation and quasimolecular excitation. The quantitative theoretical analysis of this interplay starts from an expansion of the total scattering wave function in terms of a set of suitably selected basis functions. Since we are mainly interested in the discussion of inner-shell processes, we may adopt here an independent-electron approximation[3,4] and consider, in particular, one "active" electron (or vacancy) only. Assuming the internuclear motion to proceed along classical trajectories, the motion of the active electron is described by a wave function $\psi(\mathbf{r}; t)$ obeying the time-dependent Schrödinger equation (we use atomic units unless stated otherwise)

$$\left[h(t) - i\frac{\partial}{\partial t}\bigg|_{\mathbf{r}}\right]\psi(\mathbf{r}; t) = 0 \tag{1}$$

The single-particle Hamiltonian $h(t)$ is time dependent via the dependence of the effective single-particle potential v^{eff} on the internuclear vector $\mathbf{R} = \mathbf{R}(t)$:

$$h(t) \equiv h(\mathbf{R}(t)) = -\tfrac{1}{2}\nabla_{\mathbf{r}}^2 + v^{\text{eff}}(\mathbf{r}; \mathbf{R}(t)) \tag{2}$$

The electronic coordinate vector $\mathbf{r}$ is referred to an *inertial frame of reference* (conveniently chosen to be the frame in which the center of mass of the total collision system is at rest). Use of a noninertal frame introduces additional coupling terms into the Hamiltonian,[5] which depend on the acceleration of this frame with respect to an arbitrary inertial frame. The time dependence of the internuclear vector $\mathbf{R}$ is usually calculated from a prescribed (central) interatomic potential $V = V(R)$.

We now expand the wave function $\psi(\mathbf{r}; t)$ as

$$\psi(\mathbf{r}; t) = \sum_k a_k(t)\phi_k(\mathbf{r}; \mathbf{R}(t)) \tag{3}$$

* It follows from general arguments[1] and will become apparent from the examples discussed in Section 3 that the regime of direct Coulomb excitation extends into the range $v/v_{\text{el}} \lesssim 1$ if Z_1/Z_2 is sufficiently small.

where the (not necessarily orthogonal) basis functions ϕ_k in general depend parametrically upon the internuclear vector $\mathbf{R}$. Insertion of (3) into the Schrödinger equation (1) leads in the usual way to the system of coupled equations

$$i\mathsf{N}\dot{\mathbf{a}} = \mathsf{M}\mathbf{a} \tag{4}$$

for the expansion coefficients $a_k(t)$ forming the column vector $\mathbf{a}$. The matrix N is the overlap matrix with elements $N_{kk'}(t) = \langle\phi_k(t)|\phi_{k'}(t)\rangle$, and M is the coupling matrix with elements

$$M_{kk'}(t) = \langle\phi_k(t)|h(t) - i\frac{\partial}{\partial t}\bigg|_{\mathbf{r}}|\phi_{k'}(t)\rangle \tag{5}$$

When choosing the basis set $\{\phi_k\}$ for the description of a specific collision process, one has to take into consideration the mathematical properties of the basis as well as the possibility of achieving an interpretation of the process in terms of simple excitation mechanisms.

In the regime of Coulomb excitation, target excitation and ionization are assumed to be adequately described by restricting the set $\{\phi_k\}$ to unperturbed (bound and continuum) orbitals of the target atom. When evaluating the coupling matrix (5) for this case, care must be taken to correctly fulfill the requirement that the electronic coordinates are to be referred to an inertial frame (chosen in the following to be the center-of-mass frame of the colliding nuclei; we disregard very small corrections arising from the motion of the nuclear center of mass relative to the center of mass of the total collision system). Since the target nucleus is moving with respect to the center of mass with nonuniform velocity $\dot{\mathbf{R}}_2^{(0)}(t)$ (for definition of coordinates, see Figure 1), the proper representation of the target-centered orbitals $\chi_k(\mathbf{r}_2)\, e^{-i\varepsilon_k t}$ in the center-of-mass frame reads[5]

$$\phi_k(\mathbf{r};\, t) = \chi_k(\mathbf{r} - \mathbf{R}_2^{(0)}) \times \exp\left[i\dot{\mathbf{R}}_2^{(0)} \cdot \mathbf{r} - i\left(\varepsilon_k t + \tfrac{1}{2}\int^t dt'\, \dot{\mathbf{R}}_2^{(0)2} \right) \right] \tag{6}$$

Neglecting, for simplicity, screening effects in the single-electron potential as well as in the interatomic potential, the coupling matrix elements $M_{kk'}$ corresponding to the basis functions (6) can be explicitly written as

$$M_{kk'}(t) = -Z_1\langle\chi_k(\mathbf{r})|\frac{1}{|\mathbf{r} - \mathbf{R}(t)|} - \frac{Z_2}{M_2}\frac{\mathbf{R}(t)}{R^3(t)} \cdot \mathbf{r}|\chi_{k'}(\mathbf{r})\rangle$$
$$\times \exp[-i(\varepsilon_{k'} - \varepsilon_k)t] \tag{7}$$

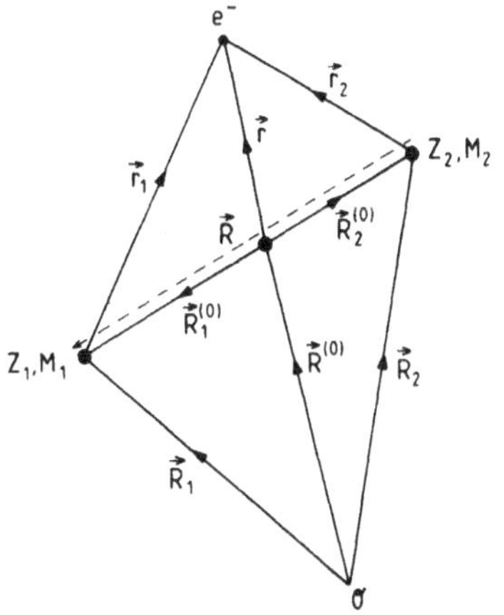

FIGURE 1. Coordinates used to describe the motion of one electron in the field of two colliding nuclei with charge number and mass Z_1, M_1 and Z_2, M_2, respectively. The position vector of the center of mass of the collision system with respect to an arbitrary origin $\mathcal{O}$ is denoted by $\mathbf{R}^{(0)}$.

where the acceleration $\ddot{\mathbf{R}}_2^{(0)}$ of the target frame with respect to the center-of-mass frame has been eliminated by means of the equation of motion for $\mathbf{R}(t)$. While the first term in (7) gives the usual matrix element for the Coulomb interaction between projectile nucleus and target electron in the target rest frame, the second term obviously reflects the perturbation of the target-centered electronic orbitals due to the recoil of the target nucleus.[6-9] The recoil term is of pure dipole character and will be small in most cases, but clearly may give significant contributions in fast collisions at very small impact parameters. In a certain sense, the matrix element (7) can be viewed as being composed of contributions from two different excitation mechanisms, which act coherently to form the total excitation probability.

In the atomic-orbital approach, the basis set $\{\phi_k\}$ to be used in expansion (3) must be augmented by projectile-centered orbitals if electron transfer channels are explicitly considered. The proper formulation of the coupled-state problem (4) for this case, including in particular contributions to the coupling matrix elements from projectile and target recoil, has been given by Fritsch[10] (see also Ref. 11).

Under conditions where the atomic-orbital aspect of a collision process still dominates, but where effects of the mutual distortion of the collision partners are no longer negligible, an expansion of the scattering wave function merely in terms of unperturbed target and projectile orbitals may be inappropriate because a large number of basis functions may be required to achieve convergence of the expansion. In this case, it is advantageous to include in the basis set $\{\phi_k\}$ functions which have, for small internuclear distances, large overlap with the "united-atom" orbitals corresponding to the field of a nucleus with charge number $Z_1 + Z_2$. By introducing such functions, one attempts to bridge the gap between the atomic-orbital picture and the molecular-orbital picture in which the united atom appears as the $R = 0$ limit of the transiently formed quasimolecule. Different approaches have been used to incorporate the united-atom aspect into the atomic-orbital expansion method. One approach[12,13] uses a triple-center expansion in which, in addition to the target and projectile orbitals, orbitals of the united atom centered about the center of mass of the collision system are introduced. In another approach, called the "pseudostate" method,[14,15] united-atom orbitals centered about either projectile or target nucleus are added to the unperturbed atomic orbitals. The pseudostate approach in fact turns out[11] to be sufficiently flexible to allow an efficient treatment even of many problems which, according to common expectation, should require a full molecular treatment.

The truly molecular description of ion–atom collisions departs from basis functions (molecular orbitals) $\tilde{\phi}_k(\mathbf{r}'; R(t))$ defined in terms of coordinates $\mathbf{r}'$ which are referred to a frame co-rotating with the internuclear axis of the collision system. In using such "body-fixed" coordinates, one emphasizes that in a slow collision the electronic wave function follows near-adiabatically the rotation of the internuclear axis. The representation of the orbitals $\tilde{\phi}_k$ in terms of "space-fixed" (nonrotating) coordinates $\mathbf{r}$ is[3,4]

$$\phi_k(\mathbf{r}; \mathbf{R}(t)) \equiv \tilde{\phi}_k(\mathbf{r}'; R(t)) = \exp[-i\theta(t) l_y]\tilde{\phi}_k(\mathbf{r}; R(t)) \tag{8}$$

where $\theta(t)$ is the angle between the z axis and the internuclear axis ($\equiv z'$ axis) at time t; l_y is the component of the electronic orbital angular momentum perpendicular to the collision plane (taken as the x–z plane). Insertion of equation (8) into (5) results in the coupling matrix elements

$$M_{kk'}(t) = M_{kk'}^{\text{pot}}(t) + M_{kk'}^{\text{dyn}}(t) \tag{9}$$

with the elements of the potential coupling matrix $\mathbf{M}^{\text{pot}}$ given by

$$M_{kk'}^{\text{pot}}(t) = \langle\tilde{\phi}_k(\mathbf{r}; R(t))|h(t)|\tilde{\phi}_{k'}(\mathbf{r}; R(t))\rangle \tag{10}$$

and those of the dynamic-coupling matrix $\mathbf{M}^{\text{dyn}}$ given by

$$M_{kk'}^{\text{dyn}}(t) = M_{kk'}^{\text{rad}}(t) + M_{kk'}^{\text{rot}}(t) \tag{11}$$

where

$$M_{kk'}^{\text{rad}}(t) = -i\dot{R}(t)\langle\tilde{\phi}_k(\mathbf{r}; R(t))|\frac{\partial}{\partial R}|\tilde{\phi}_{k'}(\mathbf{r}; R(t))\rangle \tag{12}$$

$$M_{kk'}^{\text{rot}}(t) = -\dot{\theta}(R(t))\langle\tilde{\phi}_k(\mathbf{r}; R(t))|l_y|\tilde{\phi}_{k'}(\mathbf{r}; R(t))\rangle \tag{13}$$

are the matrix elements of radial and rotational coupling, respectively. As does the recoil term of equation (7) in the atomic case, rotational coupling among molecular orbitals reflects the use of a noninertial frame of reference in defining the basis orbitals.

A preferred choice for the molecular orbitals $\tilde{\phi}_k$ is that of adiabatic orbitals $\tilde{\phi}_k^{\rm ad}$ which have the property to be exact eigenfunctions of the single-particle Hamiltonian $h(R)$ at fixed R,

$$h(R)\tilde{\phi}_k^{\rm ad}(R) = \varepsilon_k^{\rm ad}(R)\tilde{\phi}_k^{\rm ad}(R) \tag{14}$$

i.e., they diagonalize the potential coupling matrix $\mathsf{M}^{\rm pot}$. In the (orthogonal) adiabatic basis, the coupled-state problem (4) reduces to

$$i\dot{\mathbf{a}} = \mathsf{M}^{\rm dyn}\mathbf{a} \tag{15}$$

if the diagonal potential coupling is eliminated by means of an obvious phase transformation.

Any molecular basis set departing from the strictly adiabatic basis is usually called a *diabatic* basis. Diabatic orbitals have been constructed in attempts to optimize the molecular basis in such a way as to minimize the computational effort in the solution of the coupled-state problem. In general, diabatic orbitals partly diagonalize the dynamic couplings (for example, the rotational coupling in the "dynamical-state representation" of Refs. 16, 17).

In order to make the total scattering wave function $\psi(\mathbf{r}; t)$ fulfill correct boundary conditions at $t = \pm\infty$ (the time zero is taken to correspond to the distance of closest approach of the internuclear motion), the molecular basis orbitals must be required to approach in the limit $R \to \infty$ unperturbed atomic orbitals "traveling" with constant velocity relative to the center of mass of the collision system. In the case of strictly adiabatic orbitals $\tilde{\phi}_k^{\rm ad}$ (or diabatic orbitals merging into $\tilde{\phi}_k^{\rm ad}$ at $R = \infty$), this requirement is fulfilled by multiplying each $\tilde{\phi}_k^{\rm ad}$ by a "translational factor,"

$$\tilde{\phi}_k^{\rm ad}(\mathbf{r}) \to \tilde{\phi}_k^{\rm ad}(\mathbf{r}) \times \exp\left\{ i\mathbf{v}_{1,2}^{(0)} \cdot \mathbf{r} - i\left[\left[\int^t dt'\, \varepsilon_k^{\rm ad}(R(t')) + \tfrac{1}{2}\mathbf{v}_{1,2}^{(0)^2} t\right]\right]\right\} \tag{16}$$

$[\mathbf{v}_{1,2}^{(0)} \equiv \dot{\mathbf{R}}_{1,2}^{(0)}(t = \pm\infty)]$, which makes the basis asymptotically coincide with the "traveling" atomic basis (6). The assignment of translational factors to the orbitals $\tilde{\phi}_k^{\rm ad}$ is, of course, unambiguous only at $R = \infty$. For finite internuclear distances, an electron occupying a molecular orbital obviously can be found with nonvanishing probability in the vicinity of either nuclear center. In the united-atom limit ($R = 0$), in particular, the electron has completely "forgotten" to which nuclear center it was attached at $R = \infty$.

To take into account this aspect, the plane-wave factor appearing in (16) can be modified[18] by replacing the exponent $\mathbf{v}_{1,2}^{(0)} \cdot \mathbf{r}$ by $f(\mathbf{r}; \mathbf{R})\mathbf{v}_{1,2}^{(0)} \cdot \mathbf{r}$. The "switching function" $f(\mathbf{r}; \mathbf{R})$ obeys the boundary conditions $f(\mathbf{r}; R = \infty) = 1$ and $f(\mathbf{r}; R = 0) = 0$, i.e., for infinite internuclear distances the electron fully participates in the translational motion of the center to which it is attached, while in the united-atom limit no translational effect is left at all.

The quasimolecular picture of ion–atom collisions clearly admits a great variety of specific excitation mechanisms that correspond to individual electronic transitions induced by radial or rotational coupling. In inner-shell excitation, the vast number of experimental and theoretical studies undertaken in the past years have entailed a quite detailed classification of the quasimolecular mechanisms (for a recent review of the subject, see Ref. 3). We shall illustrate in the next section the different types of mechanisms by discussing pertinent examples.

It is worthwhile to consider explicitly one specific case of quasimolecular excitation mechanism which exhibits an intimate connection between the picture of Coulomb excitation and the molecular picture even for truly slow collisions. As pointed out first by Briggs,[19] excitation of quasimolecular orbitals whose binding energy is least at $R = 0$ ("energetically promoted orbitals") will take place essentially in the vicinity of the united-atom (UA) limit. By applying the Hellmann–Feynman theorem and by replacing molecular-orbital wave functions and energies by their united-atom limits, Briggs has shown that the first-order amplitude for a transition between an initial state labeled "0" and a final state labeled "f"

$$a_f = -i \int_{-\infty}^{+\infty} dt \, \langle \phi_f(\mathbf{r}; \mathbf{R}(t)) | \frac{\partial}{\partial t}\bigg|_{\mathbf{r}} | \phi_0(\mathbf{r}; \mathbf{R}(t)) \rangle$$
$$\times \exp\left\{ -i \int^{t} dt' [\varepsilon_0(R(t')) - \varepsilon_f(R(t'))] \right\} \tag{17}$$

can be rewritten as

$$a_f = - \int_{-\infty}^{+\infty} dt \, \langle \phi_f^{\mathrm{UA}}(\mathbf{r}) | \frac{Z_1}{|\mathbf{r} - \mathbf{R}_1^{(0)}(t)|} + \frac{Z_2}{|\mathbf{r} - \mathbf{R}_2^{(0)}(t)|} | \phi_0^{\mathrm{UA}}(\mathbf{r}) \rangle$$
$$\times \exp\left[-i \int^{t} dt' (\varepsilon_0^{\mathrm{UA}} - \varepsilon_f^{\mathrm{UA}}) \right] \tag{18}$$

where screening corrections in the effective molecular single-particle potential have been neglected. Equation (18) describes the excitation of a united-atom orbital formed about the center of mass, which is perturbed by the Coulomb fields of both projectile and target nucleus. The structure of the amplitude (18) resembles that of the matrix element (7) in that two different perturbations act coherently to form the total excitation probability.

To conclude this section, we mention some recent attempts to combine the atomic-orbital approach for treating ion–atom collisions with the molecular-orbital approach. Kimura and Lin[20,21] and Winter and Lane[22] have proposed and applied a matching method in which the coupled-state problem (4) is solved in a "traveling" atomic basis for internuclear distances R larger than a certain "matching distance" R_c, and in a molecular basis (without translational factor) for $R < R_c$. The solutions to equation (4) are matched twice, once in the incoming part and again in the outgoing part of the collision, to obtain the total solution. The value of R_c is determined in such a way that in the inner region the effect of electronic translational factors can be assumed to be negligible. Thereby the matching method avoids difficulties with the choice of these factors in the molecular region and emphasizes the atomic (molecular) character of distant (close) collisions.

3. IMPACT PARAMETER DEPENDENCE OF INNER-SHELL VACANCY PRODUCTION

A detailed understanding of the excitation mechanisms in ion–atom collisions would be hardly achievable if one had to rely on measurements and analyses of total cross sections only. In the field of inner-shell vacancy production, major progress towards a full understanding of the excitation mechanisms has been accomplished over the past years by determining the impact parameter dependence of this type of process from measurements of coincidences between projectile ions scattered through a fixed angle and the decay products of inner-shell vacancies (X-rays, Auger electrons). The impact parameter dependence essentially reflects the dependence of the excitation process on the minimum internuclear distance reached in a collision. Consequently, by scanning the impact parameter one may isolate excitation mechanisms that are effective in different ranges of internuclear distance.

3.1. *Asymmetric Collisions*

Let us consider first the impact parameter dependence of inner-shell vacancy production is not too slow, strongly asymmetric ($Z_1 \ll Z_2$) collisions. As mentioned in Section 2, a first-order treatment of the Coulomb interaction between projectile and target is expected to provide a good theoretical description in this case. The interest devoted in the past years to the study of Coulomb excitation in fast collisions has arisen mainly from the desire to identify various effects that are not taken into account in a standard version[23] of first-order theory, the so-called semiclassical approximation (SCA): effects of the deviation of the internuclear trajectory from the straight-line form, in particular the recoil effect exhibited by the second term in the matrix element (7); the effect of increased binding of the inner target shells due to the penetration of the projectile nucleus; relativistic effects in the electronic motion in heavy target atoms. While these effects are small in most cases, they may become strongly enhanced at very small impact parameters.[24] Clearly, the isolation of the different effects will be complicated by a partial cancellation of their contributions in the total excitation probability.

The recoil effect has been studied in detail for Ni K-shell vacancy production in p-Ni and d-Ni collisions, in which relativistic and binding effects are supposed to be small. Figure 2 shows the experimental data of Schmidt-Böcking *et al.*[24] along with the theoretical results of Rösel *et al.*[9] for the impact parameter dependence at a collision energy of 0.6 MeV/nucleon, i.e., at identical velocities of the proton and deuteron projectiles. It is seen that the calculated vacancy production rate $P_K(b)$ at small b values is significantly affected by the *destructive* interference of the (dipole) recoil amplitude with the direct excitation amplitude [cf. equation (7)]. This interference effect is expected[9] to be particularly large if $Z_1/Z_2 \simeq M_1/M_2$. The numerical results and the experimental data of Figure 2 indeed

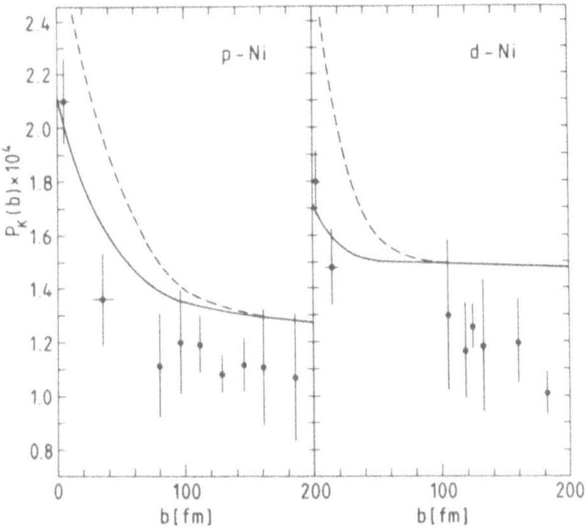

FIGURE 2. Impact parameter dependence of Ni K-shell vacancy production in 0.6 MeV/nucleon p-Ni and d-Ni collisions. Experimental data: Ref. 24. Dashed (solid) curve: SCA calculation of Ref. 9 without (including) recoil term. (From Ref. 24.)

show that the recoil effect in the case of deuteron projectiles (where the above condition is well fulfilled) leads to a much larger reduction of $P_K(b)$ than in the case of proton projectiles.

The influence of the shape of the classical trajectory $\mathbf{R}(t)$ on the probability for inner-shell excitation can be nicely illustrated by calculating the impact parameter dependence for both proton and antiproton ($\bar{p}$) projectiles, as done by Trautmann *et al.*[25] for K-shell vacancy production in Cu targets. The interatomic potential in $\bar{p}$-Cu collisions is attractive, thus allowing the projectile particles to penetrate the Cu K shell much more deeply than do proton projectiles. Figure 3 shows that $P_K(b)$ at small b values is by about a factor of 2.5 larger for $\bar{p}$-Cu collisions than for p-Cu collisions. While no experimental data exist for the cases of Figure 3, we note that data have been obtained recently for single and double ionization of helium by antiproton impact.[26] It should be mentioned here that the shape of the internuclear trajectory has a marked influence on the alignment of inner-shell vacancies following light-ion impact.[27]

The penetration of the projectile nucleus into an inner target shell will tend to enhance the effective binding energy of this shell over the unperturbed energy and thereby increase the transition energy to continuum states. Moreover, the increase in binding energy is associated with a shrinkage of the radial wave function. As long as the electronic motion can be treated nonrelativistically, the binding effect tends to reduce the excitation probability. A simple but efficient recipe for calculating the binding correction for K shells, starting from the (static) correction to the binding energy at fixed

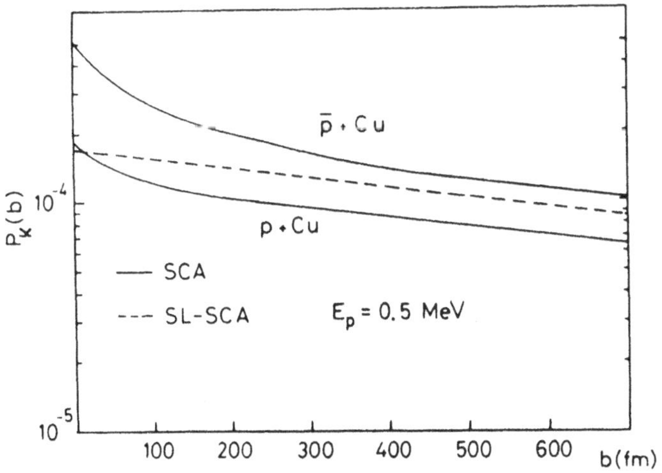

FIGURE 3. Impact parameter dependence of Cu K-shell vacancy production in 0.5-MeV p-Cu and $\bar{p}$-Cu collisions. Dashed (solid) curve: SCA calculation using straight-line trajectories (Coulomb trajectories). (From Ref. 25.)

internuclear distance, has been devised by Andersen *et al.*[28,29] Figure 4 shows that in the case of 18-MeV Ne–Ni collisions the inclusion of the binding correction into the semiclassical approximation reduces the calculated $P_K(b)$ values by more than one order of magnitude and leads to satisfactory agreement between experiment and theory.

It has been pointed out by Kleber and Unterseer[8] that the binding effect in Coulomb excitation is a dynamical effect whose dependence on collision velocity and impact parameter should be treated in a self-consistent way. To achieve this goal, Kleber and Unterseer employ a time-dependent variational principle,[30] thereby going beyond perturbation theory. The trial wave function used to describe K-shell excitation is written as

$$\begin{aligned} \psi(\mathbf{r}; t) &= C^{-1/2}(t)\{1 + \mathbf{k}(t) \cdot \mathbf{r}\} \exp(-\alpha(t) r) \\ &\equiv C^{-1/2}(t)\phi(\mathbf{r}; t) \end{aligned} \tag{19}$$

where

$$C = \pi \left(\frac{\alpha + \alpha^*}{2}\right)^{-5} \left[\left(\frac{\alpha + \alpha^*}{2}\right)^2 + \mathbf{k} \cdot \mathbf{k}^*\right] \tag{20}$$

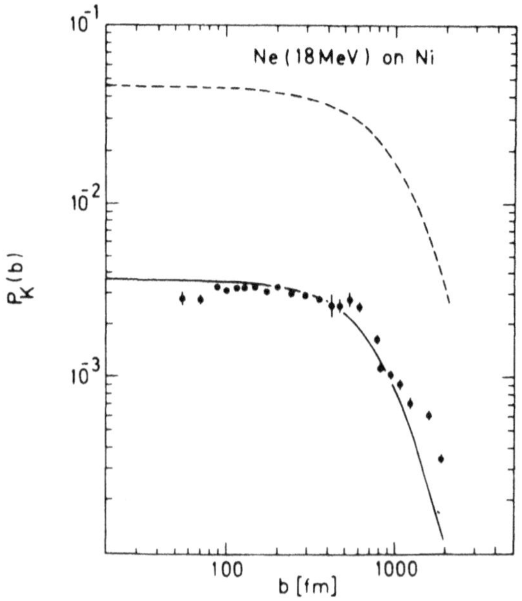

FIGURE 4. Impact parameter dependence of Ni K-shell vacancy production in 18-MeV Ne–Ni collisions. Dashed (solid) curve: SCA calculation without binding correction (including binding correction of Ref. 28). (From Ref. 24.)

The complex, time-dependent parameters $\alpha(t)$ and $\mathbf{k}(t)$ account for the time variation of the binding energy and of the dipole polarization of the K-shell wave function, respectively. The equations of motion for α and $\mathbf{k}$ are derived from the stationarity condition $\delta I = 0$ for the action integral

$$I = \int_{-\infty}^{+\infty} \mathrm{d}t \frac{\langle \phi(t)|h(t) - i\,\partial/\partial t|\phi(t)\rangle}{\langle \phi(t)|\phi(t)\rangle} \tag{21}$$

The calculated ratio of the dynamical K-shell binding-energy increase ΔE_{dyn} to the static (adiabatic) increase ΔE_{stat} at the minimum internuclear distance R_0 in head-on ($b = 0$) p-Cu collisions is shown in Figure 5 as a function of R_0 and of the ratio $\zeta = v/v_K$. The dynamic increase ΔE_{dyn} is seen to become significantly smaller than ΔE_{stat} for $\zeta > 0.2$, i.e., for collision energies beyond 0.6 MeV.

The influence of relativistic electronic motion on K-shell excitation probabilities is illustrated in Figure 6, which shows experimental data for Pb K-shell vacancy production in 135.6-MeV Cl–Pb collisions[24] along with the results of various model calculations. While the experimental data in the small-b range are by far larger than the nonrelativistic calculation including binding and trajectory effects, the agreement between experiment and theory is drastically improved if relativistic and binding effects are included in the theory[25,32] by using *united-atom* Dirac–Fock–Slater wave functions (centered about the target nucleus) as well as the associated K-shell binding energy. The use of united-atom wave functions and binding energies over the whole range of impact parameters clearly is not fully

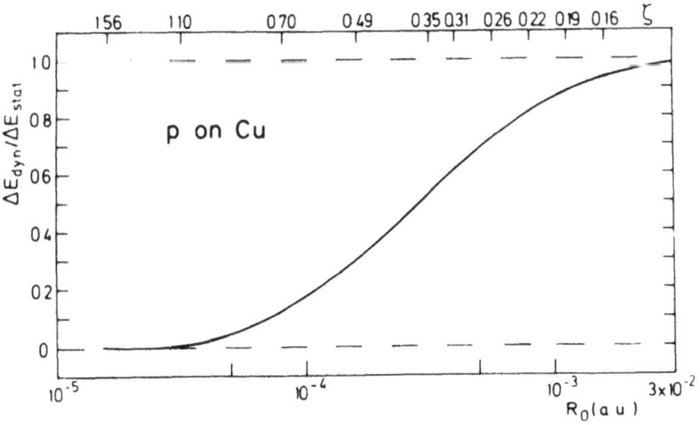

FIGURE 5. Ratio of dynamical K-shell binding-energy increase ΔE_{dyn} to static increase ΔE_{stat} at the minimum internuclear distance R_0 in head-on p-Cu collisions. The upper abscissa is defined in terms of the ratio $\zeta = v/v_K$. (From Ref. 8.)

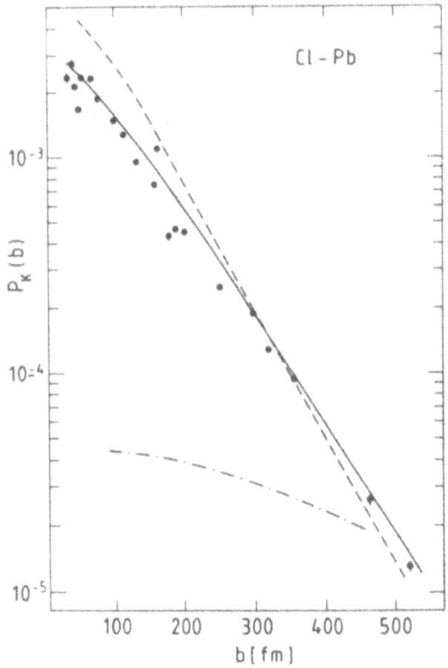

FIGURE 6. Impact parameter dependence of Pb K-shell vacancy production in 135.6-MeV Cl–Pb collisions. Experimental data: Ref. 24. Dash-dot curve: nonrelativistic SCA calculation including binding and trajectory effects (Ref. 25). Dashed curve: relativistic SCA calculation as explained in the text (Refs. 25, 32). Solid curve: relativistic SCA calculation of Ref. 33, including the binding correction of Ref. 29.

adequate, and so the agreement between experiment and theory for larger impact parameters must be considered somewhat accidental. Very good agreement between experiment and theory is achieved[33] over the whole b range if the prescription of Andersen *et al.*[29] is used to calculate the binding energy correction and an effective nuclear charge in terms of which the Dirac–Fock–Slater wave functions are evaluated. This result gives strong support to the assumption that the enhancement of the experimental data in Figure 6 over the nonrelativistic calculation in the small-b range reflect the relativistically enhanced shrinkage of the initial state wave function and the associated large probability for high-momentum components in this wave function.

Coulomb excitation of the L shell has also been a subject of investigation over the past years. The degeneracy of the L shell opens the possibility of investigating vacancy production selectively for different subshells and to study, in particular, the collision-induced alignment of the L_3 subshell. Since the latter topics will be discussed in another contribution to this volume,[27] we shall not consider L-shell Coulomb excitation any further. We merely mention that some evidence has been accumulated that a full understanding of L_3-subshell alignment within a *first-order* theory is not possible since the vacancies produced in a first interaction with the projectile nucleus may be redistributed, with substantial probability, over the different subshells by subsequent interactions.[34-37] A similar inadequacy of the first-order theory has been observed in studies of the impact parameter dependence of L-subshell vacancy production in strongly asymmetric collisions.[37-39]

3.2. Near-Symmetric, Slow Collisions

The impact parameter dependence of inner-shell vacancy production in near-symmetric, slow ion–atom collisions is expected to display a somewhat richer structure than the $P(b)$ curves pertaining to Coulomb excitation, which essentially show a monotonic decrease with increasing b. Structure in the b dependence of quasimolecular excitation may reflect localized couplings between molecular orbitals, which are effective in different ranges of the internuclear separation. The quasimolecular mechanisms of inner-shell excitation can be roughly subdivided into three classes[3]:

1. "Promotion" of electrons into the continuum;
2. Excitation due to rotational coupling among near-degenerate molecular orbitals at small internuclear distances;
3. Ionization of molecular orbitals due to direct coupling to the continuum at small internuclear distances.

The prototype case for mechanism (1) is promotion of $4f\sigma$ electrons in fairly light collision systems like Ar–Ar. From the molecular-orbital correlation diagram for this system (Figure 7), it is seen that the (diabatic) $4f\sigma$ energy curve effectively merges into the continuum when the internuclear distance falls below a "critical" value R_c of about 0.5 a.u. One may assume that in collisions for which the minimum internuclear distance R_0 is less than R_c, two $4f\sigma$ electrons are excited (or "promoted") into the continuum virtually with unit probability. The b dependence corresponding to the promotion mechanism is expected to display a step-functional behavior, with the step occurring at about that b-value for which the condition $R_0(b) = R_c$ is fulfilled. It should be realized that electron promotion into the continuum is mainly effected by the *electrostatic* potential in the quasimolecular collision complex, which causes a rapid loss of binding energy of the "promoted" orbitals when small internuclear distances are approached and eventually "squeezes" the electrons out of the complex.

If an energetically promoted molecular orbital correlates to a united-atom orbital having sufficiently large binding energy, the concept of a "critical" promotion distance R_c loses its meaning. In this case, one has to consider the couplings of the molecular orbital to higher-lying empty orbitals explicitly. The most efficient mechanism for exciting an electron out of a promoted orbital is rotational coupling among near-degenerate molecular orbitals emerging from one and the same united-atom (n, l) shell ["(n, l) rotational coupling"]. For example, in the Ar–Ar system (cf. Figure 7), $3d\sigma$

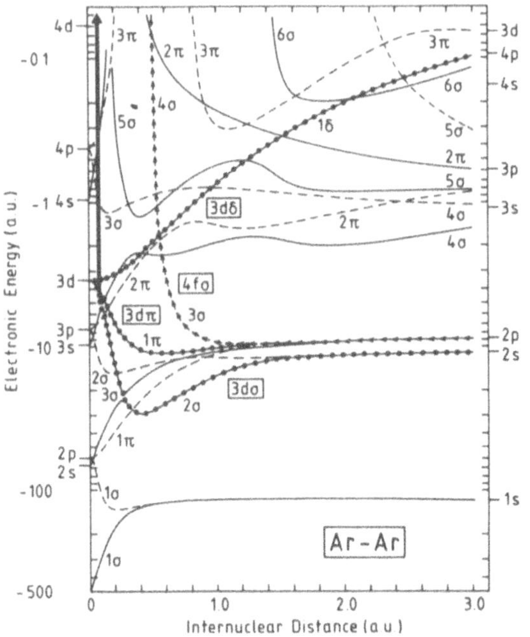

FIGURE 7. Adiabatic molecular-orbital correlation diagram for the Ar–Ar collision system, calculated from the variable-screening model. The diabatic $4f\sigma$, $3d\delta$, $3d\pi$, and $3d\sigma$ energy curves have been obtained from the adiabatic curves by ignoring the screening gaps at close pseudocrossings. The possibility of direct $3d\pi$ and $3d\sigma$ ionization is indicated by the heavy arrow. (From Ref. 3.)

and $3d\pi$ electrons may be excited via $3d$ rotational coupling into the initially empty $3d\delta$ orbital.

From simple theoretical considerations,[3] it may be expected that the impact parameter dependence of the transition probability associated with (n, l) rotational coupling exhibits, for not too low collision velocity, a double-hump structure. A so-called "kinematic" maximum at very small impact parameters (i.e., large scattering angles) arises from the rapid rotation that the internuclear axis experiences in the vicinity of the united-atom limit. For united-atom orbitals with orbital angular momentum $l = 1$, the position of the kinematic maximum is given approximately by

$$b_{\text{kin}} = Z_1 Z_2 / (\mu v^2) \tag{22}$$

where μ is the reduced mass of the collision system. The so-called "adiabatic" maximum expected to occur at larger impact parameters reflects the matching of interaction time and the characteristic time associated with the transition. For $l = 1$ united-atom orbitals, the position of the adiabatic maximum is estimated as

$$b_{ad} = (v/2\alpha)^{1/3} \tag{23}$$

where α is the expansion coefficient in a quadratic approximation to the transition energy. The b-dependence corresponding to (n, l) rotational coupling with $l > 1$ may exhibit more than one kinematic maximum as well as more than one adiabatic maximum.[3]

Excitation of promoted molecular orbitals at small internuclear distances can be accomplished not only by (n, l) rotational coupling, but also by direct coupling to loosely bound, empty orbitals and to the continuum. The physical picture underlying this excitation mechanism is exposed by the united atom ionization model outlined in Section 2. The probability of direct molecular-orbital ionization is expected to be essentially a monotonically decreasing function of impact parameter.

The impact parameter dependence of K-shell vacancy production in slow ion–atom collisions has been studied in a vast number of cases.[3,4,40] A major part of the recent activities has been devoted to the question of how far a universal description of K-shell excitation in terms of scaling laws for $2p\pi$–$2p\sigma$ rotational coupling is possible. Such scaling laws have been derived[41] on the assumption that the coupling matrix element is well approximated by its united-atom value and that the $2p\pi$–$2p\sigma$ energy splitting is adequately represented, for small internuclear distances, as αR^2. In the simplest case, where a straight-line trajectory is used to describe the internuclear motion, the excitation probability $P_K(b, v)$ for an arbitrary collision system is given by

$$P_K(b, v) = P_0(\alpha b^3/v) \tag{24}$$

where $P_0(x)$ is a universal function, and α depends on Z_1 and Z_2. This expression provides a fairly good approximation to the shape of the adiabatic maximum calculated within a full coupled-state approach (note that the description of the kinematic maximum requires a curved internuclear trajectory to be used).

The application of the scaling law for $2p\pi$–$2p\sigma$ rotational coupling in its general form[41] has led in many cases to a rather good overall description of the measured impact parameter dependence of K-shell vacancy production. Certain systematic discrepancies between experiment and theory have emerged in the range of the minumum ("valley") between kinematic and adiabatic maximum: if one fits the theoretical $P_K(b)$ curve to the experimental data in the range of the adiabatic maximum (thereby normalizing the number of incoming $2p\pi$ vacancies), the data tend to be larger than the theoretical values in the range of the valley ("filling of the valley").

In Figure 8, the b dependence of K-shell vacancy production in 363-keV Ne^+-Ne collisions is shown. It is seen that the experimental data are indeed slightly larger than the prediction of the scaling law in the range of the valley and below. Better agreement between experiment and theory is achieved if the $2p\pi$–$2p\sigma$ coupled-state problem is solved by fully taking into account the R dependence of the coupling matrix elements and molecular-orbital energies, both in a nonrelativistic calculation based on the variable-screening model[43] and a relativistic Dirac–Fock–Slater calculation.[44] It turns out that the main reason for the improved agreement is the use of realistic energy curves (instead of the quadratic approximation to these curves implied in the scaling law). In cases other than the one shown in Figure 8, in particular for higher collision velocities, it appears that the filling of the valley is partly caused by direct $2p\sigma$ ionization. However, owing to the lack of detailed calculations, the influence of direct molecular-orbital ionization in K-shell vacancy production seems to be not fully understood.

The quasimolecular mechanisms of L-shell vacancy production are obviously, because of the larger number of molecular orbitals correlating to the separated-atom L shells, more complex than the mechanisms of K-shell vacancy production. For example, L-shell excitation in the Ar–Ar collision system can be accomplished by $3d\sigma$, $3d\pi$, and $4f\sigma$ excitation (cf. Figure 7). The measured impact parameter dependence of L-shell vacancy production in Ar–Ar collisions (Figure 9) shows a steplike rise at

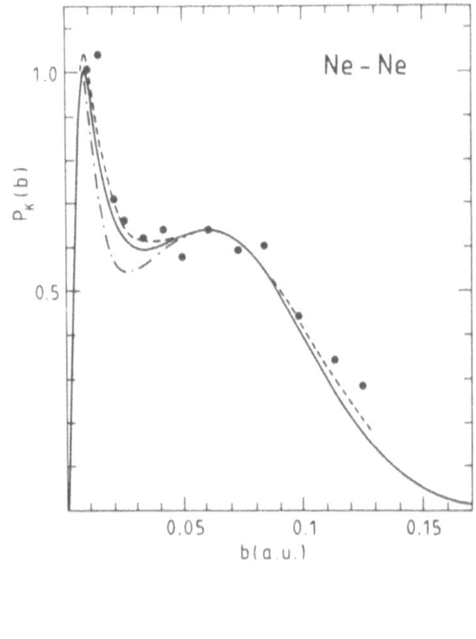

FIGURE 8. Impact parameter dependence of K-shell vacancy production in 363-keV Ne–Ne collisions. Experimental data: Auger electron data of Ref. 42. Solid curve: nonrelativistic calculation of Ref. 43, normalized to one incoming $2p\pi_x$ vacancy. Dashed curve: relativistic calculation of Ref. 44. Dash-dot curve: scaling law of Ref. 41. The dashed and dash-dot curves and the experimental data have been normalized to the solid curve at $b = 0.06$ a.u.

$b \simeq 0.45$ a.u., which may be associated with $4f\sigma$ promotion. If the $4f\sigma$ contribution is assumed to remain constant down to $b = 0$, the enhancement of the vacancy production rate over the value of 2 found in the small-b range can be quantitatively explained in terms of $3d\pi$ and $3d\sigma$ excitation via $3d\delta$-$3d\pi$-$3d\sigma$ rotational coupling.[45]

The identification of $3d$ rotational coupling as a mechanism of L-shell excitation in Ar–Ar collisions has stimulated detailed experimental and theoretical investigations aimed to clarify the role that united-atom rotational coupling plays in L-shell vacancy production in slow ion–atom collisions. Studies of the impact parameter dependence of L-shell vacancy production have revealed contributions from $3d$ rotational coupling in Ne–Ar and Ar–Kr collisions[46,47] and from $3p$ and $3d$ rotational coupling in heavy collision systems like Ag–Ag and I–Ag.[48,49] Evidence for $4f$ rotational coupling as the dominant mechanism of L-shell vacancy production in slow Kr–Kr collisions was first obtained by Shanker *et al.*[50] and subsequently confirmed in detailed studies by Shanker *et al.*[51] and Frekers *et al.*[52]

The Kr-L X-ray production rates measured independently by Shanker *et al.* and by Frekers *et al.* are in good agreement except for a discrepancy in the overall normalization by about 20%. Figure 10 shows the experimental vacancy production rates of Frekers *et al.* for a collision energy of 1.78 MeV, derived from the X-ray data by using the single-vacancy fluorescence yield. The theoretical curves in this figure refer to coupled-state calculations in which those seven adiabatic molecular orbitals were included from which the diabatic $4f\phi$, $4f\delta$, $4f\pi$, and $4f\sigma$ orbitals are obtained by "naive"

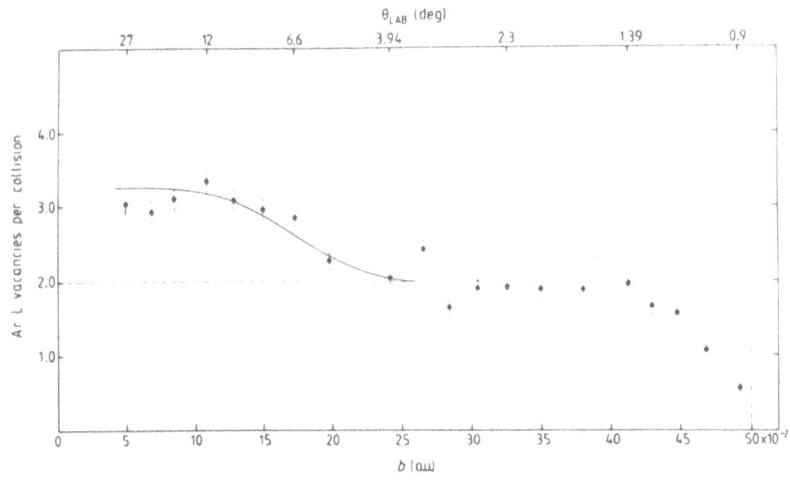

FIGURE 9. Impact parameter dependence of L-shell vacancy production in 300-keV Ar–Ar collisions. Dashed curve: theoretical prediction for $4f\sigma$ promotion. Solid curve: theoretical prediction for $4f\sigma$ promotion plus $3d$ rotational coupling. (From Ref. 45.)

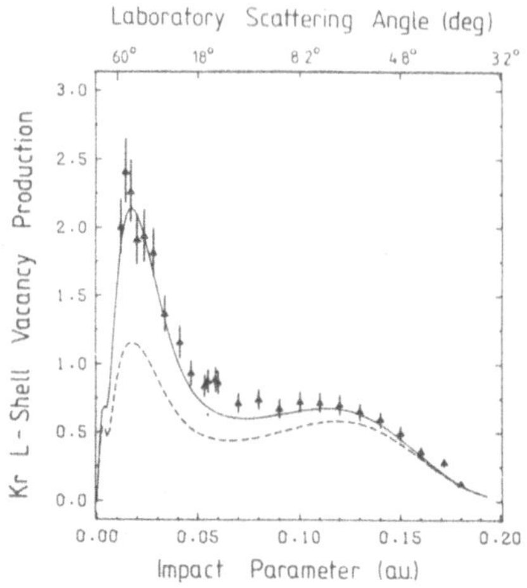

FIGURE 10. Impact parameter dependence of L-shell vacancy production in 1.78-MeV Kr–Kr collisions. Dashed curve: calculation for $4f$ rotational coupling assuming the $4f\phi$ orbital to be initially completely empty. Solid curve: calculation for $4f$ rotational coupling including normalization factor as explained in the text. (From Ref. 52.)

diabatization of avoided crossings. The initial condition in these calculations was fixed by assuming the $4f\phi$ orbital to be initially completely empty. Electronic translational factors were shown to have a very small influence only on the calculated excitation probabilities. Of major influence, however, is the choice of orbital energies, coupling matrix elements, and even of the internuclear trajectories. In Figure 11, L-shell vacancy production rates for 2-MeV Kr–Kr collisions, calculated at different stages of approximation, are shown. The adiabatic seven-state calculation using the full R dependence of the matrix elements and energies as well as trajectories generated from a Lenz–Jensen potential (solid curve) is compared to a diabatic $4f\phi$–$4f\delta$–$4f\pi$–$4f\sigma$ calculation using united-atom matrix elements (dashed curve) and

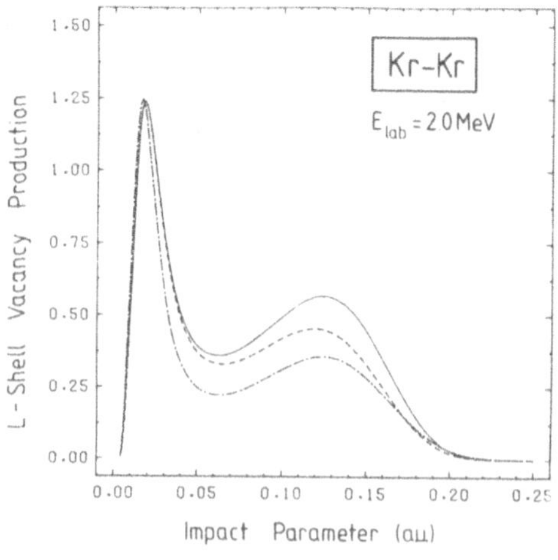

FIGURE 11. Calculated impact parameter dependence of L-shell vacancy production in 2.0-MeV Kr–Kr collisions. Solid curve: full calculation for $4f$ rotational coupling. Dashed curve: rotational-coupling matrix elements approximated by united-atom values. Dash-dot curve: approximations of Ref. 51. (From Ref. 52.)

to another diabatic calculation, in which, moreover, a cubic approximation to the R dependence of the transition energies as well as effective Coulomb trajectories have been used (dash-dot curve). The approximations underlying the latter calculation are those employed by Shanker *et al.*[50,51] in the theoretical analysis of their data.

As exemplified by the case of Figure 10, the vacancy production rates obtained from the X-ray data of Frekers *et al.*[52] by using the single-vacancy fluorescence yield are reproduced by the full coupled-state calculations only in the range of large impact parameters. The discrepancies seen at small b values can be made more apparent by plotting the ratio of experimental to theoretical vacancy production rates against minimum internuclear distance R_0 (Figure 12). The ratios for all collision energies up to 1.78 MeV approximately fall on a common curve, the values of which rise monotonically from unity at large R_0 to about 2 at small R_0. This finding is likely to reflect an increase of the dynamic L-shell fluorescence yield due to enhanced

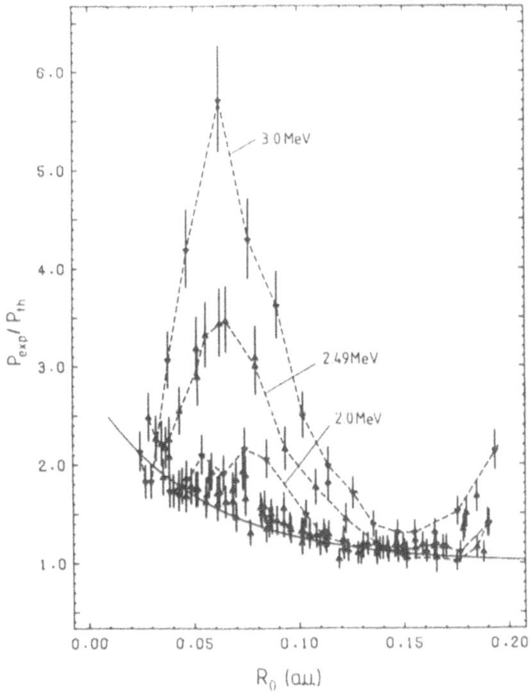

FIGURE 12. Experimental L-shell vacancy production in Kr–Kr collisions for collision energies ranging between 711 keV and 3.0 MeV, plotted as function of minimum internuclear distance R_0. The 2.0-MeV and 3.0-MeV data are from Ref. 51, the remaining data from Ref. 52. Solid curve: fit curve defining "normalization factor" as explained in the text. Dashed curves serve to guide the eye only. (From Ref. 52.)

M-shell vacancy production at small R_0 values (contributions from direct molecular-orbital ionization and rotational-coupling mechanisms not taken into account in the theoretical analysis have been estimated to be small in the energy range considered). As long as a quantitative evaluation of the fluorescence yield effect is lacking, one may assume the common curve in Figure 12 to define an R_0-dependent "normalization factor," which may be used to bring the theoretical vacancy production rates into accord with the experimental data (cf. Figure 10).

The peak displayed in Figure 12 by the curves corresponding to 2 MeV collision energy and above points to the onset of direct excitation mechanisms, but no quantitative confirmation of this assertion can be given. It appears that a complete analysis of this situation requires an appropriate treatment of the interference between rotationally induced excitation and direct molecular orbital ionization.

4. COHERENCE EFFECTS IN INNER-SHELL PROCESSES

The survey of the impact parameter dependence of inner-shell vacancy production given in the preceding section has not pointed to effects that one is inclined to call direct manifestations of the coherence of ion–atom collisions. In fact, the only indication for a coherence effect was found in K-shell Coulomb ionization in which the destructive interference between direct and recoil amplitude leads to a reduction of the ionization probability at very small impact parameters. We now consider situations where the coherence of inner-shell processes manifests itself in a direct way by giving rise to (oscillatory) interference structure in the dependence of the excitation or deexcitation probability on impact parameter, collision velocity, or energy of emitted photons or electrons. Such structure can come about through interference of different contributions to the transition amplitude, whose relative phase depends on the parameters of the collision. A survey of the various possibilities that may lead to interference structures in atomic collisions has been given recently by Niehaus.(53) The cases discussed here deal mainly with the interference of contributions to the transition amplitude from the incoming and the outgoing branch of the collision.

The investigation of interference structures in inner-shell processes has been greatly advanced in the past years through the availability(54,55) of beams of highly charged, slow ions, particularly beams of hydrogenlike or fully stripped ions in which K-shell vacancies can be carried into a heavy-ion–atom encounter. The situation prevailing in this case is illustrated in Figure 13 by means of a schematic molecular-orbital correlation diagram for the S^{15+}–Ar system, which allows the fate of a K-shell vacancy initially present in the S^{15+} ion, i.e., in the $1s\sigma$ molecular orbital, to be followed all

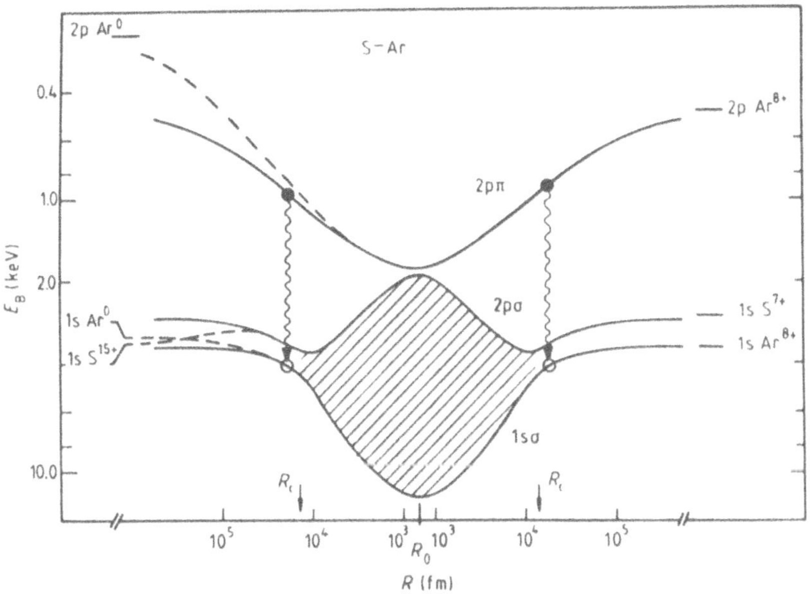

FIGURE 13. Schematic molecular-orbital correlation diagram indicating the possibility of double-passage K-K vacancy transfer via $2p\sigma$–$1s\sigma$ radial coupling and of radiative $2p\pi$–$1s\sigma$ transitions in S^{15+}-Ar collisions. The dashed curves describe the transition from the initial level ordering in the S^{15+}-Ar system to the level ordering in the charge-equilibrated S^{7+}-Ar^{8+} system. The center of the $2p\sigma$–$1s\sigma$ coupling region is denoted by R_c. The shaded area symbolizes the $2p\sigma$–$1s\sigma$ phase difference that accrues between the incoming and the outgoing passage through R_c.

through the collision. It is seen that the $1s\sigma$ vacancy can be transferred, owing to $1s\sigma$–$2p\sigma$ radial coupling on both the incoming and the outgoing branch of the collision, into the Ar K shell. This "double-passage" vacancy sharing process is to be distinguished from "single-passage" vacancy sharing in which a vacancy created, e.g., in the $2p\sigma$ orbital via $2p\pi$–$2p\sigma$ rotational coupling at small internuclear distances, is "shared" between the K shells of the collision partners owing to radial coupling on the outgoing branch of the collision. As indicated in Figure 13, a $1s\sigma$ vacancy may also be filled by radiative transitions of electrons out of higher-lying orbitals, e.g., out of the $2p\pi$ orbital. For given frequency of the emitted "molecular-orbital X rays," the essential contributions to the emission amplitude are from those points on the incoming and the outgoing branch of the collision for which the $2p\pi$–$1s\sigma$ energy difference matches that frequency.

We now discuss interference effects in double-passage vacancy sharing and molecular-orbital X-ray emission and also consider the influence of nuclear reactions on inner-shell ionization. A nuclear reaction may cause a "time delay" in the internuclear motion, which affects the relative phase between the incoming and outgoing contributions to the atomic transition amplitude.

4.1. Double-Passage Vacancy Sharing

The essential features of double-passage vacancy sharing (or of the equivalent problem of quasiresonant electron transfer) are easily derived[56] in the two-state approximation by assuming the interaction that causes the transfer to be sharply localized in the vicinity of a certain internuclear distance R_c. Applying the sudden approximation independently at each interaction region, one may express the double-passage probability P_2 in terms of the single-passage probability P_1 as

$$P_2 = 4P_1(1 - P_1)\sin^2\phi \tag{25}$$

where

$$\phi = \frac{1}{2}\int_{-t_c}^{+t_c} dt\, \Delta\varepsilon(R(t)) \tag{26}$$

and $R(\pm t_c) = R_c$. The function ϕ reflects the phase difference of the unperturbed states (with energy difference $\Delta\varepsilon$) that accrues between the inward and outward passage of the collision system through the coupling distance R_c. Taking for simplicity a straight-line trajectory to describe the internuclear motion, the dependence of ϕ on impact parameter b and collision velocity v can be written explicitly as

$$\phi(b; v) = \frac{1}{v}\int_b^{R_c} dR \frac{R}{(R^2 - b^2)^{1/2}}\Delta\varepsilon(R) \tag{27}$$

This shows that independent variation of b or v can give rise to oscillatory structure in the double-passage transfer probability (25).

There are several effects that will tend to damp or distort the simple oscillatory pattern predicted by equation (25): (1) the finite extension of the coupling region; (2) "dynamic" changes in the orbital energies and coupling matrix elements due to outer-shell processes occurring during the course of the collision; (3) coupling to higher-lying orbitals like, e.g., $2p\pi$–$2p\sigma$ rotational coupling; (4) the effect of electronic translational factors. All such effects can be taken into account in some way in numerical coupled-state calculations or in other, more refined theoretical treatments.

First experiments revealing oscillatory structure in the impact parameter dependence of K–K vacancy transfer were performed by Hagmann *et al.*,[57] who studied the F^{8+}–Ne collision system. Figure 14 shows the experimental

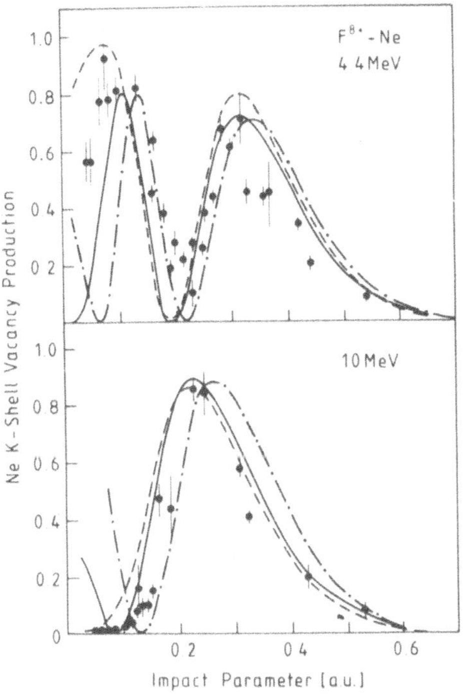

FIGURE 14. Impact parameter dependence of Ne K-shell vacancy production via K-K vacancy transfer in 4.4- and 10-MeV F^{8+}-Ne collisions. Experimental data: Ref. 57. Theoretical curves: two-state atomic-orbital calculations of Ref. 57 (solid curve) and of Ref. 58 (dashed and dash-dot curves). (From Ref. 58.)

data of Hagmann *et al.* along with the results of various model calculations. The data at 4.4 MeV collision energy display well-developed structure that is semiquantitatively reproduced by the theoretical curves. The 10-MeV data merely show a broad bump, as is expected from the v dependence of the phase ϕ [equation (27)]. The different model calculations shown in Figure 14 are all based upon an expansion of the total wave function in a space of two *atomic* orbitals (i.e., in the space of the $1s$ orbitals of the two collision partners). This space is generally believed to accommodate with sufficient accuracy the quasimolecular $1s\sigma$ and $2p\sigma$ orbitals. A closer inspection shows, however, that the theoretical transfer probabilities depend sensitively on the specific assumptions of the calculations. A detailed discussion of the use of atomic two-state expansions in the description of K-K vacancy transfer has been given by Stolterfoht.[59]

Oscillatory structure in the impact parameter dependence of K-K vacancy transfer has been observed also by Schuch *et al.*[55] in S^{15+}-Ar collisions. Figure 15 shows the experimental data at 7.9 MeV collision energy along with a variety of theoretical results including a full numerical solution of the time-dependent one-electron Schrödinger equation[60] and a multistate atomic-orbital calculation.[61] While the position of maxima and

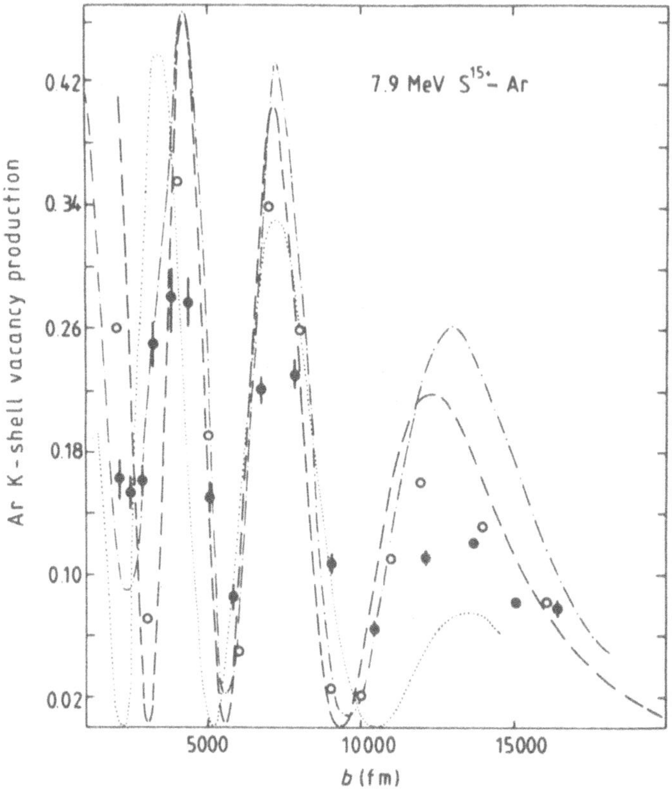

FIGURE 15. Impact parameter dependence of Ar K-shell vacancy production via K-K vacancy transfer in 7.9-MeV S^{15+}-Ar collisions. Solid circles: experimental data of Ref. 55. Open circles: theoretical result based on the direct solution of the time-dependent Schrödinger equation (Ref. 60). Dashed curve: two-state atomic orbital calculation (Refs. 58, 59). Dash-dotted curve: multistate atomic-orbital calculation (Ref. 61). Dotted curve: theoretical result based on an analytic approximation to equations (25) and (26). (From Ref. 55.)

minima in the measured transfer probabilities is quite well predicted by the calculations, the absolute values of the probabilities are not even reproduced by the most advanced calculations, i.e., the multistate atomic-orbital calculation and the direct solution of the Schrödinger equation. An explanation for these discrepancies can perhaps be found in an inadequacy of the effective potential used to describe the motion of the "active" vacancy. This potential should include the effect of dynamic changes in the screening experienced by the inner-shell electrons. For a further discussion of this point, see Schuch *et al.*(55)

The collision velocity dependence of the oscillatory structures observed in S^{15+}-Ar collisions has been analyzed by Schuch *et al.* by assuming the transfer probability to have the general form (25) and by assigning values of $(2n-1)\pi/2$ to the phase function $\phi(b)$ at the observed positions b_n of the maxima ($n = 1, 2, 3, \ldots$) and minima ($n = \frac{1}{2}, \frac{3}{2}, \frac{5}{2}, \ldots$). If plotted against impact parameter (Figure 16), the values of $v \times \phi(b_n)$ for velocities v corresponding to collision energies between 4.7 and 32 MeV are seen to fall on a common curve. This behavior is consistent with equation (27). Moreover, the slope of a continuous curve fitted through the $v \times \phi(b_n)$ points is in reasonable agreement with that calculated from equation (27) by using $2p\sigma$–$1s\sigma$ energy differences obtained from Hartree–Fock calculations or from a simple exponential interaction between the K shells of the collision partners. This leads in turn to fairly good agreement (see inset in Figure 16) between the latter energy differences and those obtained from a deconvolution of equation (27) using the $v \times \phi(b_n)$ points of Figure 16 as input.

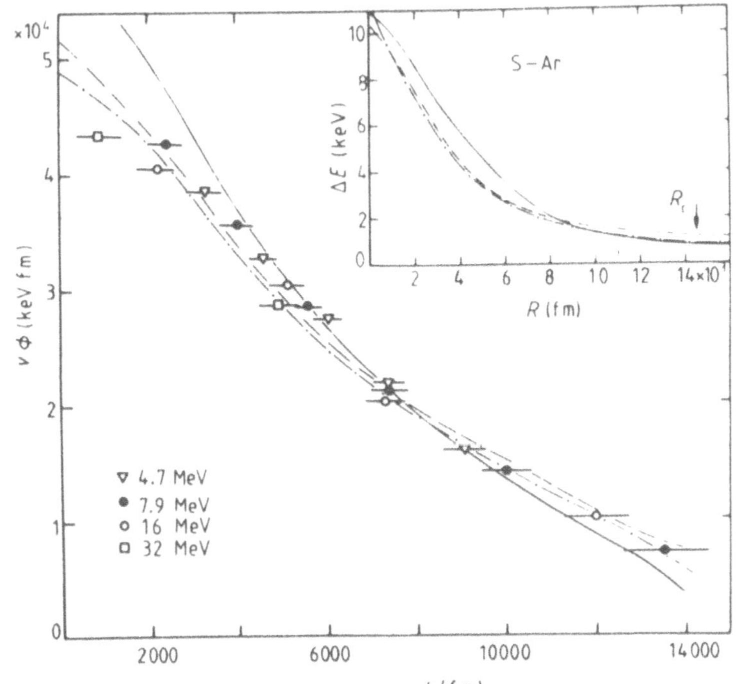

FIGURE 16. Product of collision velocity v and $2p\sigma$–$1s\sigma$ phase integral ϕ for S^{15+}–Ar collisions, plotted against impact parameter. Solid and open symbols: experimental values for the indicated collision energies. Solid curve: Hartree–Fock calculation. Dashed curve: fit curve to the experimental values. Dash-dotted curve: calculation assuming an exponential interaction between the K shells. The inset shows the corresponding $2p\sigma$–$1s\sigma$ energy differences. (From Ref. 55.)

If, instead of hydrogenlike projectile ions, completely stripped ions are used in the investigation of K-K charge transfer, the double-vacancy transfer channel competes with single-vacancy transfer. The impact parameter dependence of double K-K vacancy transfer was first investigated by Hagmann *et al.*[62] for 10-MeV F^{9+}-Ne collisions by measuring Ne-K hypersatellite Auger spectra. More recently, Schulz *et al.*[63] have studied single and double K-K vacancy transfer in 16-MeV S^{16+}-Ar collisions by observing triple coincidences between two K X rays and the scattered projectile ion. The experimental results were compared to theoretical predictions obtained by assuming completely uncorrelated behavior of the two K vacancies and by using the single-vacancy transfer probabilities measured in S^{15+}-Ar collisions as input. It was found that the observed oscillatory structure in the impact parameter dependence of the S^{16+}-Ar transfer probabilities is not as pronounced as predicted. Schulz *et al.* suspect that this finding might be associated with the dynamic change in the effective single-electron potential, which is expected to be larger in the S^{16+}-Ar system than in the S^{15+}-Ar system.

4.2. *Molecular-Orbital X Rays*

Radiative transitions between quasimolecular orbitals with time-varying energy difference $\Delta\varepsilon(R(t))$ result in the emission of continuous X-ray spectra.[64] The first-order emission amplitude for the case that a vacancy is initially present in an inner-shell molecular orbital can be schematically written as

$$D_\omega = \int_{-\infty}^{+\infty} dt\, D(R(t)) \exp[i\phi_\omega(t)] \tag{28}$$

where

$$\phi_\omega(t) = \omega t - \int_0^t dt'\, \Delta\varepsilon(R(t')) \tag{29}$$

ω is the X-ray frequency and $D(R(t))$ denotes the dipole matrix element. The largest contributions to D_ω will come from the vicinity of those times t_ω at which the phase function $\phi_\omega(t)$ is stationary, i.e., at which $\dot{\phi}_\omega(t_\omega) = 0$. This leads to the condition $\omega = \Delta\varepsilon(R_\omega)$, that is, the X-ray frequency has to match the energy difference $\Delta\varepsilon(R)$ at some value R_ω of the internuclear distance. If just one solution R_ω exists, the symmetry of the internuclear trajectory, $R(-t) = R(t)$, entails two stationary points $\pm t_\omega (t_\omega > 0)$ of the phase function $\phi_\omega(t)$. Expanding $\phi_\omega(t)$ about $\pm t_\omega$ up to second order, one obtains the stationary-phase approximation to D_ω:

$$D_\omega = D(R_\omega)\left(\frac{2\pi}{|(d/dR)\Delta\varepsilon|_{R_\omega}|\dot{R}(t_\omega)}\right)^{1/2}$$
$$\times \{\exp(-i[\phi_\omega(t_\omega) - \pi/4]) + \exp(i[\phi_\omega(t_\omega) - \pi/4])\} \qquad (30)$$

[note that $\phi(-t) = -\phi(t)$]. The two exponentials in equation (30) reflect the contributions to the emission amplitude from the incoming and the outgoing branch of the collision. The interference of these contributions gives rise to an emission probability of the form

$$I_\omega \propto |D_\omega|^2 = \frac{8\pi|D(R_\omega)|^2}{|(d/dR)\Delta\varepsilon|_{R_\omega}|\dot{R}(t_\omega)}\cos^2[\phi_\omega(t_\omega) - \pi/4] \qquad (31)$$

The phase $\phi_\omega(t_\omega)$ depends on emission frequency ω, collision velocity v, and impact parameter b, so that variation of any of these parameters may give rise to oscillatory structure in I_ω.

The "simple" stationary-phase approximation to D_ω clearly breaks down if the condition $\omega = \Delta\varepsilon(R_\omega)$ cannot be fulfilled for any real, positive R_ω ("classically forbidden case"). The emission amplitude can be continued smoothly into the classically forbidden region by applying generalized stationary-phase methods like the "uniform" Airy approximation,(65) or by adopting the "simple" Airy approximation(66,67) in which D_ω is evaluated by suitably expanding the functions $D(R(t))$ and $\phi_\omega(t)$ about $t = 0$. The Airy approximations predict an exponential decrease of I_ω in the classically forbidden region.

The experimental verification of interference structure in molecular-orbital X-ray spectra has been given by Tserruya *et al.*(68) and Schuch *et al.*,(69,70) who studied radiative transitions into the $1s\sigma$ molecular orbital in Cl^{16+}-Ar collisions. Figure 17 shows X-ray spectra taken at 2.5 MeV collision energy for different values of the impact parameter. In order to extract the R dependence of the transition energy $\Delta\varepsilon$ from the observed spectral shapes (Figure 18), Schuch *et al.* have employed two different methods: (1) an analysis of the phase difference between adjacent maxima and minima in the spectrum; (2) an analysis of the shape of the broad bump on the high-energy side of the spectrum. The first method gives $\Delta\varepsilon$ in the range of small R values (solid symbols in Figure 18). The second method assumes the shape of the broad bump to be given essentially by the square of an Airy function and allows $\Delta\varepsilon$ to be extracted in the range of larger R values (open symbols in Figure 18). It is seen from Figure 18 that the deduced energy differences agree quite well with $2p\pi$-$1s\sigma$ differences obtained from

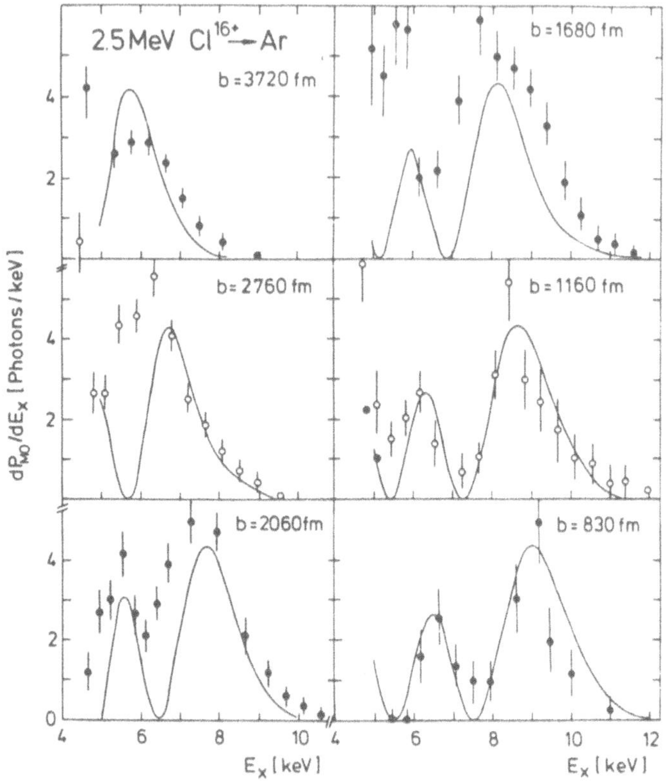

FIGURE 17. Molecular-orbital X-ray spectra emitted in 2.5-MeV Cl^{16+}-Ar collisions at different impact parameters. Solid curves: spectra calculated within the simple Airy approximation. (From Ref. 70.)

a Dirac-Fock-Slater calculation and from scaled H_2^+ energy curves, respectively. The scaled $2p\sigma$–$1s\sigma$ difference deviates considerably from the experimental points, so that one concludes that the observed X-ray spectra result mainly from $2p\pi$–$1s\sigma$ transitions. Using the experimentally determined $\Delta\varepsilon$ values, Schuch *et al.* have calculated the spectral shape of the emitted radiation from the simple Airy approximation (solid curves in Figure 17) and have found fairly good agreement between experiment and theory.

Vacancies in inner-shell molecular orbitals can be filled during a collision not only by radiative transitions but also by Auger transitions leading to continuous electron emission.[64] The identification of molecular-orbital Auger electrons is generally hampered by the large background of directly ejected electrons (δ electrons). The search for interference structure in molecular-orbital Auger electron spectra has led to conflicting results. Liesen

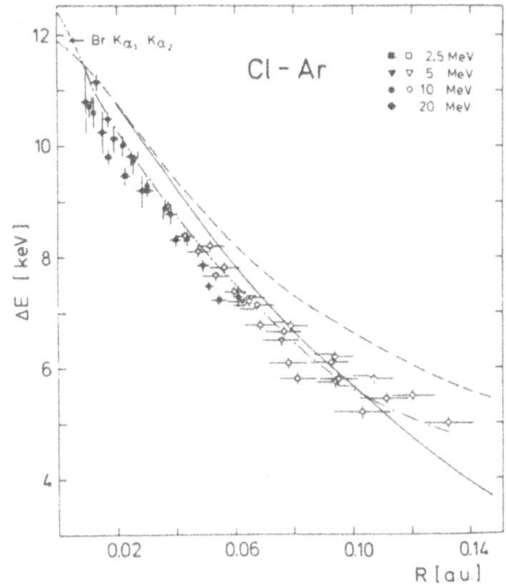

FIGURE 18. Transition energies in the Cl^{16+}-Ar collision system as function of internuclear distance. Solid and open symbols: experimental values deduced from molecular-orbital X-ray spectra as described in the text. Solid (dashed) curve: $2p\pi$–$1s\sigma$ ($2p\sigma$–$1s\sigma$) energy differences obtained from scaled H_2^+ energies. Dash-dotted curve: Dirac-Fock-Slater calculation for the $2p\pi$–$1s\sigma$ difference. (From Ref. 70.)

et al.[71] have observed oscillatory structure in the electron spectra following 450-keV Kr-Kr collisions at various, fixed projectile scattering angles (Figure 19). By analyzing their results within the simple Airy approximation,[66] they conclude that the structure is likely to be produced by Auger

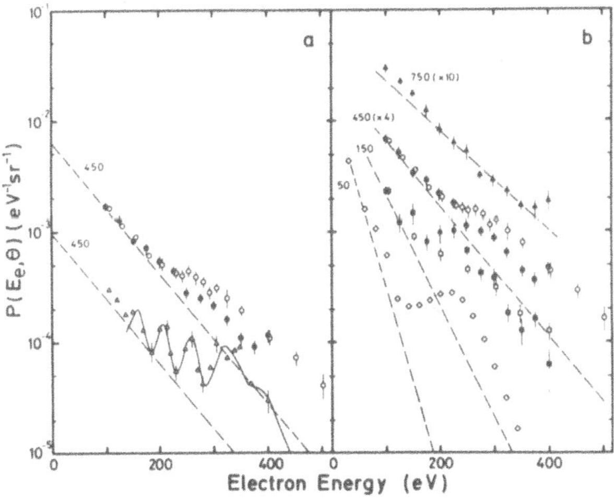

FIGURE 19. Molecular-orbital Auger electron spectra emitted in Kr-Kr collisions at a laboratory scattering angle of 15°. The numbers attached to the different data sets give the collision energy in keV. Open triangles: experimental data of Ref. 71. Open diamonds: experimental data of Ref. 72. Other symbols: experimental data of Ref. 73. The dashed lines indicate the background contribution to the spectra. The oscillating solid curve represents a fit to the data of Ref. 71, using the simple Airy approximation. (From Ref. 73.)

decay of $4p\pi$ vacancies. A recent experiment by Shergin *et al.*[73] neither was able to confirm the findings of Liesen *et al.* (Figure 19) nor did it give evidence for interference structure at collision energies other than 450 keV.

4.3. Effects of Nuclear Reactions

The examples of K-K vacancy sharing and molecular-orbital X-ray emission have shown that interference effects in inner-shell processes may be revealed by varying the relative phase between the contributions to the excitation or emission amplitude from the incoming and outgoing branches of the collision. The external parameters whose variation entails a variation of this phase are collision velocity, impact parameter and, in the case of molecular-orbital X-ray emission, also photon frequency. Another way to influence the relative phase between incoming and outgoing amplitudes consists in selecting conditions under which a time delay in the internuclear motion ("nuclear sticking") is caused by reactions of the *nuclear* particles involved in the collision.[74-77]

For a nuclear sticking time T, the first-order amplitude a for an electronic transition can be written as

$$a = -i \int_{-\infty}^{+\infty} dt\, M(R_s(t)) \exp\left[-i \int^{t} dt'\, \Delta\varepsilon(R_s(t'))\right] \tag{32}$$

Assuming, for simplicity, the nuclear scattering to be purely elastic, the time dependence of the internuclear distance is given by

$$R_s(t) = \begin{cases} R(t), & t \leq 0 \\ 0, & 0 \leq t \leq T \\ R(t-T), & t > T \end{cases} \tag{33}$$

where $R(t)$ is the time dependence for a zero-impact-parameter trajectory without sticking. By breaking the integral in equation (32) into three parts corresponding to the time intervals $[-\infty, 0]$, $[0, T]$, and $[T, +\infty]$ and assuming the coupling matrix element to vanish for $0 \leq t \leq T$, the amplitude a may be rewritten as

$$a = a_{\text{in}} + e^{-i\Delta\varepsilon_0 \cdot T} a_{\text{out}} \equiv a_{\text{in}} - e^{-i\Delta\varepsilon_0 \cdot T} a_{\text{in}}^{*} \tag{34}$$

where a_{in} and a_{out} are the contributions to the transition amplitude from the incoming and outgoing branch of the collision, respectively, for the case that no nuclear sticking occurs. The transition probability P derived from equation (34),

$$P \equiv |a|^2 = 4|a_{\text{in}}|^2 \sin^2\{\Delta\varepsilon_0 \cdot T/2 + \phi_{\text{in}}\} \tag{35}$$

[$\phi_{in} = \arg(a_{in})$], may display oscillatory behavior not only as function of the phase ϕ_{in}, but also as function of the time delay T and the united-atom transition energy $\Delta\varepsilon_0$.

The most immediate test of equation (35) is provided by measuring coincidences between elastically scattered projectile nuclei and target K X rays in the vicinity of a nuclear resonance.(78–83) Figure 20 shows results for elastic p-^{138}Ba scattering at $\theta_{lab} = 172°$ in the vicinity of the isobaric analog resonance at $E_p = 10.00$ MeV. The K-shell ionization probability P_K exhibits a rapid variation as a function of proton energy when the resonance is crossed. A large variation of P_K is indeed expected since a value of about 0.54 for the phase shift $\Delta\varepsilon_0 \cdot T$ is estimated from the united-atom K-shell ionization energy and the width Γ of the scattering resonance ($T = 1/\Gamma$).

Effects of a nuclear resonance have been found not only in target ionization, but also in electron capture. Scheurer *et al.*(84) observed a fluctuation of the capture probability in the vicinity of the nuclear scattering resonances at 462 keV proton energy in p-^{12}C collisions and at 1058 keV in p-^{14}N collisions. More recently, Horsdal *et al.* (85) have studied the probability for electron capture across the 1955-keV resonance in p-^{20}Ne collisions. Their results are shown in Figure 21.

The possibility to study time-delay effects on K-shell ionization in *deep-inelastic* nuclear reactions has been pointed out by Anholt(86) and by

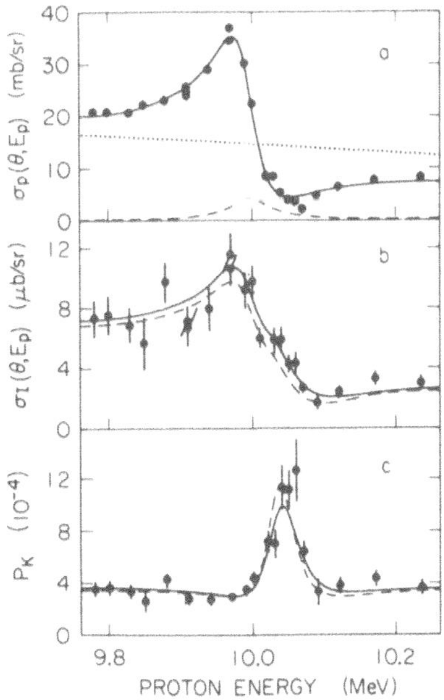

FIGURE 20. Effect of nuclear time delay on K-shell ionization across the 10.00-MeV isobaric analog resonance in elastic p-^{138}Ba scattering. (a) Elastic scattering cross section at $\theta_{lab} = 172°$, plotted as function of proton energy. Solid curve: best fit to the data. Dotted curve: contribution of elastic background scattering. Dashed curve: upper limit to contribution of compound-elastic scattering. (b) Differential ionization cross section. Solid (dashed) curve: fit to the data without (with) inclusion of compound-elastic contribution in the nuclear scattering amplitude. (c) Ionization probability. Meaning of solid and dashed curves is the same as in case (b). (From Ref. 83.)

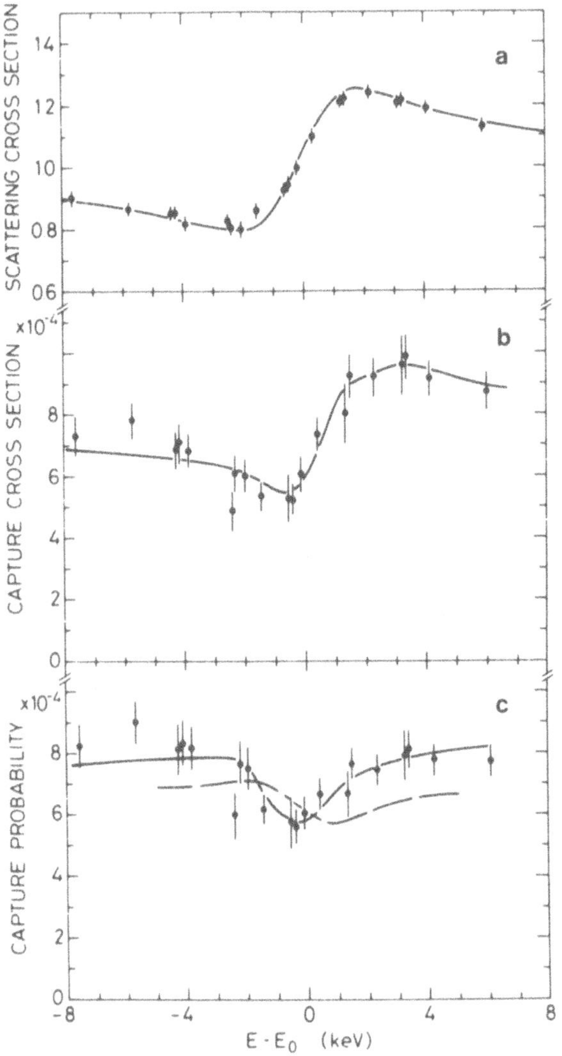

FIGURE 21. Effect of nuclear time delay on electron capture across the 1.955-MeV resonance in elastic p-^{20}Ne scattering. (a) Elastic scattering cross section at $\theta_{lab} = 30°$, normalized to the Rutherford cross section. The resonance energy is denoted by E_0. Solid curve: best fit to the data. (b) Differential capture cross section, normalized to the Rutherford cross section. Solid curve: model calculation as described in Ref. 85. (c) Capture probability. Solid and dashed curves: model calculations as described in Ref. 85. (From Ref. 85.)

Müller *et al.*[87] In such reactions, the two colliding heavy nuclei are supposed to form a composite system that rotates for a certain time interval during which the internuclear distance remains almost constant. The associated time delay is estimated to be of order 10^{-21} s for collision systems like Xe–Pb. Assuming K-shell ionization in the heavier collision partner to proceed mainly via $1s\sigma$ excitation, the phase shift $\Delta\varepsilon_0 \cdot T$ relevant to Pb K-shell ionization in Xe–Pb collision is estimated to be larger than 0.5. Attempts to identify time delay effects in Xe–Pb and Xe–Th collisions[88] have failed owing to the large background of K X rays caused by internal conversion of nuclear γ-ray transitions in the targetlike fragments.

More favorable conditions for the observation of time-delay effects in deep-inelastic collisions apparently exist in the (symmetric) U–U system in which $2p\sigma$ excitation is expected to be the dominant mechanism of K-shell ionization. Since the probability of $2p\sigma$ excitation is much larger than that of $1s\sigma$ excitation, the background due to internal conversion in U–U collisions should be relatively much smaller than that in Xe–Pb and Xe–Th collisions. Stoller *et al.*[89,90] have measured the probability of K-shell ionization in 7.5 MeV/nucleon U–U collisions as function of the Q value of the nuclear reaction. Upon subtracting from the total probability a computed contribution of internal conversion, they have obtained the purely atomic ionization probability P_K. By comparing the Q dependence of P_K with the calculated dependence of this quantity upon the time delay T, Stoller *et al.* deduced a semiempirical relationship between T and Q. The time delay in the deep-inelastic region is found to be about 1×10^{-21} s, a value that is consistent with the result of an analysis of positron spectra emitted in U–U collisions.[91]

In principle, time-delay effects on inner-shell ionization should be observable in the energy distribution of the ejected electrons ("δ electrons").[92,93] According to equation (35), this distribution should exhibit oscillatory behavior for $T > 0$. Senger *et al.*[94] have measured δ-electron spectra in deep-inelastic 8.6-MeV/nucleon U–Au collisions resulting in fission of the uranium projectiles. In such ternary nuclear reactions, the background of conversion electrons is significantly reduced owing to the strong dependence of the conversion coefficients on the nuclear charge number. Figure 22 shows preliminary results of Senger *et al.* for different ranges of the total kinetic-energy loss. It is seen that with increasing energy loss, the δ-electron spectra tend to develop structure that is reproduced by model calculations assuming a certain distribution of the time delay about a mean value of order 10^{-21} s.

Kirsch *et al.*[95] have pointed out that the time delay due to a nuclear reaction may influence the interference pattern in molecular-orbital X-ray emission. For radiative transitions into the $1s\sigma$ orbital in 5.8-MeV/nucleon Pb–Pb collisions, a shift in the position of the highest maximum in the

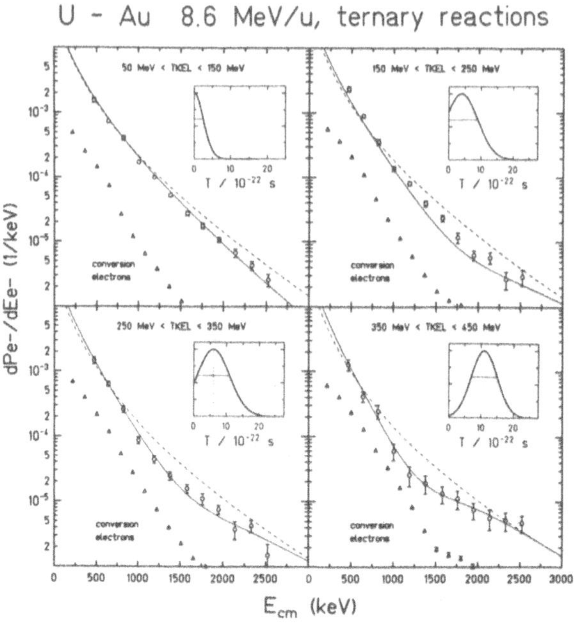

FIGURE 22. Spectra of δ electrons accompanying 8.6 MeV/nucleon U–Au collisions leading to ternary nuclear reactions in the indicated ranges of the total kinetic-energy loss (TKEL). The contribution of conversion electrons is already subtracted in the δ electron spectra. Dashed curves: spectra calculated on the assumption of no time delay in the internuclear motion. Solid curves: spectra calculated for a distribution of nuclear contact times T as shown in the insets. (From Ref. 94.)

X-ray spectrum by about 25 keV is estimated for a time delay of 10^{-21} s. At present, however, no experimental tests of this prediction are available.

5. EFFECTS OF EXTERNAL FIELDS IN ION–ATOM COLLISIONS

External (electric or magnetic) fields may influence atomic systems through the exchange of energy and momentum. In the past years, a great deal of effort has been devoted to the study of atomic collision processes proceeding in the presence of (static or time-dependent) external fields.[96–99] The effect of an external field on the collision dynamics is usually very small since the electric or magnetic interaction is much smaller than the characteristic energies of the collision system. This holds in particular for inner-shell processes in ion–atom collisions, and accordingly it is not surprising that such processes have been studied so far only theoretically. Field effects in highly excited systems like Rydberg atoms are, however, easily accessible to experimental investigation.[96]

In the present context, it is appropriate to discuss examples in which the external field serves to reveal coherence properties of atomic collisions. One such example is that of Rydberg states in fast ions, which are exposed to an electric field. In this case, the field is not applied *during* the collision in which the highly excited ionic states are formed, but it influences the evolution of these states *after* the ion-atom encounter is terminated. In order to illustrate the effect of external magnetic fields on ion-atom collisions, we discuss some recent theoretical work dealing with charge exchange in the presence of very strong (astrophysically relevant) fields.

5.1. Coherent Evolution of Rydberg States in Electric Fields

The influence of an external electric field on Rydberg states may be demonstrated by considering the time evolution of hydrogenlike states with unperturbed wave function

$$\psi_{nlm}^{(Z)}(t) = \phi_{nlm}^{(Z)} \exp(-iE_n^{(Z)}t) \tag{36}$$

[$E_n^{(Z)} = -Z^2/(2n^2)$; Z is the effective nuclear charge] in a static, uniform field of strength F. To first order in F, the perturbed atomic Hamiltonian is diagonalized by the "parabolic" or "Stark" representation $\phi_{nkm}^{(Z)}$ labeled by the "electric" quantum number $k \equiv n_1 - n_2$, where $n_1 = 0, 1, \ldots, n - |m| - 1$ and $n_2 = n - n_1 - |m| - 1$ are parabolic quantum numbers.[100] The associated energies are

$$E_{nk}^{(Z)} = E_n^{(Z)} + \frac{3}{2}\frac{F}{Z}nk \tag{37}$$

Assume now that the field F starts to act on the unperturbed states at $t = 0$. The wave function $\psi_{nm}^{(Z)}(t)$ evolving from $\psi_{nlm}^{(Z)}(t = 0) \equiv \phi_{nlm}^{(Z)}$ is obtained by expanding $\phi_{nlm}^{(Z)}$ in terms of Stark states $\phi_{nkm}^{(Z)}$,

$$\phi_{nlm}^{(Z)} = \sum_k \left(\begin{matrix} (n-1)/2 & (n-1)/2 \\ (m+k)/2 & (m-k)/2 \end{matrix} \middle| \begin{matrix} l \\ m \end{matrix} \right) \phi_{nkm}^{(Z)} \tag{38}$$

where, for the sake of simplicity, the axis of quantization has been assumed to coincide with the field direction; the expansion coefficients denoted by bracket symbols are Clebsch-Gordan coefficients.[101] Recognizing that the time evolution of the individual Stark states is determined by phase factors $\exp(-iE_{nk}^{(Z)}t)$, one obtains for $t > 0$

$$\psi_{nm}^{(Z)}(t) = \sum_k \left(\begin{matrix} (n-1)/2 & (n-1)/2 \\ (m+k)/2 & (m-k)/2 \end{matrix} \middle| \begin{matrix} l \\ m \end{matrix} \right) \times \phi_{nkm}^{(Z)} \exp\left[-i\left(E_n^{(Z)} + \frac{3}{2}\frac{F}{Z}nk\right)t\right] \tag{39}$$

If the field is switched off at $t = t_0$, the field-dependent parts of the phase factors in equation (39) become "frozen in" at their values for $t = t_0$. The wave function for $t > t_0$ then reads

$$\psi_{nm}^{(Z)}(t) = \left\{ \sum_k \begin{pmatrix} (n-1)/2 & (n-1)/2 & l \\ (m+k)/2 & (m-k)/2 & m \end{pmatrix} \phi_{nkm}^{(Z)} \exp\left(-i\frac{3}{2}\frac{F}{Z} nkt_0 \right) \right\}$$
$$\times \exp(-iE_n^{(Z)} t) \tag{40}$$

The probability $P_{n'l'm'}^{(Z)}$ to find, for $t \geq t_0$, the unperturbed state $\psi_{n'l'm'}^{(Z)}(t)$ in $\psi_{nm}^{(Z)}(t)$ is given by

$$P_{n'l'm'}^{(Z)} \equiv |\langle \psi_{n'l'm'}^{(Z)}(t) | \psi_{nm}^{(Z)}(t) \rangle|^2$$
$$= \delta_{nn'}\delta_{mm'} \sum_{\kappa = 0, \pm 1, \ldots, \pm \kappa_0} \left(\sum_K C_{K+\kappa}^{ll'} C_{K-\kappa}^{ll'*} \right) \exp\left(-i\frac{3Fnt_0}{Z}\kappa \right) \tag{41}$$

where $\kappa_0 = n - |m| - 1$ and

$$C_k^{ll'} \equiv \begin{pmatrix} (n-1)/2 & (n-1)/2 & l \\ (m+k)/2 & (m-k)/2 & m \end{pmatrix} \begin{pmatrix} (n-1)/2 & (n-1)/2 & l' \\ (m+k)/2 & (m-k)/2 & m \end{pmatrix}$$

This probability is periodic in the field strength F and in the "dwell time" t_0, with the corresponding fundamental periods given by

$$\Delta F = \frac{2\pi}{3} \cdot \frac{Z}{nt_0}, \qquad \Delta t = \frac{2\pi}{3} \cdot \frac{Z}{nF} \tag{42}$$

The oscillating F or t_0 dependence of $P_{n'l'm'}^{(Z)}$ reflects what one calls "Stark beats," i.e., the interference of coherently developing Stark states whose phases are "detuned" owing to the action of the electric field.

The prototype system on which Stark beats have been studied is the hydrogen $n = 2$ system.[102] This system may be experimentally prepared in the (metastable) $2s$ substate by waiting after beam-foil or beam-gas excitation until the $2p$ state has completely decayed via radiative $2p \to 1s$ transitions. If the $2s$ state is exposed to an electric field, the $2p$ state becomes admixed again in the total wave function ("coherent regeneration"). The probability $P^{(1)}_{2p,m=0}$ to find the $2p$, $m = 0$ substate is, according to equation (41), given by

$$P^{(1)}_{2p,m=0} = \sin^2(3Ft_0) \tag{43}$$

The predicted t_0 dependence of $P^{(1)}_{2p,m=0}$ is observed experimentally[102] by scanning the intensity of the Lyman-α radiation emitted perpendicular to the beam axis as function of the distance that the hydrogen atoms have traveled in the field.

A much more complicated situation prevails in the case of high-n Rydberg states formed in interactions of fast, highly charged ions with foil or gas targets. In general, a rather large manifold of states is expected to become excited in these interactions. A delicate question is to what extent different states $\phi^{(Z)}_{nlm}$ are excited coherently (in the case of partially coherent excitation, the complete theoretical description of Stark beats requires a density matrix description). Even if one may consider the states $\phi^{(Z)}_{nlm}$ separately, equation (41) shows that the high principal quantum numbers may give rise to high-order harmonics in the probability $P^{(Z)}_{n'l'm'}$ by which the beat pattern may be smeared out. Moreover, in the case of Rydberg states, $P^{(Z)}_{n'l'm'}$ cannot be measured individually for the quantum numbers n', l', m'. What one can measure is either delayed X-ray emission or electron spectra following field ionization of Rydberg states. Both methods provide an average of the probability $P^{(Z)}_{n'l'm'}$ over a certain n', l', m' range (for the discussion of field ionization experiments, the quantum number n' is to be replaced by the energy of the continuum final state).

In view of these points, it was somewhat surprising that electric-field-induced structure has been observed both in delayed X-ray spectra[103-105] and in the electron yield due to field ionization of Rydberg states.[106-110] By means of the field-ionization method, systematic studies have been performed of the dependence of the field-induced structure on projectile charge number and charge state as well as on the type of target (foil or gas target).

A typical setup[108] for the field-ionization experiments is shown in Figure 23. The "deflector field" is a weak, static electric field oriented perpendicular to the ion beam axis. This field serves two purposes: (1) it deflects continuum electrons ("cusp electrons") traveling with the beam velocity out of the beam direction; (2) it influences the Rydberg states in the ion beam. These states are field-ionized in the electric field of a tandem electron spectrometer. The spectrometer field is considerably stronger than the deflector field. Accordingly, the main contribution to the field ionization yield is from Rydberg electrons which are only weakly affected by the deflector field.

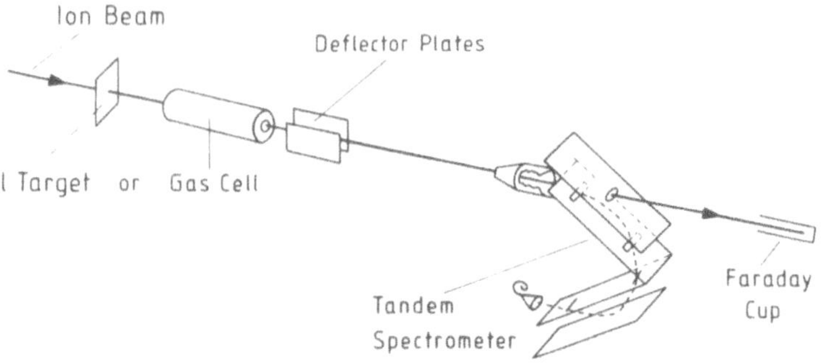

FIGURE 23. Schematic diagram of the setup used in field-ionization experiments on high Rydberg states.

Figure 24 shows electron spectra[108] following the interaction of an 85-MeV Ne^{6+} beam with a helium gas target. The spectrum for zero deflector field F_D displays the "cusp peak" at an electron energy corresponding to the beam velocity. The line structure superimposed on the tails of the cusp results from excited Ne^{6+} configurations that autoionize in the helium target. A shoulderlike structure shows up at an energy that is about 300 eV higher

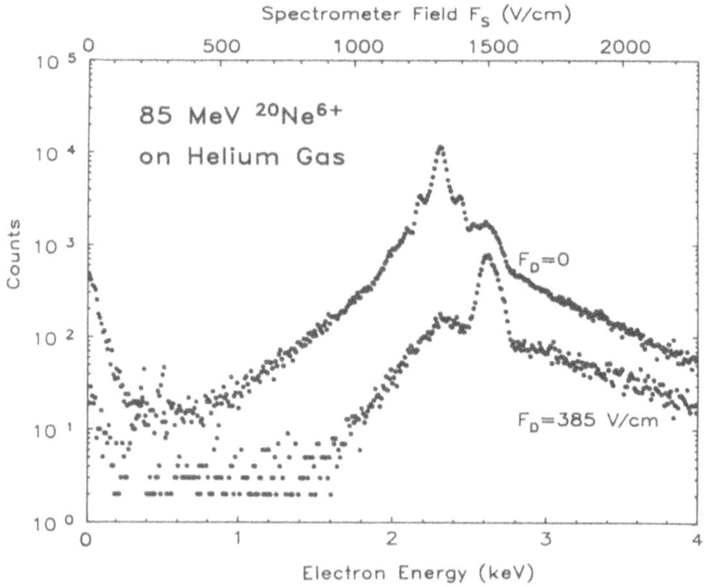

FIGURE 24. Electron spectra following collisions of 85-MeV Ne^{6+} ions with a helium gas target, taken for zero deflector field F_D and for $F_D = 385$ V/cm, respectively. (From Ref. 108.)

than the cusp energy. For $F_D = 385$ V/cm, the cusp peak has almost disappeared, while the shoulder has developed into a pronounced peak. The latter peak ("Rydberg peak") is associated with electrons resulting from field ionization of Rydberg states in the first stage of the spectrometer. These electrons are expected to appear at an "apparent" energy larger than the cusp energy since field ionization takes place only after the Rydberg ions have traveled a certain distance in the spectrometer. Fixing the spectrometer field at the value corresponding to the Rydberg peak, one can measure the Rydberg electron yield as function of the deflector field.

Figure 25 shows the (background-subtracted) Rydberg electron yield as function of F_D for interactions of 72 MeV O^{3+} ions with a helium gas target.(109,110) A marked oscillatory structure is observed in the yield. The corresponding Fourier power spectrum (Figure 26) reveals a fundamental frequency of 0.52 $(V/cm)^{-1}$ as well as the presence of higher harmonics. Results similar to those shown in Figures 25 and 26 have been obtained(109,110) for a variety of projectile ions in different charge states interacting with foil or gas targets.

The universal appearance of structure in the field ionization yield makes an interpretation in terms of Stark beats tempting.(106,108,111) However, as compared with the "ideal" Stark beat experiment on the hydrogen $n = 2$ system, the situation in field ionization experiments on Rydberg states is

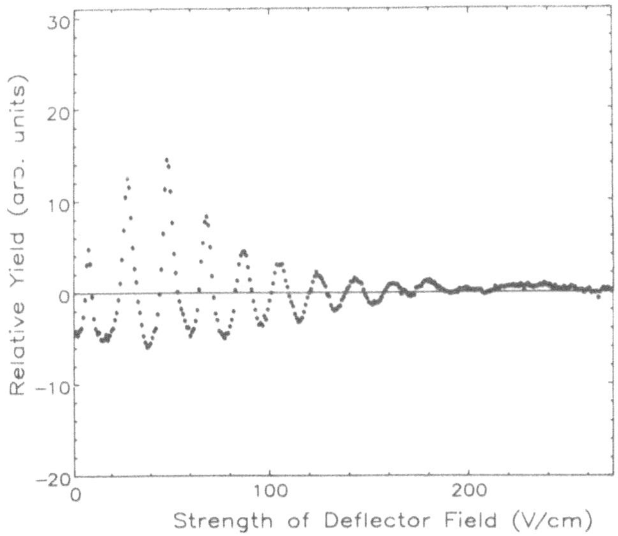

FIGURE 25. Background-subtracted Rydberg electron yield as function of deflector field strength for interactions of 72-MeV O^{3+} ions with a helium gas target. (From Ref. 109.)

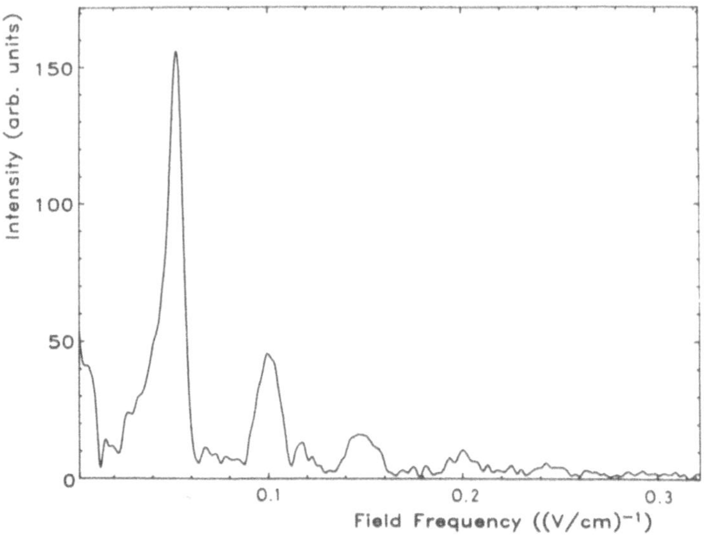

FIGURE 26. Fourier power spectrum corresponding to the yield curve of Figure 25. (From Ref. 109.)

obscured by the incomplete knowledge of the influence that the electron spectrometer has on the observed electron yield. A detailed study[109] of the field distribution and the electron trajectories in the spectrometer indicates that in experiments using the setup of Figure 23, only Rydberg electrons that have been ionized in a sharply localized spatial region will fully traverse the spectrometer and finally be recorded. This spatial localization implies, via the strong sensitivity of the field ionization probability on the principal quantum number n, that (independent of how the primary n-distribution of the Rydberg states looks like) only a narrow band of n values contributes to the observed yield. For example, for the case shown in Figure 25, one may estimate that a few n values close to 70 give major contributions to the yield.

The n selectivity of the electron spectrometer appears to be crucial to the occurrence of pronounced structure in the field ionization yield. If just one n-value contributes, the theory sketched above predicts a Fourier spectrum with a single fundamental period given by equation (42). The experimentally observed periods indeed agree in many cases with the periods derived from the field strength at the point of ionization and the related n value. This finding gives good support to the assumption that the structure seen in the field-ionization yield reflects Stark beats in high-n Rydberg states.

There remains the question whether the analysis of the beat pattern observed in field-ionization experiments allows conclusions to be drawn[108,111] on the production mechanisms of Rydberg states in fast ion beams interacting with foil or gas targets. An important clue to the understanding of these mechanisms is the knowledge of the l-distribution which, according to equation (41), is hidden in the weight factors of the different harmonics corresponding to a given n value. The experiments indicate that the weight of higher harmonics is usually small. Unfortunately, because of the complexity of the weight factors, one cannot immediately conclude from this behavior that the l-distribution has a simple shape, e.g., that low or high l values are preferentially populated.

5.2. *Ion–Atom Collisions in Strong Magnetic Fields*

The interest devoted to atomic collision processes in strong magnetic fields has arisen mainly from astrophysical motivations.[112,113] The field strengths estimated to be present in degenerate astrophysical objects range from 10^2–10^5 T in white-dwarf stars to 10^7–10^9 T in neutron stars. Field strengths of this order of magnitude, while far from being accessible in terrestrial laboratories, are challenging from a theoretical point of view because the interaction energies associated with them are comparable to or larger than binding energies of ground or low-lying excited states in simple atomic systems like the hydrogen atom. The large distortion of the atomic states by the magnetic field is expected to give rise to qualitatively new phenomena.

Recent theoretical work on ion–atom collisions in strong magnetic fields has concentrated on charge exchange in one-electron collision systems.

Symmetric resonant charge exchange in a strong magnetic field was first considered by Bivona *et al.*[114,115] within a two-state atomic-orbital approach using straight-line internuclear trajectories directed parallel to the field. Electronic translational factors as well as the cyclotron motion of the center of mass of the collision system have been ignored. It was shown that the probability $P(b, v; B)$ for an electron initially bound to center 2 to be transferred to center 1 can be expressed as

$$P(b, v; B) = \sin^2 \eta \tag{44}$$

where

$$\eta = \frac{1}{v}\int_{-\infty}^{+\infty} dz \frac{V_{12} - SV_{22}}{1 - S^2} \tag{45}$$

$$V_{12} = \int d\mathbf{r}\, \tilde{\phi}_1^* V e^{-ig} \tilde{\phi}_2 \tag{46}$$

$$V_{22} = \int d\mathbf{r}\, \tilde{\phi}_2^* V \tilde{\phi}_2 \tag{47}$$

$$S = \int d\mathbf{r}\, \tilde{\phi}_2^* \, e^{ig} \tilde{\phi}_1 \tag{48}$$

and V is the electrostatic interaction in the initial channel. The functions $\tilde{\phi}_1$ and $\tilde{\phi}_2$ are "magnetically dressed" wave functions, i.e., eigenfunctions of the atomic Hamiltonian including the coupling to the magnetic field **B**. An essential ingredient of equations (46) and (48) is the phase factor e^{ig}, where

$$g = \frac{1}{2c} (\mathbf{B} \times \mathbf{b}) \cdot \mathbf{r} \tag{49}$$

(**b** is the two-dimensional impact parameter vector). This phase expresses the difference in gauge of the magnetic vector potentials referred to centers 1 and 2, respectively. The factor e^{ig} appears in equations (46) and (48) formally in the same way as does the electronic translational factor (which has been ignored here). In general, the effect of an additional (oscillating) phase factor in the matrix elements is to reduce the transition probability.

Bivona *et al.*(114,115) have calculated the total cross section for resonant charge transfer in H^+-H and Rb^+-Rb collisions. They find a substantial reduction of the cross section through the influence of the magnetic field. For example, in H^+-H collision at $v = 0.22$ a.u., a field $B = 0.5B_0$ ($B_0 = 2.35 \times 10^5$ T $\triangleq$ 137.04 a.u. is the "critical" field strength) results in a cross section that is about 34% smaller than the cross section in the field-free case. Similarly, in Rb^+-Rb collisions at $v = 0.033$ a.u., a field $B = 0.05B_0$ reduces the cross section by about 43%. According to the analysis of Bivona *et al.*, the reduction is mainly caused by the phase factor e^{ig}, while the "dressing" of the atomic wave functions has only a minor effect on the cross section.

The influence of an external magnetic field on charge exchange in asymmetric ion–atom collisions has been investigated by Grosdanov and McDowell(116) and by Bivona and McDowell.(117)

Grosdanov and McDowell(116) derive the classical equations of motion for the three-body problem in a uniform magnetic field and employ the classical-trajectory Monte Carlo method in a study of He^{2+}-H(1s) collisions for impact energies ranging between 25 and 75 keV/nucleon. The principal result is that for the field strengths considered ($B \leq 0.2B_0$), the total capture cross section is enhanced over the field-free cross section. The maximum enhancement is found to be about 17%.

In the work of Bivona and McDowell,(117) the $He^+(2lm)$ substate population through charge exchange in He^{2+}-H collisions has been investigated within an atomic-orbital approach including the H(1s), $He^+(2s)$, $He^+(2p_0)$, and $He^+(2p_{\pm 1})$ states. Straight-line internuclear trajectories lying parallel to the field direction are assumed, and the cyclotron motion of the center of mass is ignored. In the B range considered ($B \leqslant 0.2B_0$), unperturbed atomic wave functions are used, while the effect of the field on the energy levels is taken into account to first order in the magnetic interaction. Electronic translational factors are included in the calculation. In the nondiagonal coupling and overlap matrix elements connecting the H(1s) state with the He$^+$ states, the phase g of equation (49) then conspires with the translational-factor phase to give an additional phase factor of the form

$$\exp\left\{i\left[\frac{1}{2c}(\mathbf{B}\times\mathbf{b})-\mathbf{v}\right]\cdot\mathbf{r}\right\} \tag{50}$$

The term in square brackets can be viewed as a field- and impact-parameter-dependent effective velocity $\mathbf{v}_{\rm eff} = \mathbf{v} - (1/2c)(\mathbf{B}\times\mathbf{b})$. If $\mathbf{B} \parallel \mathbf{v}$, as is assumed by Bivona and McDowell, the field-dependent component of $\mathbf{v}_{\rm eff}$ is perpendicular to the collision velocity $\mathbf{v}$. The results of Bivona and McDowell show that for velocities $v \leqslant 1$ a.u., the magnetic field has a major influence on the capture cross section both through the phase factor (50) and through the field-induced removal of the $2s$–$2p$ degeneracy in the final He$^+$ ion. The redistribution of flux into the different final substates leads in particular to a strong reduction of the $2p_{+1}$ capture probability and to a corresponding enhancement of the $2p_{-1}$ probability (Figure 27).

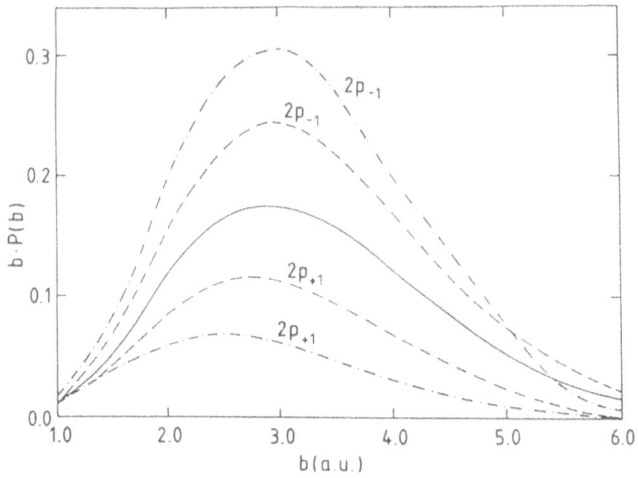

FIGURE 27. Impact parameter dependence of the linearly weighted capture probability into the $2p_{\pm1}$ states in He^{++}-H(1s) collisions at $v = 1.0$ a.u. Solid curve: $B = 0$. Broken curves: $B = 0.1B_0$. Dash-dotted curves: $B = 0.2B_0$. (From Ref. 117.)

The quasimolecular treatment of ion–atom collisions in the presence of a strong magnetic field requires as a first step the calculation of magnetically dressed, one-electron molecular orbitals for arbitrary values of the internuclear distance and of the angle θ between internuclear axis and field direction. The pertinent Hamiltonian reads[118]

$$H = -\tfrac{1}{2}\nabla^2 - \left(\frac{Z_1}{r_1} + \frac{Z_2}{r_2}\right) + \frac{B}{2c}(\cos\theta\, l_z + \sin\theta\, l_x)$$

$$+ \frac{B^2}{8c^2}\{(x^2 + y^2) - \sin^2\theta(x^2 - z^2) - \sin 2\theta\, xz\} \qquad (51)$$

(the coordinates are chosen such that the z axis coincides with the internuclear axis and the y axis is perpendicular to the field direction). The

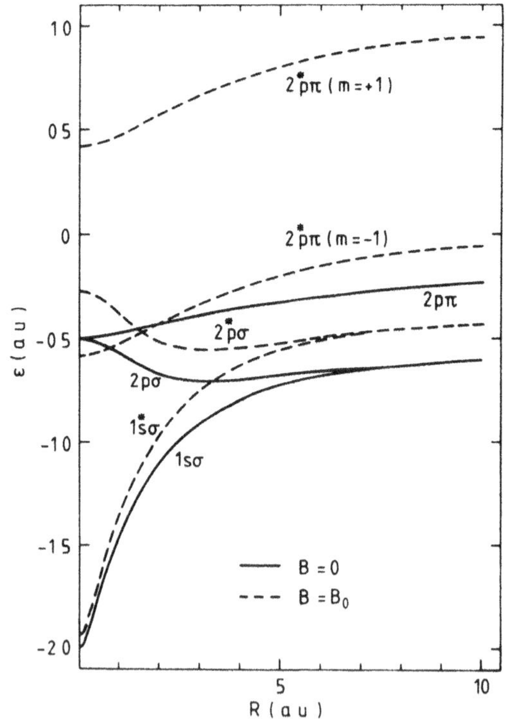

FIGURE 28. Energy curves of the lowest orbitals of the H_2^+ quasimolecule for $B = 0$ and $B = B_0$ ($\theta = 0$). (From Ref. 118.)

diagonalization of H in a suitably chosen basis set is complicated by the fact that for $\theta \neq 0$ the angular momentum projection l_z is no longer a good quantum number.

Calculations of energy surfaces and wave functions associated with the Hamiltonian (51) have been initiated by Wille,[118-120] who used basis functions of the Hylleraas type. Figure 28 displays the lowest energy curves of the magnetically dressed H_2^+ quasimolecule for $\theta = 0$ and $B = B_0$ in comparison with the field-free case. The dressed energy surfaces $\varepsilon^*_{1s\sigma}(R, \theta)$ and $\varepsilon^*_{2p\sigma}(R, \theta)$ have been used[120] in a calculation of probabilities and cross sections for resonant charge transfer in H^+–$H(1s)$ collisions. The transfer probability is given by

$$P(b, v; B) = \sin^2\left\{\int_0^{+\infty} dt[\varepsilon^*_{2p\sigma}(R(t), \theta(t)) - \varepsilon^*_{1s\sigma}(R(t), \theta(t))]\right\} \quad (52)$$

Note that the magnetic field enters this probability only through the dressing of the molecular-orbital energies. An additional field-dependent phase factor as it shows up in the probability derived within the two-state atomic-orbital

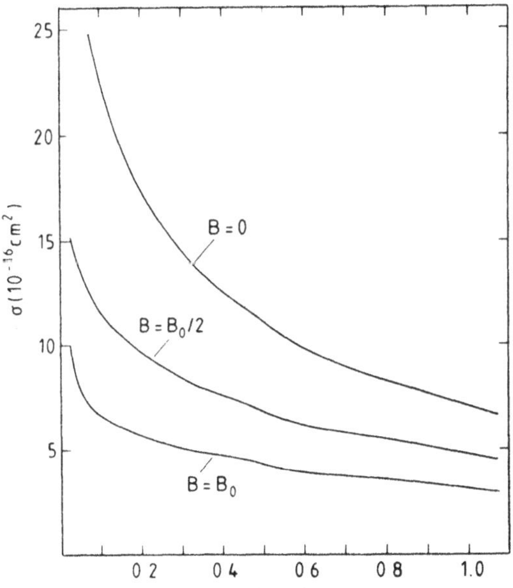

FIGURE 29. Collision velocity dependence of the total cross section for resonant charge transfer in H^+–H collisions, calculated for different values of the magnetic field strength B. (From Ref. 120.)

approach [cf. equations (44) to (49)] does not arise in the molecular treatment since in this case all wave functions are referred to one and the same frame of coordinates. Figure 29 shows the transfer cross section for different values of the magnetic field strength, calculated with straight-line internuclear trajectories. It is seen that the cross section decreases with increasing field strength. For $B = B_0$, the cross section is smaller than the one for $B = 0$ by about a factor of 2.5. This result qualitatively confirms the two-state atomic-orbital calculations of Bivona *et al.*,[114,115] which also gave a field-induced decrease of the transfer cross section.

REFERENCES

1. D. H. Madison and E. Merzbacher, in *Atomic Inner-Shell Processes* edited by B. Crasemann (Academic, New York, 1975), Vol. I, pp. 1-72.
2. P. Richard, in *Atomic Inner-Shell Processes*, edited by B. Crasemann (Academic, New York, 1975), Vol. I, pp. 73-158.
3. U. Wille and R. Hippler, Mechanisms of Inner-Shell Vacancy Production in Slow Ion-Atom Collisions, *Phys. Rep.* **132**, 129-260 (1986).
4. J. S. Briggs, The Excitation of Inner Shells in Slow Atomic Collisions, *Rep. Prog. Phys.* **39**, 217-289 (1976).
5. G. B. Schmid, Transformation Theory and Translational Factors in Inelastic Atomic Collisions, *Phys. Rev. A* **15**, 1459-1468 (1977).
6. G. Ciocchetti, A. Molinari, and R. Malvano, A Method to Determine the Lifetime of a Nuclear State by Measuring the *K*-Shell Ionization Probability, *Nuovo Cimento* **29**, 1262-1274 (1963).
7. P. A. Amundsen, On the Role of Dipole Transitions in the Semiclassical Approximation for *K*-Shell Ionization, *J. Phys. B: At. Mol. Phys.* **11**, 3197-3220 (1978).
8. M. Kleber and K. Unterseer, Dynamical Treatment of Binding, Polarization and Recoil in Asymmetric Ion-Atom Collisions, *Z. Phys. A* **292**, 311-318 (1979).
9. F. Rösel, D. Trautmann, and G. Baur, Recoil Effects in Atomic Inner Shell Ionization, *Nucl. Instrum. Methods* **192**, 43-52 (1982).
10. W. Fritsch, Atomic Orbital Expansion Description for Slow Ion-Atom Collisions: A Curved-Line Trajectory Study, *J. Phys. B: At. Mol. Phys.* **15**, L389-L393 (1982).
11. W. Fritsch and C. D. Lin, in *Electronic and Atomic Collisions*, edited by J. Eichler, I. V. Hertel, and N. Stolterfoht (North-Holland, Amsterdam, 1984), pp. 331-342.
12. D. G. M. Anderson, M. J. Antal, and M. B. McElroy, The Triple-Center Expansion for Charge Exchange in Atomic Scattering Theory, *J. Phys. B: At. Mol. Phys.* **7**, L118-L121 (1974).
13. T. G. Winter and C. D. Lin, Triple-Center Treatment of Ionization in *p*-H Collisions, *Phys. Rev. A* **29**, 3071-3077 (1984).
14. I. M. Cheshire, D. F. Gallaher, and A. J. Taylor, Coupled-State Calculations of Proton-Hydrogen Scattering with a Pseudo-State Expansion, *J. Phys. B: At. Mol. Phys.* **3**, 813-832 (1970).
15. W. Fritsch and C. D. Lin, Close-Coupling Calculations for Inelastic Processes in Intermediate Energy Ion-Atom Collisions, *J. Phys. B: At. Mol. Phys.* **15**, 1255-1268 (1982).

16. H. NAKAMURA AND M. NAMIKI, Semiclassical Theory of Rotationally Induced Non-adiabatic Transitions, *Phys. Rev. A* **24**, 2963-2974 (1981).
17. H. NAKAMURA, Dynamical-State Representation and Nonadiabatic Electronic Transitions in Atomic Collisions, *Phys. Rev. A* **26**, 3125-3136 (1982).
18. S. B. SCHNEIDERMAN AND A. RUSSEK, Velocity-Dependent Orbitals in Proton-on-Hydrogen-Atom Collisions, *Phys. Rev.* **181**, 311-321 (1969).
19. J. S. BRIGGS, Perturbed United-Atom Description of Direct Ionization in Slow Heavy Ion-Atom Collisions, *J. Phys. B: At. Mol. Phys.* **8**, L485-L488 (1975).
20. M. KIMURA AND C. D. LIN, Unified Treatment of Slow Atom-Atom and Ion-Atom Collisions, *Phys. Rev. A* **31**, 590-592 (1985).
21. M. KIMURA AND C. D. LIN, Charge Transfer and Excitation Processes in p-He Collisions Studied Using a Unified Atomic-Orbital-Molecular-Orbital Matching Method, *Phys. Rev. A* **34**, 176-184 (1986).
22. T. G. WINTER AND N. F. LANE, "R-Matrix" Approach to Electron Transfer in α-H Collisions, *Phys. Rev. A* **31**, 2698-2701 (1985).
23. J. BANG AND J. M. HANSTEEN, Coulomb Deflection Effects on Ionization and Pair-Production Phenomena, *Mat. Fys. Medd. Dan. Vid. Selsk.* **31**(13), 1-43 (1959).
24. H. SCHMIDT-BÖCKING, K. E. STIEBING, W. SCHADT, N. LÖCHTER, G. GRUBER, S. KELBCH, AND K. BETHGE, Trajectory, Binding and Relativistic Effects in the K-Shell Ionization Process Investigated by Particle-Photon Coincidences, *Nucl. Instrum. Methods* **192**, 71-77 (1982).
25. D. TRAUTMANN, F. RÖSEL, AND G. BAUR, Accurate Calculation of Inner-Shell Ionization, *Nucl. Instrum. Methods* **214**, 21-27 (1983).
26. L. H. ANDERSEN, P. HVELPLUND, H. KNUDSEN, S. P. MØLLER, K. ELSENER, K.-G. RENSFELT, AND E. UGGERHØJ, Single and Double Ionization of Helium by Fast Antiproton and Proton Impact, *Phys. Rev. Lett.* **57**, 2147-2150 (1986).
27. W. JITSCHIN, this volume, Chapter 6.
28. J. U. ANDERSEN, E. LAEGSGAARD, M. LUND, AND C. D. MOAK, Z_1-Scaling for Impact-Parameter Dependence of Inner-Shell Ionization by Heavy Ions, *Nucl. Instrum. Methods* **132**, 507-515 (1976).
29. J. U. ANDERSEN, E. LAEGSGAARD, AND M. LUND, Impact-Parameter Dependence of K-Shell Ionization, *Nucl. Instrum. Methods* **192**, 79-101 (1982).
30. M. KLEBER AND J. ZWIEGEL, Variational Method for Electron Excitation in Ion-Atom Collisions, *Z. Phys. A* **280**, 137-142 (1977).
31. K. UNTERSEER AND M. KLEBER, Dynamics of Inner-Shell Electrons in Ion-Atom Collisions, *Nucl. Instrum. Methods* **192**, 35-41 (1982).
32. A. JAKOB, D. TRAUTMANN, F. RÖSEL, AND G. BAUR, Wave Function Effects in Inner-Shell Ionization, *Nucl. Instrum. Methods* **232** [B4], 218-226 (1984).
33. J. SEIDEL, S. RÖHL, R. LOREK, S. HUCHLER, AND M. DOST, K-Shell Ionization by ^{16}O and ^{32}S Ions: Reduced-Velocity Dependence of the Binding Effect, *Phys. Rev. A* **32**, 2142-2149 (1985).
34. L. SARKADI AND T. MUKOYAMA, Higher Order Processes in L-Shell Ionization, *Nucl. Instrum. Methods* **232** [*B*4], 296-302 (1984).
35. L. SARKADI, L_3-Subshell Alignment Calculations in the Second-Order Born Approximation for Light- and Heavy-Ion Impact on Au, *J. Phys. B: At. Mol. Phys.* **19**, 2519-2530 (1986).
36. L. SARKADI, Coupled-State Analysis of the Heavy-Ion-Induced L_3-Subshell Alignment, *J. Phys. B: At. Mol. Phys.* **19**, L755-L759 (1986).
37. S. ZEHENDNER, G. B. BAPTISTA, R. DÖRNER, E. JUSTINIANO, J. KONRAD, H. SCHMIDT-BÖCKING, AND R. SCHUCH, Impact Parameter Dependence of the L-Subshell Ionization Probabilities and of the L_3-Subshell Alignment Tensor Components in 4 MeV p-Sm Collisions, *Z. Phys. D At. Mol. Clusters* **4**, 243-247 (1987).

38. K. DEXHEIMER, J. ULLRICH, K. E. STIEBING, W. SCHADT, S. KELBCH, C. KELBCH, R. SCHUCH, S. ZEHENDNER, AND H. SCHMIDT-BÖCKING, Experimental Evidence for Vacancy Sharing between *L* Subshells in Very Asymmetric Ion-Atom Collisions, *J. Phys. B: At. Mol. Phys.* **19**, 3083-3090 (1986).
39. H. SCHMIDT-BÖCKING, J. ULLRICH, R. DÖRNER, K. DEXHEIMER, S. KELBCH, V. DANGENDORF, R. SCHUCH, S. ZEHENDNER, S. HAGMANN, AND G. B. BAPTISTA, *L*-Subshell Ionization Probabilities in Very Asymmetric Ion-Atom Collisions, *Nucl. Instrum. Methods B* **24/25**, 64-68 (1987).
40. W. E. MEYERHOF AND K. TAULBJERG, *K*-Shell Ionization in Heavy-Ion Collisions, *Ann. Rev. Nucl. Sci.* **27**, 279-332 (1977).
41. K. TAULBJERG, J. S. BRIGGS, AND J. VAABEN, Inner-Shell Excitation in Heteronuclear Collisions III. A Scaling Law for Differential and Total Cross Sections for *K*-Shell Excitation, *J. Phys. B: At. Mol. Phys.* **9**, 1351-1371 (1976).
42. N. LUZ, S. SACKMANN, AND H. O. LUTZ, Impact-Parameter Dependence of *K*-Shell Excitation in Slow Ion-Atom Collisions, *J. Phys. B: At. Mol. Phys.* **12**, 1973-1993 (1979).
43. R. FREKERS AND H. SCHULZE, private communication (1986).
44. B. THIES, B. FRICKE, W.-D. SEPP, AND H. HARTUNG, New Calculations of $P(b)$ Curves for $1s\sigma$ Excitation in Low-Z ($Z \leq 10$) Ion-Atom Scattering, *J. Phys. B: At. Mol. Phys.* **19**, L617-L621 (1986).
45. R. SHANKER, R. BILAU, R. HIPPLER, U. WILLE, AND H. O. LUTZ, Impact Parameter Dependence of Ar *L*-Shell Excitation in Slow Ar-Ar Collisions, *J. Phys. B: At. Mol. Phys.* **14**, 997-1007 (1981).
46. R. SHANKER, R. HIPPLER, U. WILLE, R. BILAU, AND H. O. LUTZ, $3d\sigma$ Excitation in Slow Ne-Ar Collisions, *Z. Phys. A Atoms Nuclei* **313**, 281-288 (1983).
47. R. BILAU, W. R. MCMURRAY, U. WILLE, R. SHANKER, R. HIPPLER, AND H. O. LUTZ, $3d\sigma$ Excitation in Slow Ar-Kr Collisions, *Z. Phys. D At. Mol. Clusters* **1**, 39-46 (1986).
48. G. PRESSER, J. STÄHLER, R. WERNER, AND U. WILLE, Impact Parameter Dependence of *L*-Shell Excitation in 17 to 40 MeV Ag + Ag and Ag + Cs Collisions, *J. Phys. B: At. Mol. Phys.* **16**, 197-214 (1983).
49. E. MORENZONI, M. NESSI, P. BÜRGY, CH. STOLLER, AND W. WÖLFLI, Impact Parameter Dependence of *L*- and *K*-Shell Excitation in I-Ag Collisions, *Z. Phys. A At. Nuclei* **311**, 7-17 (1983).
50. R. SHANKER, R. HIPPLER, U. WILLE, R. BILAU, AND H. O. LUTZ, Evidence for Rotationally Induced $4f\sigma$ Excitation in Slow Kr-Xe and Kr-Kr Collisions, *J. Phys. B: At. Mol. Phys.* **15**, L495-L500 (1982).
51. R. SHANKER, U. WILLE, R. BILAU, R. HIPPLER, W. R. MCMURRAY, AND H. O. LUTZ, $4f\sigma$ Excitation in Slow Kr-Kr and Kr-Xe Collisions, *J. Phys. B: At. Mol. Phys.* **17**, 1353-1371 (1984).
52. R. FREKERS, H. SCHULZE, U. WILLE, AND B. CLEFF, Impact Parameter Dependence of *L*-Shell Vacancy Production in Slow Kr-Kr and Xe-Xe Collisions, *Z. Phys. D At. Mol. Clusters* **6**, 131-144 (1987).
53. A. NIEHAUS, in *Atomic Inner-Shell Physics*, edited by B. Crasemann (Plenum, New York, 1985), pp. 377-416.
54. P. H. MOKLER, D. H. H. HOFFMANN, W. A. SCHÖNFELDT, D. MAOR, W. E. MEYERHOF, AND Z. STACHURA, Atomic Collision Studies at Moderate Projectile Velocities Using Highly Charged, Decelerated Heavy Ions from the GSI-Unilac, *Nucl. Instrum. Methods* **232** [B4], 34-39 (1984).
55. R. SCHUCH, H. INGWERSEN, E. JUSTINIANO, H. SCHMIDT-BÖCKING, M. SCHULZ, AND F. ZIEGLER, Interference Effects in *K* Vacancy Transfer of Hydrogen-Like S Ions Colliding with Ar, *J. Phys. B: At. Mol. Phys.* **17**, 2319-2338 (1984).

56. W. LICHTEN, Resonant Charge Exchange in Atomic Collisions. II. Further Applications and Extensions to the Quasi-Resonant Case, *Phys. Rev.* **139**, A27–A34 (1965).
57. S. HAGMANN, C. L. COCKE, J. R. MCDONALD, P. RICHARD, H. SCHMIDT-BÖCKING, AND R. SCHUCH, Quasiresonant Charge Transfer in Inner-Shell Excitation: Impact Parameter Dependence of *K*-Vacancy Creation in $F^{q+} \rightarrow$ Ne Collisions, *Phys. Rev. A* **25**, 1918–1929 (1982).
58. N. STOLTERFOHT, Model Calculations of Charge Exchange between *K* Shells in Near-Symmetric Ion–Atom Collisions, *J. Phys. B: At. Mol. Phys.* **16**, 2385–2397 (1986).
59. N. STOLTERFOHT, in *Progress in Atomic Spectroscopy, Part D*, edited by H. J. Beyer and H. Kleinpoppen (Plenum, New York, 1987), pp. 415–476.
60. N. GRÜN, private communication to R. Schuch *et al.* (as cited in Ref. 55).
61. W. FRITSCH AND C. D. LIN, Close-Coupling Study of *K*-Shell Vacancy Production in Near-Symmetric Collisions, *Phys. Rev. A* **31**, 1164–1167 (1985).
62. S. HAGMANN, C. L. COCKE, P. RICHARD, A. SKUTLARTZ, S. KELBCH, H. SCHMIDT-BÖCKING, AND R. SCHUCH, in *Electronic and Atomic Collisions*, edited by J. Eichler, I. V. Hertel, and N. Stolterfoht (North-Holland, Amsterdam, 1984), pp. 385–392.
63. M. SCHULZ, E. JUSTINIANO, J. KONRAD, R. SCHUCH, AND A. SALIN, *K*-Shell to *K*-Shell Charge Transfer in Collisions of Bare, Decelerated S Ions with Ar, *J. Phys. B: At. Mol. Phys.* **20**, 2057–2073 (1987).
64. R. ANHOLT, X-Rays from Quasimolecules, *Rev. Mod. Phys.* **57**, 995–1053 (1985).
65. W. FRITSCH AND U. WILLE, Theory of Collision Broadening in Molecular-Orbital X-Ray Spectra, *Jpn. J. Appl. Phys. Suppl.* **17-2**, 387–392 (1978).
66. W. FRITSCH AND U. WILLE, On the Theory of Molecular-Orbital X-Rays, *J. Phys. B: At. Mol. Phys.* **12**, L335–L340 (1979).
67. A. Z. DEVDARIANI, U. N. OSTROSKII, AND A. NIEHAUS, *Analytical Formulae for Spectral Shapes Arising from Emission during Atom–Atom Collisions, J. Phys. B: At. Mol. Phys.* **18**, L161–L166 (1985).
68. I. TSERRUYA, R. SCHUCH, H. SCHMIDT-BÖCKING, J. BARRETTE, WANG DA-HAI, B. M. JOHNSON, K. W. JONES, AND M. MERON, Interference Effects in the Quasimolecular *K* X-Ray Production Probability for 10 MeV Cl^{16+}-Ar Collisions, *Phys. Rev. Lett.* **50**, 30–33 (1983).
69. R. SCHUCH, H. SCHMIDT-BÖCKING, I. TSERRUYA, B. M. JOHNSON, K. W. JONES, AND M. MERON, X-Ray Spectroscopy of Cl–Ar Quasi-Molecular Orbitals from $1s\sigma$–$2p\pi$ Transitions, *Z. Phys. A At. Nuclei* **320**, 185–189 (1985).
70. R. SCHUCH, R. HOFFMANN, B. M. JOHNSON, K. W. JONES, M. MERON, H. SCHMIDT-BÖCKING, AND I. TSERRUYA, Quasimolecular X-Ray Spectroscopy for Slow Cl^{16+}-Ar Collisions, *Phys. Rev. A* **37** (1988), in press.
71. D. LIESEN, A. N. ZINOVIEV, AND F. W. SARIS, Oscillations in Coincident Molecular-Orbital Auger Spectra Emitted in 450-keV Kr^{+4}-Kr Collisions, *Phys. Rev. Lett.* **47**, 1392–1395 (1981).
72. V. V. AFROSIMOV, G. G. MESKHI, N. N. TSAREV, AND A. P. SHERGIN, Auger Spectroscopy of Quasimolecules, *Zh. Eksp. Teor. Fiz.* **84**, 454–465 (1983) [*Sov. Phys. JETP* **57**, 263–269 (1983)].
73. A. P. SHERGIN, R. STÖTZEL, Z. ROLLER, R. BILAU, AND H. O. LUTZ, Impact-Energy Dependence of Quasimolecular Auger Emission in Kr–Kr Collisions, *Phys. Rev. A* **34**, 4490–4492 (1986).
74. G. CIOCCHETTI AND A. MOLINARI, *K* Electron Shell Ionization and Nuclear Reactions, *Nuovo Cimento* **40B**, 69–86 (1965).
75. W. E. MEYERHOF, in *Electronic and Atomic Collisions*, edited by J. Eichler, I. V. Hertel, and N. Stolterfoht (North-Holland, Amsterdam, 1984), pp. 31–49.

76. W. E. MEYERHOF AND J.-F. CHEMIN, Nuclear Reaction Effects on Atomic Inner-Shell Ionization, *Adv. At. Mol. Phys.* **20**, 173-239 (1985).
77. R. ANHOLT, in *Atomic Inner-Shell Physics*, edited by B. Crasemann (Plenum, New York, 1985), pp. 581-625.
78. J. S. BLAIR, P. DYER, K. A. SNOVER, AND T. A. TRAINOR, Nuclear "Time-Delay" and X-Ray-Proton Coincidences near a Nuclear Scattering Resonance, *Phys. Rev. Lett.* **41**, 1712-1715 (1978).
79. W. DUINKER, J. VAN ECK, AND A. NIEHAUS, Experimental Evidence for the Influence of Inner-Shell Ionization on Resonant Nuclear Scattering, *Phys. Rev. Lett.* **45**, 2102-2105 (1980).
80. J. F. CHEMIN, R. ANHOLT, CH. STOLLER, W. E. MEYERHOF, AND P. A. AMUNDSEN, Measurement of ^{88}Sr K-Shell Ionization Probability across the Nuclear Elastic-Scattering Resonance at 5060 keV, *Phys. Rev. A* **24**, 1218-1222 (1981).
81. J. F. CHEMIN, W. E. MEYERHOF, R. ANHOLT, J. D. MOLITORIS, AND CH. STOLLER, Measurement of the K-Shell Ionization Probability across a Wide Resonance: ^{88}Sr (p, p_0) at 6.06 MeV, *Phys. Rev. A* **26**, 1239-1242 (1982).
82. M. DOST, R. LOREK, S. RÖHL, J. SEIDEL, AND W. KOENIG, K-Shell Ionization Probability in Elastic Proton Scattering on ^{138}Ba through an $f_{7/2}$ Isobaric-Analog Resonance, *Phys. Rev. A* **32**, 2077-2087 (1985).
83. D. W. SPOONER, CH. STOLLER, J.-F. CHEMIN, W. E. MEYERHOF, J.-N. SCHEURER, AND X.-Y. XU, Large Resonance Effect in K-Shell Ionization Probability in Elastic Proton Backscattering on ^{138}Ba, *Phys. Rev. Lett.* **58**, 341-344 (1987).
84. J. N. SCHEURER, O. K. BAKER, AND W. E. MEYERHOF, Large Angle Scattering and Nuclear Resonance Effect in Electron Capture in H^+-C and H^+-N Collisions, *J. Phys. B: At. Mol. Phys.* **18**, L85-L89 (1985).
85. E. HORSDAL, B. JENSEN, AND K. O. NIELSEN, Experimental Study of Charge Transfer near a Nuclear-Scattering Resonance, *Phys. Rev. Lett.* **57**, 675-678 (1986).
86. R. ANHOLT, Electronic $1s\sigma$ Vacancy Production in Deep-Inelastic Nuclear Reactions, *Phys. Lett.* **88B**, 262-264 (1979).
87. U. MÜLLER, J. REINHARDT, G. SOFF, B. MÜLLER, AND W. GREINER, K-Vacancy Production in Deep-Inelastic Nuclear Reactions, *Z. Phys. A At. Nuclei* **297**, 357-358 (1980).
88. CH. STOLLER, J. F. CHEMIN, R. ANHOLT, W. E. MEYERHOF, AND W. WÖLFLI, X-Ray Study of the Deep Inelastic Reactions 1160 MeV ^{136}Xe + Pb and ^{136}Xe + ^{232}Th, *Z. Phys. A At. Nuclei* **310**, 9-17 (1983).
89. CH. STOLLER, M. NESSI, E. MORENZONI, W. WÖLFLI, W. E. MEYERHOF, J. D. MOLITORIS, E. GROSSE, AND CH. MICHEL, Nuclear Reaction Times in the Deep-Inelastic U + U Collision Deduced from K-Shell Ionization Probabilities, *Phys. Rev. Lett.* **53**, 1329-1332 (1984).
90. J. D. MOLITORIS, Nuclear Time Delay Effects on K-Vacancy Production in Deep-Inelastic U + U Collisions, Ph.D. thesis, Stanford University (1986).
91. H. BACKE, P. SENGER, W. BONIN, E. KANKELEIT, M. KRÄMER, R. KRIEG, V. METAG, N. TRAUTMANN, AND J. B. WILHELMY, Estimates of the Nuclear Time Delay in Dissipative U + U and U + Cm Collisions Derived from the Shape of Positron and δ-Ray Spectra, *Phys. Rev. Lett.* **50**, 1838-1841 (1983).
92. J. REINHARDT, B. MÜLLER, W. GREINER, AND G. SOFF, Delta Electrons: An Atomic Clock for Deep-Inelastic Collisions, *Z. Phys. A* **292**, 211-212 (1979).
93. G. SOFF, J. REINHARDT, B. MÜLLER, AND W. GREINER, Delta-Electron Emission in Deep-Inelastic Heavy-Ion Collisions, *Phys. Rev. Lett.* **43**, 1981-1984 (1979).
94. P. SENGER, H. BACKE, M. BEGEMANN-BLAICH, H. BOKEMEYER, P. GLÄSSEL, D. v. HARRACH, M. KLÜVER, W. KONEN, K. POPPENSIEKER, K. STIEBING, J. STROTH, AND K. WALLENWEIN, Nuclear Contact Times in Dissipative Heavy Ion Collisions Measured via δ-Ray Spectroscopy, Preprint No. GSI-86-51 (1986).

95. J. KIRSCH, B. MÜLLER, AND W. GREINER, MO-Radiation Interference Phenomena as a "Clock" for Nuclear Reaction Times, *Z. Phys. D At. Mol. Clusters* **1**, 47–50 (1986).
96. *Atomic and Molecular Physics Close to Ionization Thresholds in High Fields*, edited by J. P. Connerade, J. C. Gay, and S. Liberman, *J. Phys. (Paris)* **43**, *C*2 (1982).
97. G. FERRANTE, in *Fundamental Processes in Energetic Atomic Collisions*, edited by H. O. Lutz, J. S. Briggs, and H. Kleinpoppen (Plenum, New York, 1983), pp. 585–610.
98. G. FERRANTE, in *Fundamental Processes in Atomic Collision Physics*, edited by H. Kleinpoppen, J. S. Briggs, and H. O. Lutz (Plenum, New York, 1985), pp. 343–395.
99. M. R. C. MCDOWELL AND M. ZARCONE, Scattering in Strong Magnetic Fields, *Adv. At. Mol. Phys.* **21**, 255–304 (1985).
100. H. A. BETHE AND E. E. SALPETER, *Quantum Mechanics of One- and Two-Electron Atoms* (Academic, New York, 1957).
101. D. PARK, *Relation between the Parabolic and Spherical Eigenfunctions of Hydrogen, Z. Phys.* **159**, 155–157 (1960).
102. H. J. ANDRÄ, in *Progress in Atomic Spectroscopy*, Part B, edited by W. Hanle and H. Kleinpoppen (Plenum, New York, 1979), pp. 829–953.
103. E. H. PINNINGTON, H. G. BERRY, J. DESESQUELLES, AND J. L. SUBTIL, Stark Effect Modulations in He II and Li III, *Nucl. Instrum. Methods* **110**, 315–320 (1973).
104. C. F. MOORE, W. J. BRAITHWAITE, AND D. L. MATTHEWS, Electric Field Induced Structure in X-Ray Yields from Highly Ionized Oxygen Atoms, *Phys. Lett.* **47A**, 353–354 (1974).
105. P. RICHARD, C. L. COCKE, S. J. CZUCHLEWSKI, K. A. JAMISON, R. L. KAUFFMAN, AND C. W. WOODS, Observation of Stark Beats in the H-Like *K* X Rays of Oxygen and Fluorine, *Phys. Lett.* **47A**, 355–356 (1974).
106. Z. VAGER, E. P. KANTER, D. SCHNEIDER, AND D. S. GEMMELL, Coherent Stark Levels in Beam-Foil-Excited Fast Rydberg Atoms, *Phys. Rev. Lett.* **50**, 954–957 (1983).
107. E. P. KANTER, D. SCHNEIDER, Z. VAGER, D. S. GEMMELL, B. J. ZABRANSKY, GU YUAN-ZHUANG, P. ARCUNI, P. M. KOCH, D. R. MARIANI, AND W. VAN DE WATER, Ionization of Fast Foil-Excited Ion Beams in Electromagnetic Fields, *Phys. Rev. A* **29**, 583–594 (1984).
108. D. SCHNEIDER, W. ZEITZ, R. KOWALLIK, G. SCHIWIETZ, T. SCHNEIDER, N. STOLTERFOHT, AND U. WILLE, Effects of External Electric Fields on High Rydberg States Formed in Foil and Gas Interactions of 85-MeV Ne^{6+} Ions, *Phys. Rev. A* **34**, 169–175 (1986).
109. R. KOWALLIK, Messungen von Quanteninterferenzen an schnellen Rydbergionen in elektrischen Feldern, Diploma thesis, Free University of Berlin (1986).
110. W. ZEITZ, R. KOWALLIK, AND D. SCHNEIDER, unpublished.
111. T. SEIDEMAN, M. SHAPIRO, AND Z. VAGER, Theory of the Coherent Decay of High-Lying Rydberg States in Beam-Foil Encounters, *Phys. Rev. A* **35**, 87–102 (1987).
112. R. H. GARSTANG, Atoms in High Magnetic Fields, *Rep. Prog. Phys.* **40**, 105–154 (1977).
113. D. MOSS, Magnetic Fields in Stars, *Phys. Rep.* **140**, 1–74 (1986).
114. S. BIVONA, B. SPAGNOLO, AND G. FERRANTE, in *Electronic and Atomic Collisions* (*Abstracts of XIII ICPEAC*) edited by J. Eichler, W. Fritsch, I. V. Hertel, N. Stolterfoht, and U. Wille (North-Holland, Amsterdam, 1983), p. 692.
115. S. BIVONA, B. SPAGNOLO, AND G. FERRANTE, Charge Transfer in the Presence of a Strong Magnetic Field, *J. Phys. B: At. Mol. Phys.* **17**, 1093–1106 (1984).
116. T. P. GROSDANOV AND M. R. C. MCDOWELL, A Classical Trajectory Monte Carlo Study of Collisions of He^{2+} with H in a Strong Magnetic Field, *J. Phys. B: At. Mol. Phys.* **18**, 921–931 (1985).
117. S. BIVONA AND M. R. C. MCDOWELL, Asymmetric Charge Exchange in a Strong Static Magnetic Field, *J. Phys. B: At. Mol. Phys.* **20**, 1541–1554 (1987).

118. U. WILLE, in *Fundamental Processes in Atomic Collision Physics*, edited by H. Kleinpoppen, J. S. Briggs, and H. O. Lutz (Plenum, New York, 1985) pp. 719–727.
119. U. WILLE, Vibrational and Rotational Properties of the H_2^+ Molecular Ion in a Strong Magnetic Field, *J. Phys. B: At. Mol. Phys.* **20**, L417–L422 (1987).
120. U. WILLE, Resonant Charge Transfer in Slow H^+–H Collisions in the Presence of a Strong Magnetic Field, *Phys. Lett.* **125A**, 52–56.

CHAPTER 8

Coherence in Grazing Ion–Surface Collisions

H. Winter and R. Zimny

1. INTRODUCTION

Coherence in atomic collisions is closely related to symmetry and selection of parameters that specify the collision process like impact parameter, plane of scattering, and final target states. In general, coincidence methods are used in the experiments to achieve these conditions, and in most cases a large degree of coherence in collisional excited terms is observed. A number of studies on this topic are discussed throughout this volume. This type of (time-consuming) experiment provides detailed information on the interaction processes and exhibit sensitive tests of theoretical approaches.

In ion–solid interactions coherence was first observed after beam-foil excitation[1] where an "alignment" in the spatial distribution of orbital angular momenta of electronic terms excited by a thin carbon foil was found. Symmetry is closely related to the nature of anisotropy so that the alignment is a consequence of the axial symmetry in the "classical" beam-foil geometry. Berry and co-workers[2] subsequently showed that the excitation process has to be located at the surface; by tilting the foil with respect to the beam axis, which means a violation of the axial symmetry in the collision,[3] an "orientation" of excited terms occurs.

Since this orientation monotonically increases with the angle of tilt of the foil, the largest anisotropy is obtained at angles that are not feasible in foil transmission experiments. Consequently Andrä[4] scattered fast ions from a solid surface at grazing incidence and observed a large anisotropy in excited terms. In Figure 1 we have sketched the typical geometry of grazing ion–surface collisions.

H. Winter and R. Zimny • Institute for Nuclear Physics, University of Münster, D-4400 Münster, West Germany.

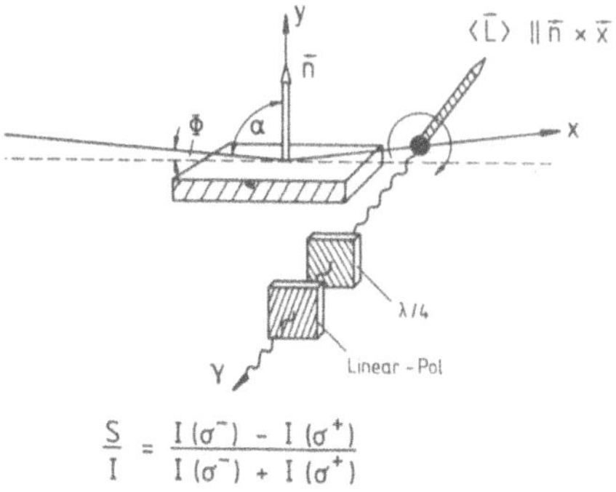

FIGURE 1. Geometry in grazing ion-surface collisions.

The coherence of excited terms obtained by ion-surface interaction is probed by the detection of elliptically polarized light emitted in transitions from excited terms. Since the anisotropy mainly results in an orientation, the polarized light is basically described by the Stokes parameter S/I, which refers to circular polarization by $S/I = [I(\sigma^-) - I(\sigma^+)]/[I(\sigma^-) + I(\sigma^+)]$, where $I(\sigma^-)$ and $I(\sigma^+)$ are the intensity of light with negative and positive helicity, respectively.

That a fairly large coherence can be obtained in these collisions is demonstrated by data for the interaction of 230-keV Ar^+ ions with a Si(111) surface at a grazing angle $\Phi_{in} \approx 0.2°$.[5] In Figure 2 a spectral scan of $I(\sigma^-)$, $I(\sigma^+)$ over the Ar II $4s'\,^2D$–$4p'\,^2F$ multiplet transition is shown. The pronounced difference between both intensities corresponds to $S/I = (95 \pm 1)\%$, which is to our knowledge the highest polarization observed after the excitation by a direct collision.

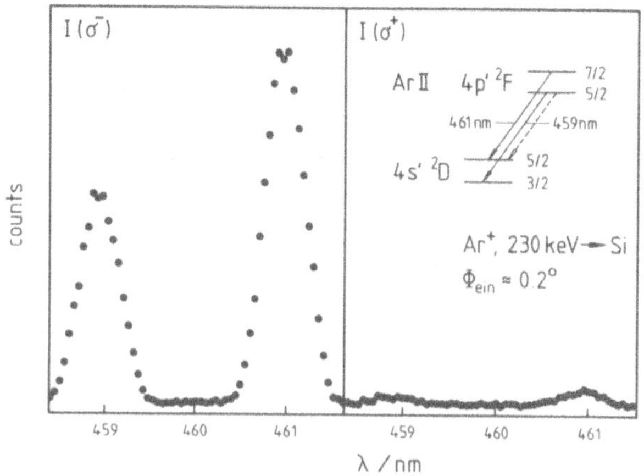

FIGURE 2. Intensities $I(\sigma^-)$ and $I(\sigma^+)$ of the Ar II $4s'\,^2D$–$4p'\,^2F$ multiplet transitions after the interaction of 230-keV Ar^+ ions with a Si(111) surface at $\Phi_{in} \approx 0.2°$.

This large polarization can be used to perform high-resolution atomic spectroscopy and to produce polarized beams.[31] In this chapter we will concentrate on the discussion of mechanisms that cause the large coherence observed after grazing ion-surface scattering. Such a discussion needs a detailed study of the interaction processes at surfaces in general so that we first have to consider aspects of scattering and charge exchange at surfaces before we discuss polarization phenomena.

We will learn that grazing ion-surface collisions take place in an extreme geometry that contains some interesting features with respect to ion-surface interaction and collision physics in general. Adopting well-established concepts in the description of atom-surface scattering to the geometry of grazing ion-surface interactions results in some attractive features with respect to experiment and theory, which also gives a description of coherence phenomena.

2. PLANAR CHANNELING OF IONS AT A SOLID SURFACE

Grazing scattering of ions with solid surfaces can be described in terms of a planar channeling in the collective potential of screened surface atoms. The geometry of this process is sketched in Figure 3. A crucial consequence of the interaction with the solid in this extreme geometry, i.e., grazing angles of the order of (a few tenths of) a degree, is a pronounced difference in motion parallel and normal to the surface.

At a grazing angle Φ we have at an energy of the projectiles E_0 for the motion parallel to the surface $E_x = E_0 \cos^2 \Phi \approx E_0$, whereas the motion normal to the surface is given by $E_y = E_0 \sin^2 \Phi$ so that, e.g., for $\Phi = 0.2°$ E_x and E_y differ by five orders of magnitude ($\sin^2 \Phi \approx 10^{-5}$). This property of the scattering implies interesting aspects with respect to the ion-surface

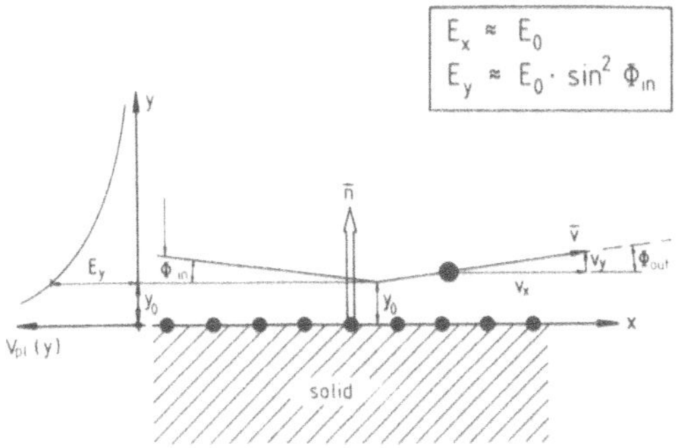

FIGURE 3. Scattering of ions in the collective planar potential at a solid surface.

interaction, since at projectile energies in the keV domain the scattered atoms escape from the surface at eV energies.

The trajectory of scattered projectiles results from planar channeling in the collective potential[6]:

$$V_{pl} = 2\pi Z_1 Z_2 e^2 a_F N_s \cdot f_{pl}(y/a_F)$$

where the screening is taken into account by the Moliere approximation[7]

$$f_{pl} = 1.17 \exp\left(-0.3 \frac{y}{a_F}\right) + 0.46 \exp\left(-1.2 \frac{y}{a_F}\right) + 0.02 \exp\left(-6 \frac{y}{a_F}\right)$$

and

$$a_F = 0.885 a_0 (Z_1^{1/2} + Z_2^{1/2})^{-2/3}$$

is the "Firsov screening parameter"[8] for a projectile–target combination Z_1, Z_2; $a_0 = 0.0529$ nm is the Bohr radius and N_s is the planar density of atoms at the surface.

During the process of scattering in this potential the energy $E_y = E_0 \sin^2 \Phi$ determines the distance of closest approach y_0 according to $V_{pl}(y_0) = E_y$. In the case $y_0 > a_F$, collisions with individual surface atoms play a minor role, and the conditions for channeling are met below critical energies

$$E_{yc} = 2\pi Z_1 Z_2 e^2 a_F N_s$$

and critical angles

$$\Phi_c = (E_{yc}/E_0)^{1/2}$$

In Figure 4 we plot the planar potential $V_{pl}(y)$ for the interaction of H^+ ions with a Ni(111) surface and indicate for a given E_y a distance of closest approach y_0. In general this distance exceeds a_F so that for an ideally flat surface a well-defined backscattered beam should be observed.

A distribution of 200-keV H^+ ions scattered off a Si(111) surface is shown in Figure 5 and may serve as a demonstration that we are able to closely approach the ideal conditions in an actual experiment. The data in Figure 5 display the detector signal in dependence on the angle of scattering; the two peaks stem from part of the direct beam that passes the target without interaction and the scattered beam. At an angle of scattering of about 0.2° we achieve a half width in the distribution of 0.26°. This width has to be considered as an upper limit, since no deconvolution with respect to the divergence of the projectile beam and the length of the target is performed.

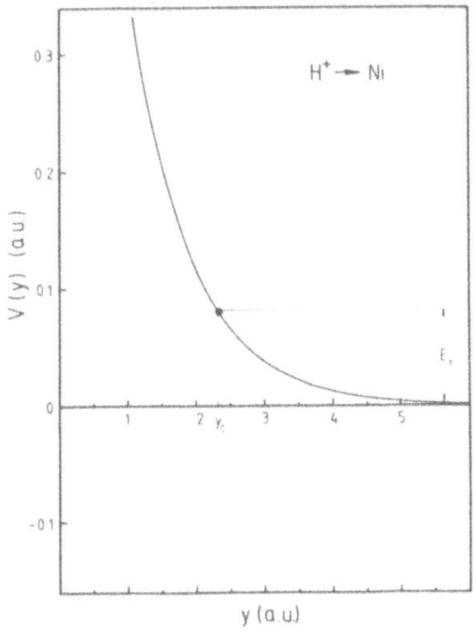

FIGURE 4. Planar potential for H^+-Ni(111).

The experiment clearly shows that grazing ion-surface collisions are characterized by well-defined trajectories. These trajectories are located in a plane of scattering given by the direction of the incoming beam and the surface normal. Therefore these ion-surface collisions take place in a well-defined plane of scattering, and in contrast to ion-atom collisions no further selection of this plane has to be performed by time-consuming coincidence methods.

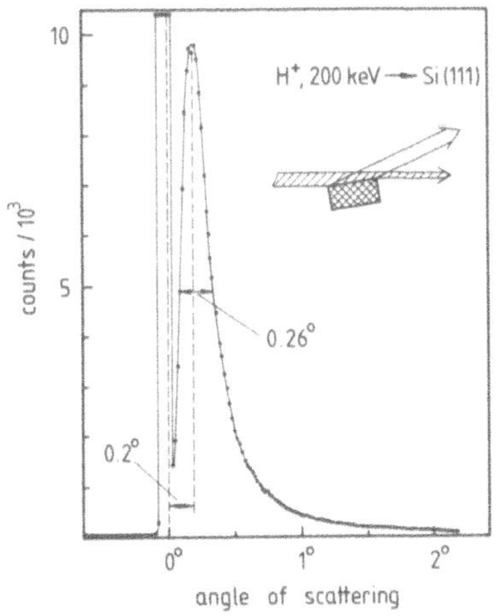

FIGURE 5. Scattering distribution of 200-keV H^+ ions after the interaction with a Si(111) surface.

Ohtsuki *et al.*[9] pointed out that the interaction of the scattered ion with its own image charge adds an attractive component to the planar potential V_{pl} which modifies the effective potential. This effective potential is attractive down to small distances to the surface and shows a pronounced repulsive behavior close to the surface. As a result the distance of closest approach y_0 varies only slightly with E_y, whereas for a pure planar potential V_{pl} this variation is dramatic. An indication for this type of effective potential may be seen in experiments with projectiles of low energies, where we find significantly broader scattering distributions than at higher energies (see Figure 5). However, further experimental tests are needed to probe the effects of the image charge on the scattering process.

The planar channeling at surfaces implies no penetration of the projectiles into the bulk of the solid. As a consequence the energy of backscattered ions remains rather well defined. Relative energy loss and straggling are observed to amount to about 1% or even less.[20]

3. ELECTRONIC TRANSFER PROCESSES

In the preceding section we discussed the trajectory of a projectile scattered at grazing incidence and found that the velocity component normal to the surface is significantly smaller than the velocity of the projectile. We neglect for the moment the (high) velocity component of the motion along the surface and describe the electronic processes by the well-established approaches of atom–solid interaction at hyperthermal energies.[10]

In Figure 6 we consider the perturbation of an atom by a metallic surface at a distance R. The image charges of electron and ion core cause an additional potential term H_p in the Hamiltonian

$$H = H_a + H_p = H_a + Z/D - 1/4d \tag{1}$$

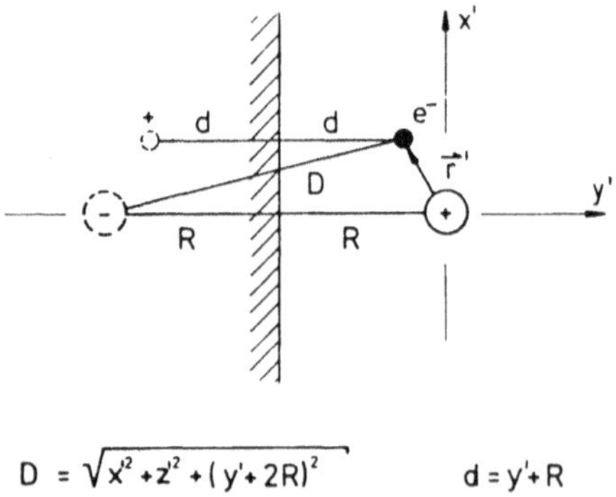

$$D = \sqrt{x'^2 + z'^2 + (y' + 2R)^2} \qquad d = y' + R$$

FIGURE 6. Atom–metal interaction.

where H_a represents the Hamiltonian of the unperturbed atom, Z is the charge of the ion core, and D and d are given in the figure.

Expanding the perturbation H_p in powers of $1/R$, we find

$$H_p = \frac{2Z-1}{4R} + \frac{(1-Z)}{4R^2} y' + O(1/R^3) \tag{2}$$

For sufficiently large distances ($R \gg y'$) we deduce from equation (2) an energy shift of an atomic level

$$\Delta E_a = \langle \psi_a | H_p | \psi_a \rangle \approx (2Z-1)/4R \tag{3}$$

and a dipole electric image field directed along y, y'

$$F_y = (1-Z)/4R^2 \tag{4}$$

Thus the presence of a metal surface in the vicinity of atoms ($Z = 1$) and ions ($Z \geqslant 2$) leads to a positive shift in energy, whereas the energy of affinity levels of negative ions ($Z \leqslant 0$) is lowered. The electric image field given by equation (4) vanishes for neutral atoms because of screening of the ion core by the valence electron, but occurs for each sort of ion ($Z \neq 1$) in the vicinity of the surface.

Well-established electron transfer processes at surfaces are sketched in Figure 7, indicating the relevance of the binding energy of an atomic term E_B with respect to work function W and Fermi energy ε_F of the solid.

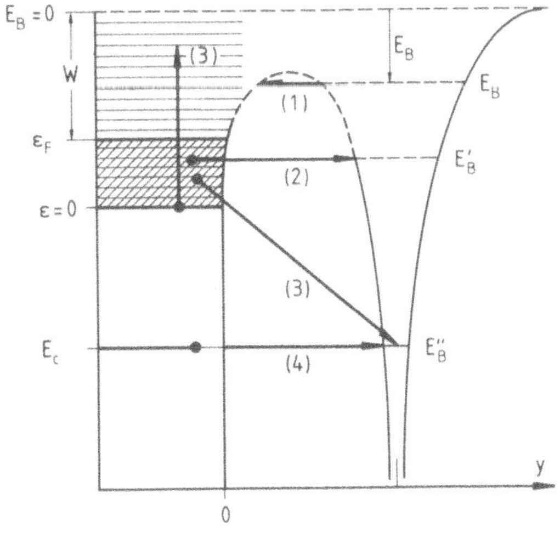

FIGURE 7. Electron transfer processes at surfaces.

Depending on these energies, resonant ionization (1), resonant neutralization (2), Auger-type processes (3), and core interactions (4) will occur. Furthermore, violent atomic collisions with single surface atoms may play a role.[10a]

We limit our discussion to one-electron resonant transfer processes, which we describe by a Pauli master equation[11]

$$\frac{d}{dt} P_n = \sum_{\mathbf{k}} [P_{\mathbf{k}} \cdot W_{\mathbf{k}n}(y) - P_n W_{n\mathbf{k}}(y)] \tag{5}$$

where P_n is the occupation probability of an atomic state $|n\rangle$, $P_{\mathbf{k}}$ the occupation of an electronic state $|\mathbf{k}\rangle$ of the solid, and

$$W_{\mathbf{k}n}(y) = W_{n\mathbf{k}}(y) = 2\pi|\langle\mathbf{k}|V(y)|n\rangle|^2\, \delta(E_n - E_{\mathbf{k}}) \tag{6}$$

is the transfer rate given by a "golden-rule" type expression;[10] the δ function in equation (6) accounts for static resonance between both states. An implicit assumption of this concept of resonance tunneling ("under-barrier transition") between bound states is the existence of a potential barrier between solid and atomic level $|n\rangle$.

Assuming thermodynamical equilibrium for the conduction electrons at all times $P_{\mathbf{k}}$ is given by the Fermi–Dirac distribution $f_{\mathbf{k}}$, and we write equation (5) by

$$\frac{d}{dt} P_n = \Delta_n(y)[\Delta_n^0(y)/\Delta_n(y) - P_n] \tag{7}$$

with the total transition rate

$$\Delta_n(y) = \sum_{\mathbf{k}} W_{n\mathbf{k}}(y) \tag{8}$$

and the neutralization rate

$$\Delta_n^0(y) = \sum_{\mathbf{k}} f_{\mathbf{k}} W_{n\mathbf{k}}(y) \tag{9}$$

In equation (7) the term $P_n^e = \Delta_n^0/\Delta_n$ is the probability of occupation of an atomic state in equilibrium ($dP_n/dt = 0$). For a solution of equation (7) we assume a classical trajectory with $y = y_0 + v_y|t|$. Since the final state formation takes place at the outgoing part of the trajectory,[12] we get for the final population of an atomic state

$$\begin{aligned} P_n(\infty) &= \int_{y_0}^{\infty} dy\, P_n^e(y) \frac{\Delta_n(y)}{v_y} \exp\left[-\frac{1}{v_y}\int_y^{\infty} \Delta_n(y')\, dy'\right] \\ &= \int_{y_0}^{\infty} dy\, P_n^e(y) \cdot F_n(y) \end{aligned} \tag{10}$$

Thus $P_n(\infty)$ is given by a weighting procedure of the equilibrium occupation probability $P_n^e(y)$ with the function $F_n(y)$ over the exit part of the trajectory. The factor $\Delta_n(y)\,dy/v_y$ of $F_n(y)$ represents the probability that a transition between solid and projectile with velocity v_y occurs between y and $y + dy$, whereas the second factor gives the probability that this atomic state will survive on the remaining part of the trajectory. So far the spin degree of freedom of electrons (or more generally the degeneracy of atomic levels) has been disregarded. To approximate the effect of the spin-degeneracy we define from equations (8) and (9) the ionization rate

$$\Delta_n^-(y) = \Delta_n(y) - \Delta_n^0(y) \tag{10a}$$

to write equation (7) in the form

$$\frac{d}{dt} P_n = \Delta_n^0(y)(1 - P_n) - \Delta_n^-(y) P_n \tag{10b}$$

For neutralization the final atomic level is twofold degenerate (s level) and the initial $|k\rangle$ states are doubly occupied. In conclusion $\Delta_n^0(y)$ has to be multiplied by 2, whereas the ionization rate remains unchanged, and we obtain

$$P_n(\infty) = \int_{y_0}^{\infty} N_n^e(y) F_n^*(y) \tag{10c}$$

with the "equilibrium" neutral fraction

$$N_n^e(y) = 2P_n^e(y)/[1 + P_n^e(y)] \tag{10d}$$

and the modified weighing function

$$F_n^*(y) = \frac{\Delta_n^0(y) + \Delta_n(y)}{v_y} \exp\left\{-\frac{1}{v_y}\int_y^{\infty} dy'[\Delta_n^0(y') + \Delta_n(y')]\right\} \tag{10e}$$

For small neutralization rates the neutralization probability $P_n(\infty)$ increases by a factor 2 compared to "spinless electrons," whereas for large rates the correct limit is achieved. A more detailed discussion of this intricate problem of level degeneracies is given in Refs. 19b, 12a, and 12b.

Since the wave functions of $|n\rangle$ and $|\mathbf{k}\rangle$ have exponential tails, the same holds for $W_{\mathbf{k}n}$ in equation (6) so that the transition rate can be well approximated at larger distances by

$$\Delta_n(y) = \Delta_0 \exp(-y/y_w) \tag{11}$$

where y_w is the "interaction length."

So $\Delta_n(y)$ is large close to the surface and exponentially decays with y. As a consequence the two factors of $F_n(y)$ are opposite in their dependences on y so that a pronounced maximum is to be expected around some distance y_s from the surface. Thus y_s can be considered as "distance of formation" of an atomic state.

To illustrate this important aspect of the atom–solid interaction we display in Figure 8 $F_n(y)$ for the scattering of 100-eV Na atoms at $\Phi_{\text{in}} = \Phi_{\text{out}} = 60°$ on a W(110) surface.[44] Calculations of Remy[45] yield $y_w = 1.1a_0$ and $\Delta_0 = 2.9 \times 10^{18}\,\text{s}^{-1}$ so that we get with $v_y = 2.7 \times 10^{14}\,a_0/s$ the dependence of $F(y_n)$ shown in the figure. The formation of the final Na $3s$ term is well localized at a distance $y_s = 10.3a_0$, which results from the maximum $F_n(y_s)$ at

$$y_s = y_w \ln(\Delta_0 y_w / v_y) \tag{12}$$

For the discussion of coherence in ion–surface scattering one has to point out that y_s represents a well-defined effective impact parameter, and collisions with surfaces are thus characterized by a selection of that quantity. Furthermore we notice that y_s is significantly larger than the distance of closest approach y_0, which has some consequences for the interaction process. We will show in the next section that y_s can be roughly estimated by $y_s \approx \langle r_{nl}\rangle + d/2$, with r_{nl} i.e., the mean radius of a $|nl\rangle$ state and d is the interplanar spacing in the solid.[18]

Finally we have to point out that the transitions between atom and solid cause a Heisenberg broadening of the atomic level of Lorentzian type of half-width $\Gamma(y) = \Delta_n(y)$. At the distance of formation y_s we get from equations (11) and (12) $\Delta_n(y_s) = v_y/y_w = 2.45 \times 10^{14}\,\text{s}^{-1}$, which yields $\Gamma(y_s) = 0.16\,\text{eV}$. This broadening of level is ignored in equation (6).

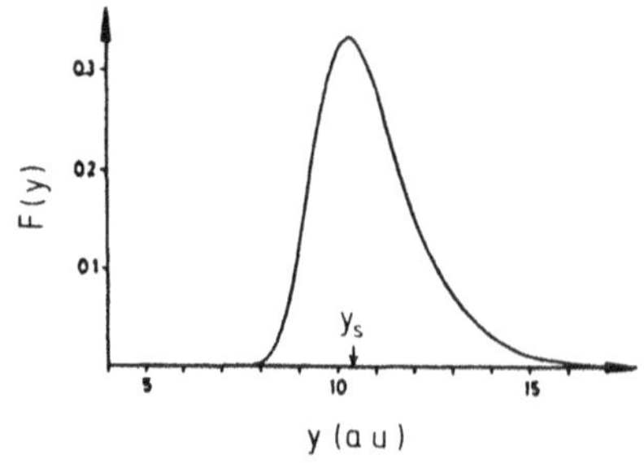

FIGURE 8. $F_n(y)$ given by Ref. 44.

3.1. PCI-Stark Effect: Distance of Formation of Manifolds in He II

In the preceding section we have shown that a simple theoretical concept of the dynamic ion-surface interaction results in a well-defined distance of formation y_s. We will now discuss an actual experimental determination of this quantity, where we use the orientation of orbital angular momenta of excited terms modified by the electric image field of ions as our probe.

Fröhling[13] and Graser and Varelas[14] observed that the orientation of the He II $n = 4$ manifold after ion-surface scattering is significantly influenced by the time the ion spent in close vicinity of the surface. The same observation is found in highly ionized *N*,[15] whereas in almost all other atomic and ionic terms no dependence or modification of that type is reported.

These experimental findings are consistently interpreted in terms of a postcollision interaction (PCI) Stark effect, which modifies the collision induced anisotropy of excited terms. Zimny *et al.*[15] first pointed out that the (long-range) electric image field at a metallic surface given by equation (4) is responsible for the mixing of opposite parity states. Since this electric dipole field does not occur for neutral atoms ($Z = 1$), this Stark mixing can only be observed in terms of ions. Furthermore the opposite parity levels have to be almost degenerate to cause an appreciable mixing via linear Stark effect; i.e., we have to investigate manifolds of hydrogenlike ions.

The concept of the experiment[17] is shown in Figure 9. We select by a narrow slit of $0.1 \times 10\ \text{mm}^2$ 200-keV He^+ ions scattered at an angle of exit Φ_{out}. Variation of Φ_{out} modifies the vertical velocity $v_y = v \sin \Phi_{out}$ and the interaction time with the electric image field which drops off with a R^{-2} dependence [see equation (4)].

The experimental data for the dependence of S/I of the He II $n' = 3 - n = 4$ manifold transition on Φ_{out} after the interaction of He^+ ions with a copper surface at $\Phi_{in} \approx 0.5°$ are displayed in Figure 10. Since the electric

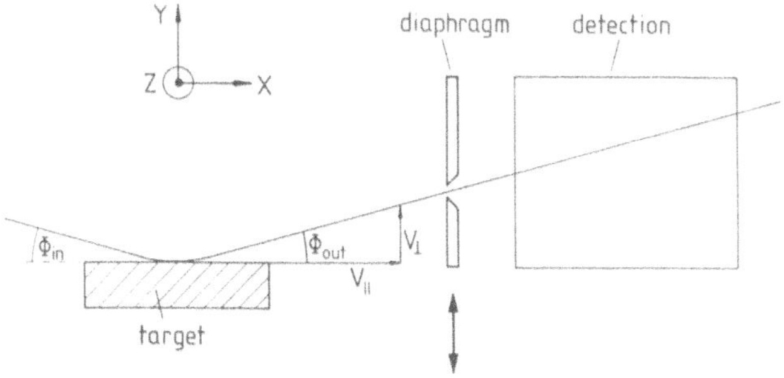

FIGURE 9. Sketch of the experimental setup for the study of PCI-Stark effect.

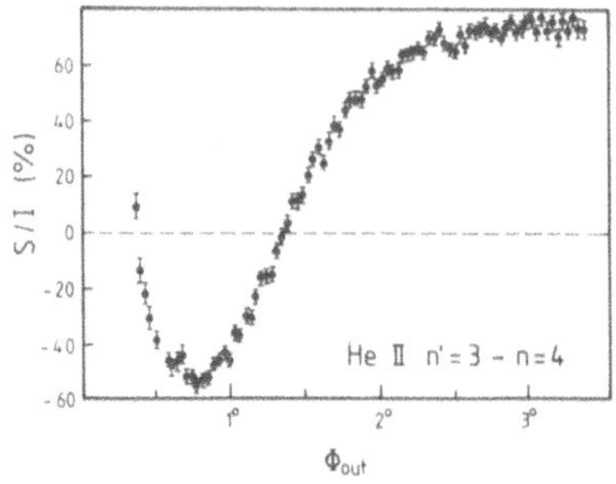

FIGURE 10. Dependence of S/I on Φ_{out} for the fluorescence light emitted in the He II $n' = 3$, $n = 4$ transitions.

image field is directed along y this field causes a mixing of opposite parity levels ($\Delta l = \pm 1$) with magnetic quantum numbers $\Delta m = \pm 1$ so that the initial collision-induced orientation is coherently modified.

In case Φ_{out} and v_y are sufficiently large, the interaction time in the electric field is too short to cause an appreciable mixing; this holds for our experimental data in Figure 10 down to angles Φ_{out} of about 2.5°. A further reduction of Φ_{out} results in times that allow some mixing of levels within the $n = 4$ manifold, so that S/I decreases. At $\Phi_{out} = \Phi_0 = 1.35°$ we observe a crossover of S/I and for smaller angles a minimum and subsequent increase of S/I.

For a theoretical treatment of the PCI-Stark effect we consider the nonrelativistic Hamiltonian for a hydrogenlike ion in an electric field $\mathbf{F}$

$$H = H_0 + H_1 = p^2/2 - Z/r + \mathbf{r} \cdot \mathbf{F} \tag{13}$$

With the orbital angular momentum operator $\mathbf{L}$ and $E_0 = -Z^2/2n^2$ we define the "Runge–Lenz vector"[16]

$$\mathbf{A} = (-E_0)^{-1/2}[\tfrac{1}{2}(\mathbf{p} \times \mathbf{L} - \mathbf{L} \times \mathbf{p}) - Z\mathbf{r}/r] \tag{14}$$

with the commutators

$$\mathbf{A} \times \mathbf{A} = i\mathbf{L}, \qquad \mathbf{L} \times \mathbf{A} = i\mathbf{A} \tag{15}$$

In the unperturbed atom $\mathbf{L}$ and $\mathbf{A}$ are constants of motion, i.e., $[\mathbf{L}, H_0] = [\mathbf{A}, H_0] = 0$. For hydrogenlike atoms we have for the dipole operator within an n manifold[16]

$$\mathbf{d} = -\mathbf{r} = \frac{3}{2}\frac{n}{Z}\mathbf{A} \tag{16}$$

so that we get the equations of motion in the Heisenberg representation

$$\frac{d\mathbf{L}}{dt} = i[H_1, \mathbf{L}] = \mathbf{A} \times \boldsymbol{\omega}$$
$$\frac{d\mathbf{A}}{dt} = i[H_1, \mathbf{A}] = \mathbf{L} \times \boldsymbol{\omega} \tag{17}$$

with

$$\boldsymbol{\omega} = \frac{3}{2}\frac{n}{Z}\mathbf{F} \tag{18}$$

From the coupled equations (17) we deduce for the projection of **L** with respect to the z axis

$$\langle L_z\rangle(\psi) = \langle L_z\rangle(0)\cos\psi - \langle A_x\rangle(0)\sin\psi \tag{19}$$

where the Stark phase ψ is given by

$$\psi = \int_{t_1}^{t_2} \boldsymbol{\omega}\mathbf{e}_y\, dt = \frac{3}{2}\frac{n}{Z}\int_{t_1}^{t_2} F_y\, dt \tag{20}$$

Application of this formalism to surface scattering implies that the interaction of the valence electron with the dipole image field starts at distances y_s where the final formations of states (manifolds) take place. The electric field F_y acts on the ion during escape from the surface with $v_y = v\sin\Phi_{\text{out}}$ so that

$$\psi = \frac{3}{2}\frac{n}{Z}\frac{1}{v_y}\int_{y_s}^{\infty} F_y\, dy \tag{21}$$

Taking F_y as given by equation (4), y_s is the only unknown quantity in equation (21) so that the determination of the phase ψ allows one to deduce the distance of formation y_s with respect to the image plane. The precision of this procedure is limited because of experimental reasons (finite angular resolution, etc.), the averaging on a complete manifold, and the distribution of distances given by $F_n(y)$. We neglect in equation (19) $\langle A_x\rangle$ and make use of the correlation between S/I and $\langle L_z\rangle$ with respect to the phase ψ.[17]

With these approximations we get for the angle at crossover Φ_0 the phase $\psi = \pi/2$ and from equation (21)

$$y_s = \frac{3}{4\pi}\frac{n}{Z}(Z-1)\frac{1}{v\sin\Phi_0} \tag{22}$$

TABLE I
Angle of Exit Φ_0 where S/I Vanishes for Transitions Starting from Different Manifolds $n = 4, 5, 6$ and Distance of Formation y_s Computed from Equation (22)

	Φ_0 (deg)	y_s (nm)
$n = 4$	(1.35 ± 0.2)	(0.75 ± 0.1)
$n = 5$	(1.0 ± 0.2)	$(1.25 + 0.3/-0.2)$
$n = 6$	(0.85 ± 0.2)	$(1.75 + 0.6/-0.5)$

In Table I we give the results of our evaluation for the $n = 4, 5, 6$ manifolds in He II.

An interesting aspect with respect to y_s follows from a discussion of the effective potential in the rest frame of an atom that is a distance R apart from a surface. Limiting our discussion to the y' direction we get from equation (1) the potential

$$V_e = V_c + \frac{Z}{2R + y'} - \frac{1}{4(R + y')} \tag{23}$$

which is plotted in Figure 11 for $R = 4a_0$ in comparison to the pure Coulomb potential (solid line). We notice that the interaction of the valence electron with its own image leads to a saddle point structure of V_e between surface

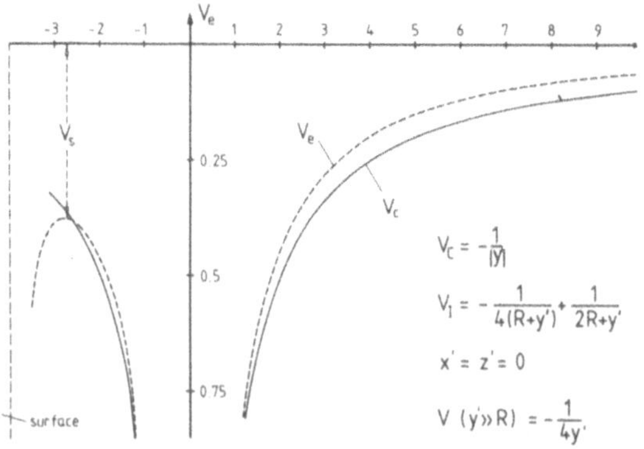

FIGURE 11. Effective potential of an atom at a distance $R = 4$ a.u. from the surface.

and ion core, which results in a lower limit of binding energies for stable terms at a given distance R.

Taking the energy at the saddle point V_s as a limit, we find for He II terms ($Z = 2$) $y_s = 2.058/V_s$. This (static) estimate for y_s agrees astonishingly well with our data in Figure 12, where we plot the experimental distances of formation y_s as a function of the binding energies of some He II manifolds. From the good agreement we deduce

$$y_s = \sqrt{2Z}/|E_B| = n^2(2/Z)^{3/2} \tag{23a}$$

as an estimate for the distance of formation, where $|E_B| < W + \varepsilon_F$ and the image plane is located 3–4 a_0 from the topmost layer of the surface.[17a]

At distances $R < y_s$, captured electrons will reionize with high probability, since "over-barrier transitions" are already classically allowed and expected to be highly effective in view of the long dwell times of the projectiles in the vicinity of the surface. Hence, the conditions for resonance tunneling are only met if the distance y_s given by equation (12) is comparable to the result of equation (23a).

We complete our discussion of our data with respect to the PCI-Stark effect by data for the Li II $2s\,^3S$–$2p\,^3P$ emission, which are shown in Figure 13. For the Li II ion we expect the presence of an electric image field; the spacings in energy between opposite parity levels, however, are too large to observe an appreciable Stark mixing by the image field, so that no dependence of S/I on Φ_{out} is observed. The same observation holds for hydrogen,[17] because the neutral H I atom is not affected by an electric image field.

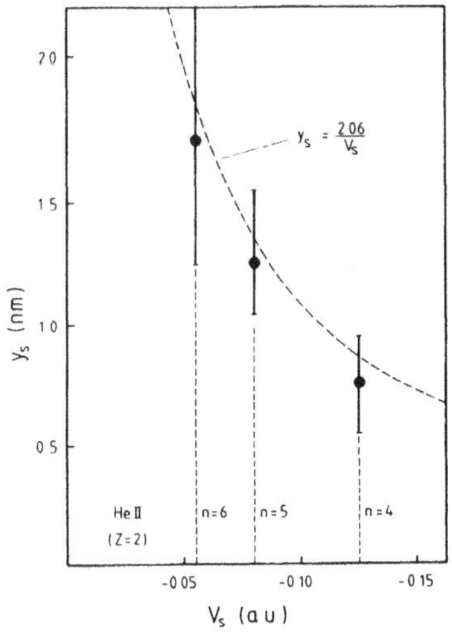

FIGURE 12. Distance of formation y_s versus binding/threshold energy V_s. Comparison of our data for He II $n = 4, 5, 6$ with the estimate $y_s = 2.06/V_s$.

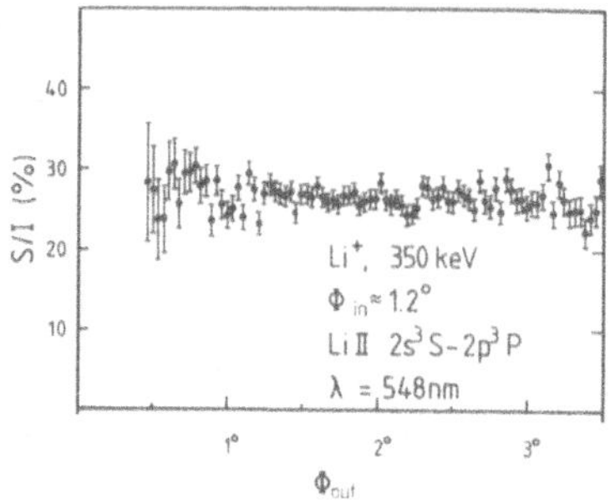

FIGURE 13. Dependence of S/I on Φ_{out} for the fluorescence light emitted in the Li II $2\,^3S$-$2\,^3P$ transition.

3.2. *Resonant Charge Exchange in Grazing Ion–Surface Collisions*

For a discussion of resonant charge exchange in grazing ion–surface collisions we consider as a typical example the distributions of charge states after the interaction of K^+ and N^+ ions with a Ni(111) surface. In Figure 14 the level schemes of some relevant levels in the neutral K I and N I atoms with binding energies E_B in comparison to the work function W and Fermi energy ε_F of Ni are displayed.

For very small velocities v_y we get from equation (10) $P_n(\infty) = \theta(|E_B| - W)$ so that the charge state distribution is determined by the equilibrium occupation probability P_n^e. Thus we expect from this concept for K completely ionized and for N completely neutralized projectiles after the surface scattering with small parallel velocities v_x.

In Figure 15 we show the dependence of the neutral charge state fraction on the velocity v_x after the interaction of N^+ ions at $\Phi_{in} \approx 0.5°$ with a Ni(111)-surface. For small velocities v_x we indeed observe the expected

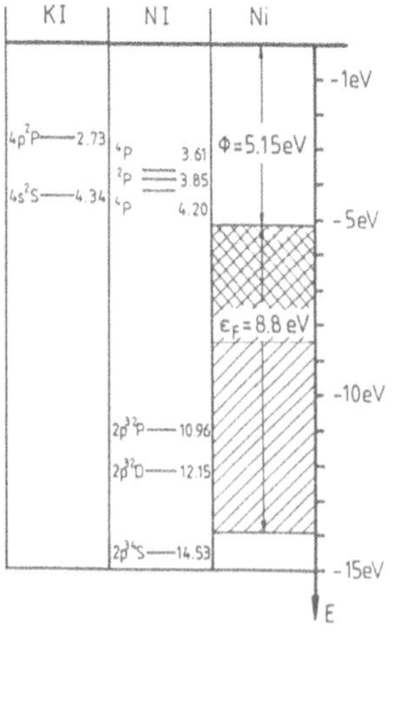

FIGURE 14. Level schemes of K I and N I and electronic structure of a Ni surface.

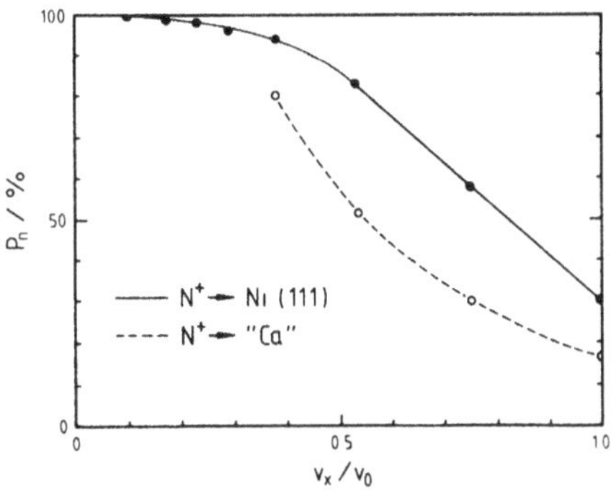

FIGURE 15. Neutral charge state fraction in dependence on the parallel velocity v_x after the interaction of N^+ ions with a Ni(111) surface at $\Phi_{in} \approx 0.5°$ (solid line) and with a Ca-coated surface (dashed line).

result: at projectile energies of some keV 99.6% of the backscattered atoms are neutral. Also the experimental findings with K^+ ions at comparable velocities fully agree with the model, since only 0.2% of the projectiles are neutralized at an energy of 15 keV.

When we increase the velocity of the projectiles and v_x, respectively, we notice in Figure 15 a significant drop in the neutral fraction. This observation indicates that we have to incorporate the parallel velocity v_x in the model of charge exchange at surfaces. But before we discuss this aspect, we demonstrate the effect of the quality of the surface on this type of experiments. The dashed line in Figure 15 connects data that were taken with a target coated with some layers of Ca of rather poor surface quality. The charge state fractions clearly deviate from the data for clean Ni(111) and compare well with fractions after beam-foil, which indicates other mechanisms of interaction.

The binding energies of excited terms that decay via fluorescence in the visible spectral region are generally smaller than the work functions of pure metals, so that we expect according to the model no excitation of those terms via surface scattering at small velocities v_x. In Figure 16 we show the excitation function of Na I $3\,^2P$ by a plot of the normalized intensity of the Na I $3\,^2S$-$3\,^2P$, $\lambda = 589$-nm emission versus the projectile energy E_0. The intensity of light is low, and at energies E_0 below some keV (which was not accessible in our experiment) no excitation of the $3\,^2P$ terms seems to occur. At higher E_0 an increasing population of $3\,^2P$ is observed, which also indicates the necessity of incorporating v_x into the model of charge exchange at surfaces.

3.3. *Incorporation of the Parallel Velocity into the Model of Resonant Charge Exchange at Surfaces*

For incorporating the parallel velocity into the "static" model we assume that the electronic structure of the solid can be well approximated

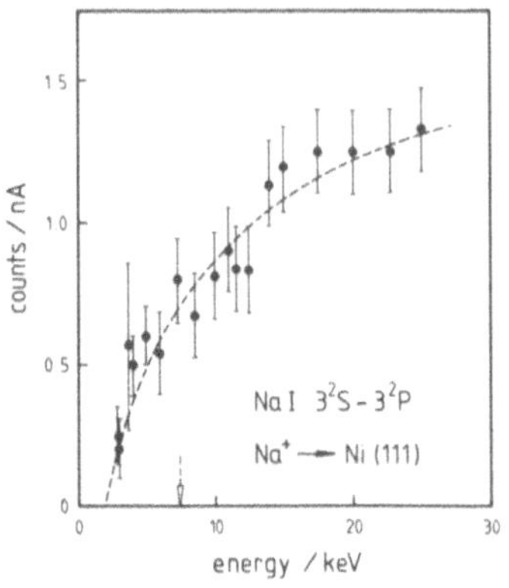

FIGURE 16. Normalized intensity of the Na I $3\,^2S$–$3\,^2P$ emission versus projectile energy.

by the model of a semi-infinite free electron gas (planar "jellium" model). In this case we have invariance of the electronic structure with respect to a translation parallel to the surface. As a consequence we find by Galilei transformation that in the rest frame of the projectile with parallel velocity v_x, $P_{\mathbf{k}}$ appears to be modified.

With the relative momentum $q = mv_x$ we have to replace in the rest frame of the atom $P_{\mathbf{k}} = f_{\mathbf{k}}$ by $P_{\mathbf{k}} = f_{\mathbf{k}+\mathbf{q}}$.(19) This transformation is simply performed in momentum space where the occupied electronic states in the conduction band of the solid are represented by a "Fermi" sphere of radius k_F. By pure geometrical arguments we obtain distribution functions in momentum and energy representation as indicated in Figure 17.

The effect of v_x on the charge exchange at surfaces can be nicely illustrated by the modification of the Fermi–Dirac distribution of the target

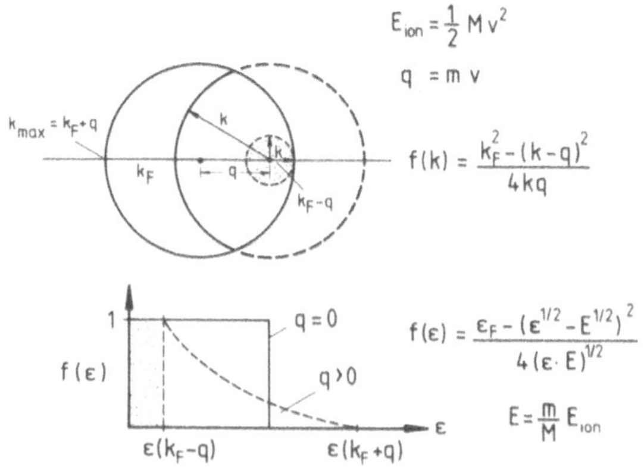

FIGURE 17. Computation of modified density of states of occupied conduction band states in the rest frame of the moving atom.

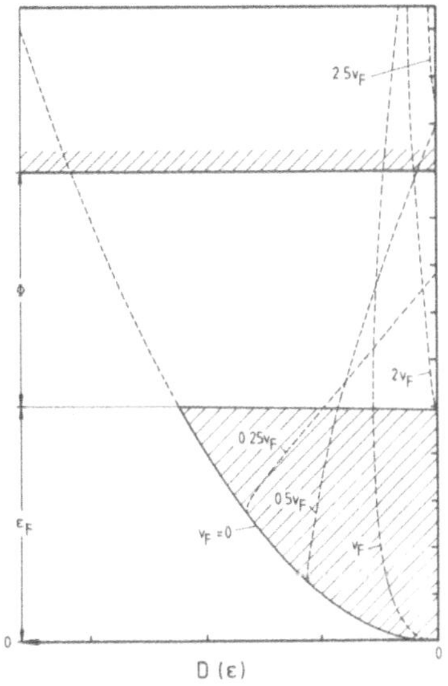

FIGURE 18. Density of states of a free electron gas in a rest frame moving with velocity v_x with respect to the solid.

electrons in the rest frame of the atom. In Figure 18 we have sketched the effective distribution of bulk density of states for $v_x = 0$ and finite parallel velocities v_x. Similar to the effect of very high temperatures of the solid the parallel motion results in a virtual population of electronic states above the Fermi edge so that with increasing v_x occupied states of the solid come into resonance with atomic levels with $|E_B| < W$; i.e., we achieve a "kinematic resonance."

In Figure 19 the atom conduction band interaction for $|E_B| < W$ is displayed. For $v_x = 0$ ($q = 0$) no occupied states of the solid are in resonance

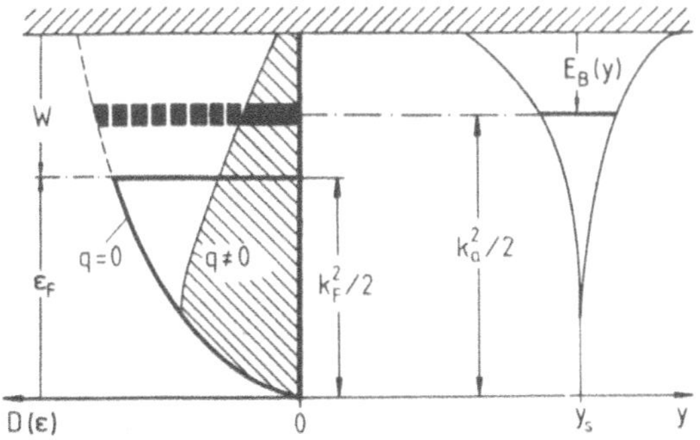

FIGURE 19. Atom-solid interaction for $|E_B| < W$ and zero and finite parallel velocity v_x.

with the atomic level, whereas at $q \neq 0$, occupied as well as unoccupied states of the solid as indicated by the solid and broken bars in the figure are in resonance with the atomic level. For a calculation of the probability of equilibrium occupation P_n^e of an atomic state after charge exchange at surfaces we have to weight the resonant transitions between solid and atom by the matrix element $W_{\mathbf{k}n}(y_s)$ given in equation (6).

According to van Wunnik *et al.*[19] one can nicely visualize the calculation of the charge exchange in momentum space by considering the overlap of the occupied electron states, represented by the Fermi sphere of radius k_F, with the surface of an atomic resonance sphere of radius k_a as defined in Figure 19. The resonance surface represents all electronic states of the solid which are in resonance with the atomic state, where the overlap with the Fermi sphere yields the fraction of resonant occupied states of the solid. For $|E_B| < W$ we have $k_a > k_F$, so that at $q = k_a - k_F$, the resonance surface of the atom starts to overlap with the Fermi sphere.

To account for the matrix elements, we compute as a representative example $W_{\mathbf{k}n}$ for a hydrogenic $2s$ wave function at a distance $y = y_s$ and plot in the $(k_y, k_\parallel)$ plane of Figure 20 lines of constant value of the matrix elements. $W_{\mathbf{k}n}$ increases between consecutive lines by a factor of 4 in the direction of the poles of the resonance sphere. Computing P_n^e by equations (8) and (9) is then performed by a summation over the surface of the resonance sphere including $W_{\mathbf{k}n}(y_s)$. Since the contributions of the matrix elements dominate in the pole region, one expects for $k_a > k_F$ for all velocities v_x a small population of an atomic state. For $k_a < k_F$ and $q \leqslant (k_F - k_a)$ the resonance sphere overlaps with the whole Fermi sphere and a very pronounced neutralization should occur.

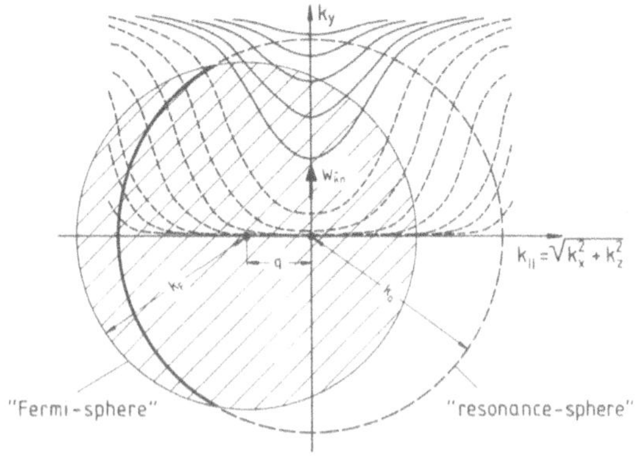

FIGURE 20. Resonant charge exchange at surfaces represented in momentum space. The lines represent constant values of the matrix elements $W_{\mathbf{k}n}$ (y_s) for an atomic $2s$ state.

3.4. Experimental Tests of the Model

The model of charge exchange incorporating the parallel velocity outlined in the previous sections was successfully applied to describe the formation of negative ions at cesiated surfaces by van Wunnik, Brako, Makoshi, Newns, Los, and Geerlings.(19) We refer to the literature for a detailed discussion of the experiments. It turns out that this model also provides a good basis to understand charge exchange processes at low and medium energy grazing ion-surface interaction, since the parallel velocity is equivalent to the velocity of the projectiles which is sufficiently large to cause appreciable effects.

A convincing demonstration of the ability of the model to describe charge exchange over a large domain of projectile velocities is given by recent experiments of Nienhaus *et al.*(19a) and corresponding model calculations by Zimny.(19b) Investigating the neutral fraction after the interaction of K-ions with a aluminum surface at grazing incidence results in a characteristic dependence on projectile velocity (Figure 21).

Since $|E_B(y_s)| < W$ (see also Figure 14 for K-Ni), we expect from the model a fully ionized beam at very small velocities, and an increase of the neutral fraction due to the "kinematic resonance" and a subsequent decrease at increasing velocities. The solid and dashed lines represent quantitative treatments(19b) with and without incorporation of the electronic spin (see Section 3) which show that charge transfer into groundterms of alkali atoms at grazing ion-surface scattering is well understood.

For a discussion of experiments with respect to the formation of excited terms in Na I after the interaction of fast Na^+ ions with a Ni(111) surface

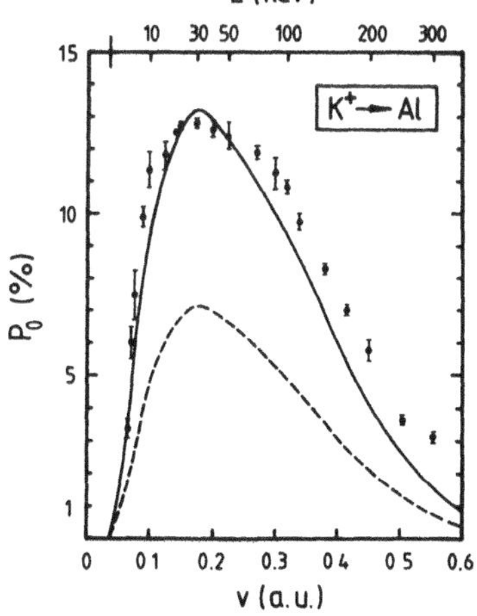

FIGURE 21. Neutral fraction P_0 after K^+-Al(111) scattering at a grazing angle of about 0.5°. Upper scale: projectile energy in keV; lower scale: projectile velocity in atomic units (2.2 mm/ns).

we have sketched the $3s$, $3p$, and $4d$ terms and their level shifts and distances of formation in Figure 22.

Describing the excitation of Na I $3p$ we have $|E_B| < W$, and an energy E_0 larger than 8 keV is needed to resonantly populate this term. This energy is indicated by an arrow in Figure 16 and clearly deviates from the experimental findings. Assuming a lower work function of the Ni target—which was not measured during the experiments—brings the theoretical estimate in agreement with the data. However, by this lower work function the observed neutralization of the Na I $3s$ ground term ($n^0 \approx 25\%$ for 10 keV $\leqslant E_0 \leqslant$ 30 keV) is significantly too low. So we are left at the moment with the speculation that the lower onset of the intensity of the Na I resonance line may be due to a "broadening" of the resonance or an incomplete description of a real electronic structure of a surface and of the atomic level shift.

According to the model we expect for terms with lower binding energies E_B a higher parallel velocity v_x to reach the threshold of excitation. This threshold is about 20 keV for Na I $4d$ excitation, whereas only 8 keV is sufficient for populating the $3p$ term. A spectrum taken after the interaction of Na^+ ions with a Na(111) surface at $E_0 = 28$ keV which is slightly above the theoretical threshold for $4d$ excitation is shown in Figure 23. The data clearly show the $3s$–$3p$ resonance transition, but only a very small contribution from $3p$–$4d$, which agrees with the estimate based on the model.

Increasing the projectile energy, i.e., parallel velocity v_x, should lead to a relative increase of the intensity of the $3p$–$4d$ transition. This feature is indeed found in experiments of Fröhling,[20] who observed a decrease of the intensity ratio $I(3s\text{–}3p)/I(3p\text{–}4d)$ from about 300–400 at $E_0 = 100$ keV to about 30 at 300 keV.

Charge exchange in atom–surface interaction as sketched in Figure 20 implies that for $k_a > k_F$ the solid acts as a rather effective "ionizer," i.e.,

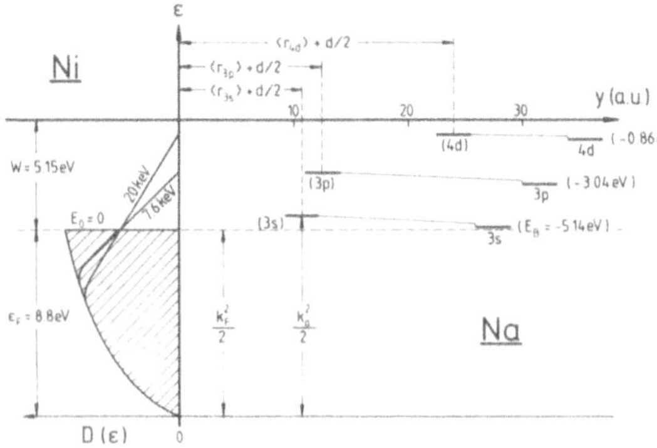

FIGURE 22. Terms of Na atom at a Ni(111) surface.

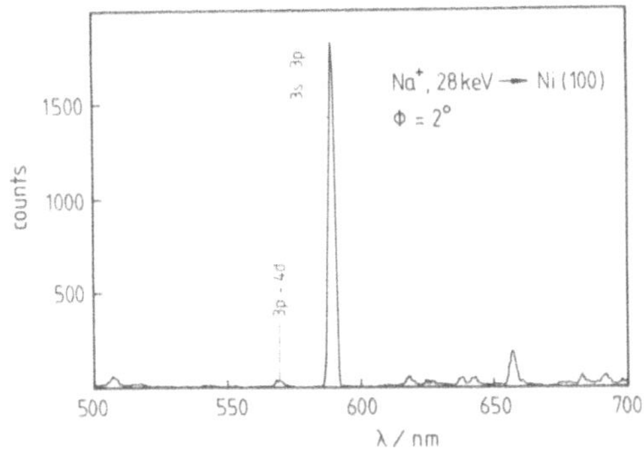

FIGURE 23. Spectrum after the interaction of 28-keV Na^+ ions with a Ni(111) surface.

suppresses the formation of excited terms. As can be visualized from that figure, this property of the surface will be strongly enhanced for an increase of k_a at a given k_F. As a result we expect a pronounced decrease in population of terms with increasing principal quantum numbers and momentum k_a, respectively.

A Balmer spectrum of Neumann *et al.*[21] after the interaction of 4-keV H_2^+ ions with clean Ni(111) is displayed in Figure 24. The authors clearly found a deviation from the Jackson-Schiff scaling law,[22] which holds in most ion-atom and beam-foil collisions and interpret their data by n^{-4}–n^{-6} scalings. Similar observations were also made by Leung, Tolk, Heiland, and co-workers[23] and by Fröhling *et al.*[20] for some alkali atoms.

That a modified scaling seems also to hold for highly excited Rydberg terms after surface interaction can be extracted from electron spectra which were taken in the direction of the specularly reflected ions. The spectra

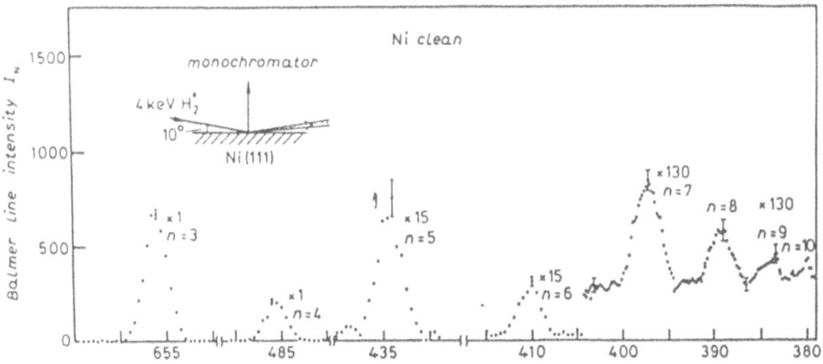

FIGURE 24. Balmer spectrum of Neumann *et al.* (Ref. 21) after the interaction of 4-keV H_2^+ ions with Ni(111).

show at an energy which corresponds to the ion velocity the well known "cusp" structure,[24] which is, however, significantly broadened in comparison to ion–atom or beam–foil interactions. In Figures 25a and 25b two electron spectra of Strohmeier[25] are shown after the interaction of 150-keV protons with a He gas target and a Si(111) surface, respectively. The data clearly exhibit a broadening of the cusp after ion–surface scattering. Evaluating both data sets by a procedure given by Lucas *et al.*[26] yields for the ion–atom collision the expected n^{-3} scaling, whereas the surface interaction is described by the exponent—3.72 ± 0.06. The population of Rydberg levels at the limit of ionization by ion–surface interaction exhibits a first example of a clear deviation from the Jackson-Schiff scaling law; i.e., the cross section in the rest frame of the atom $d\sigma/d\varepsilon$ vanishes at the ionization threshold.

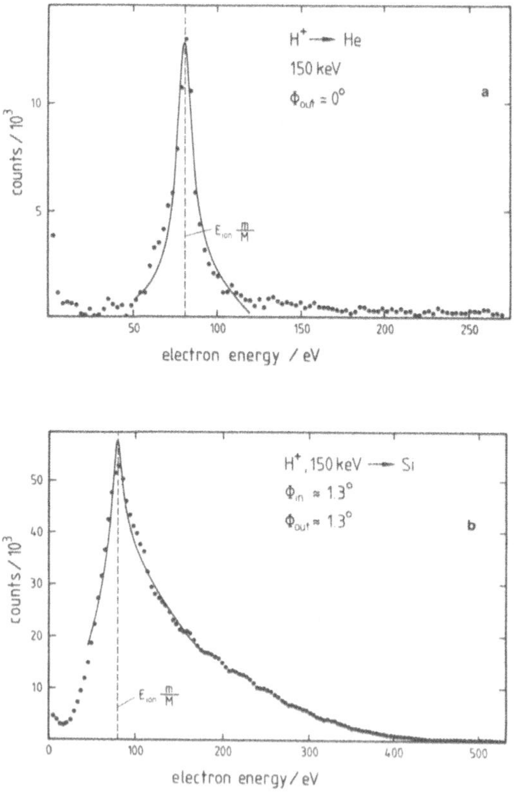

FIGURE 25. (a) Spectrum of electrons emitted at $\Phi_{out} = 0°$ after interaction of 150-keV protons with He. (b) Spectrum of electrons emitted at $\Phi_{out} = 1.3°$ after interaction of 150-keV protons with a Si(111) surface at $\Phi_{in} = 1.3°$.

The formation of Rydberg levels as observed in the previous experiment can hardly be understood by a contribution of valence band electrons, since the parallel velocity $v_x = 2.5v_0$ is too large to obtain any overlap between Fermi and resonance sphere. Therefore one has to consider other mechanisms that contribute to the excitation process, whereas resonant transfer processes only proceed via ionization into empty states of the conduction band.

With respect to other mechanisms of excitation we have to recall that the distance of final state formation y_s clearly exceeds in most cases the distance of closest approach y_0. In general y_0 is sufficiently small to cause a strong interaction between solid and ion core and to achieve some equilibrium with respect to charge state and excitation. On the exit part of the trajectories this distribution of states is changed by resonant charge exchange processes where the final state formation takes place at larger distances y_s.

Thus the formation of final states in ions or atoms are the result of a number of consecutive capture and loss processes so that the identity of the projectile is completely lost during the interaction with the surface. Varying the charge state of the projectile consequently has no effect on the charge states of the scattered particles.[27] Further indications for those interaction processes are given by the large orientation of orbital angular momenta in singly and multiply charged ions and the population of multiply excited terms after ion-surface scattering.

4. COHERENCE IN GRAZING ION-SURFACE COLLISIONS

In our introductory remarks we already showed that a large orientation of orbital angular momenta of excited terms is achieved after grazing ion-surface scattering. We will now discuss this unique feature of collisional excitation by ion-solid interaction in more detail.

For describing the coherence of excited terms we use the concept of the density operator,[28] which is given at the time t_s of the final state formation by

$$\rho(t_s) = \rho_L(t_s) \otimes \frac{1_s}{2S+1} \otimes \frac{1_I}{2I+1} \tag{24}$$

where we assume that the interaction is a pure orbital angular momentum mechanism[29] and is not affected by electronic and nuclear spins S and I. Expansion of ρ in the basis of irreducible tensor operators T_q^k[28]

$$\rho_L = \sum_{kq} {}^L\rho_q^k T_q^k \tag{25}$$

where the components with $k = 0, 1, 2$ can be associated with "population," "orientation," and "alignment" of magnetic sublevels of terms with orbital angular momenta L. Fine and hyperfine interactions with splittings $\hbar\omega_{JF_1F_2}$ yield the time dependences for $[(LS)JI]F$ terms with lifetime τ

$$^L\rho_q^k(t \gg t_s) = \sum_{JF_1F_2} \frac{(2J+1)^2(2F_1+1)(2F_2+1)}{(2I+1)(2S+1)} \begin{Bmatrix} J & J & k \\ L & L & S \end{Bmatrix}^2 \begin{Bmatrix} F_1 & F_2 & k \\ J & J & I \end{Bmatrix}^2 \, {}^L\rho_q^k(t_s)$$

$$\times \exp(-i\omega_{JF_1F_2}t - t/\tau) \tag{26}$$

This transfer of anisotropy between orbital angular momenta and electronic/nuclear spins given by equation (26) can be used to perform high-resolution atomic spectroscopy[30] and to produce spin-polarized beams after ion–surface scattering.[31]

The anisotropic sublevel populations of excited terms are generally probed by the polarization of the fluorescence light emitted in transitions between terms with orbital angular momenta L, L_0. Since after grazing ion–surface scattering the orientation of orbital angular momenta plays a dominant role with respect to the anisotropy obtained, we limit our discussion to the normalized Stokes parameter

$$\frac{S}{I} = \frac{I(\sigma^-) - I(\sigma^+)}{I(\sigma^+) + I(\sigma^+)} \tag{27}$$

where $I(\sigma^-)$, $I(\sigma^+)$ are the intensities of light with negative and positive helicity, respectively. S/I (and the Stokes parameter describing the linear polarization fraction) is directly related to the tensorial components of the density operator. We obtain by taking into account the depolarization due to a finite electronic spin S ($I = 0$ or $\omega_{JF_1F_2} \cdot \tau \ll 1$) for unresolved multiplet transitions $(LS)J$, $(L_0S)J_0$ observed in a direction normal to the plane of scattering (z axis)[28]

$$\frac{S}{I} = \frac{\begin{Bmatrix} 1 & 1 & 1 \\ L & L & L_0 \end{Bmatrix} \sum_J (2J+1)^2 \begin{Bmatrix} L & L & 1 \\ J & J & S \end{Bmatrix} {}^L\rho_0^1}{\sqrt{\frac{2}{3}} \sum_J (2J+1)^2 \left[\begin{Bmatrix} 1 & 1 & 0 \\ L & L & L_0 \end{Bmatrix} \begin{Bmatrix} L & L & 0 \\ J & J & S \end{Bmatrix}^2 {}^L\rho_0^0 + (1/\sqrt{3}) \begin{Bmatrix} 1 & 1 & 2 \\ L & L & L_0 \end{Bmatrix}^2 {}^L\rho_0^2\right]} \tag{28}$$

In most cases the second term in the denominator of equation (28) is about an order of magnitude smaller than the first one, so that S/I is well approximated by a proportionality with ${}^L\rho_0^1/{}^L\rho_0^0$.

In quite a few experiments the orientation of terms in atoms and ions covering the Periodic Table has been investigated for various kinds of targets and experimental conditions. At present one can obtain by grazing ion-surface scattering at conditions that will be specified below

$$L = 1\ (p \text{ states}): \qquad \rho_0^1/\rho_0^0 \leqslant 0.5$$

$$\left.\begin{array}{l} L = 2\ (d \text{ states}): \\ L = 3\ (f \text{ states}): \end{array}\right. \qquad \rho_0^1/\rho_0^0 \leqslant 0.85$$

which is a fairly large anisotropy with respect to the "simple" process of scattering. According to equation (28) the magnitude of S/I observed in the experiments depends on the degree of depolarization caused by electronic and nuclear spins, i.e., on the type of multiplet transition.

Already the first experiments clearly showed that the anisotropy obtained after ion-surface scattering depends on the "quality" of the surface. Since the vacuum and the polishing of polycrystalline samples were poor, the data are strongly affected by deposition of adsorbates and defects of the surface. By sputtering of adsorbates and surface atoms these poor experimental conditions could be partially compensated, especially in experiments with high fluxes of projectiles. Achieving some equilibrium between deposition/removal of adsorbates and creation/removal of defects, experimental data as shown in Figure 26 for 300-keV Ar^+ ions on Cu and graphite at $\Phi \approx 1°$ are obtained in early investigations of this phenomenon.[29]

Comparing these data with the experimental results already displayed in Figure 2 at UHV conditions and with a highly polished Si(111) surface ($S/I \approx 95\%$) clearly demonstrates the effect of a clean and well-defined

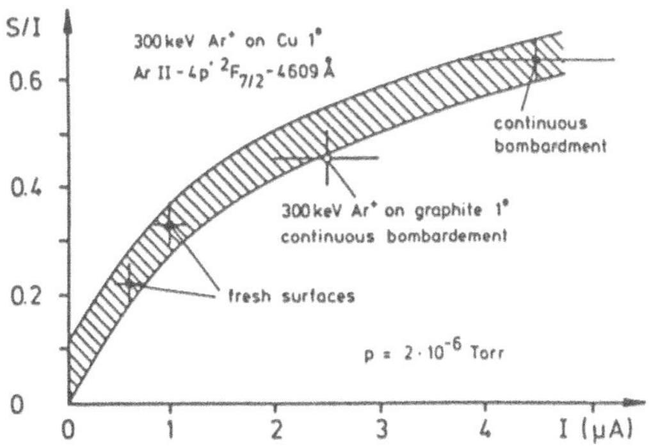

FIGURE 26. Dependence of S/I of the Ar II $4s'\,^2D$–$4p'\,^2F$ emission on the ion current of the projectiles (Ref. 29).

surface. For a direct comparison it seems to be more appropriate to make use of the polarization fraction $I(\sigma^-)/I(\sigma^+)$ instead of S/I: from the data in Figure 26 fractions of about 2-4 are obtained, whereas improved conditions yield fractions larger than about 40. An explicit study of the change of orientation of this Ar II $4p'\,^2$F term due to adsorbates has been performed by Church, Lee, and Lenoir,[32] and for terms in Li by Hermann *et al.*[32a]

Relevant for the study of the interaction mechanisms and of applications is the dependence of the orientation on the energy of the projectiles E_0. Typical examples are given in Figure 27, where we have shown S/I of the fluorescence light of the He I $2\,^3S$-$3\,^3P$ and Na I $3\,^2S$-$3\,^2P$ transitions as a function of E_0. The dependence on energy shown in Figure 27 is given by a saturation curve. As expected from arguments of symmetry, S/I seems to vanish for $E_0 \approx 0$, and saturation is obtained at energies of about 5 keV/amu; i.e., $v_x \approx v_0/2$. Similar dependences have also been observed by us for N, by Tolk *et al.*[33] for H, and Lenoir *et al.*[33a] for Ar.

Extending the energy of the projectiles into the MeV regime still yields a significant amount of orientation of excited terms as shown for two terms of N III in Figure 28. The largest energy in these experiments was $E_0 =$ 12 MeV, which corresponds to $v_x \approx 6v_0$.

In summarizing these observations we describe the dependence of S/I on the (parallel) velocity $v \approx v_x$ by the scheme given in Figure 29, which we have consequently divided into three sections. The first part is given by an increase of S/I from zero at $v_x \approx 0$ to a value of saturation at $v_x \approx v_0/2$, which is of the order of the Fermi velocity v_F of the metals used as targets. Above this velocity S/I is about constant and for velocities $v_x > 2v_F$ resonant transfer processes discussed in Section 3 are assumed to be of minor importance.

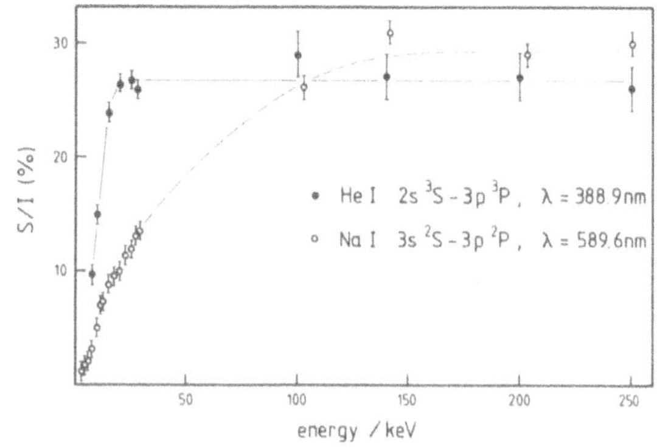

FIGURE 27. Dependence of S/I on projectile energy of light emitted in the He I $2\,^3S$-$3\,^3P$ and Na I $3\,^2S$-$3\,^2P$ transitions.

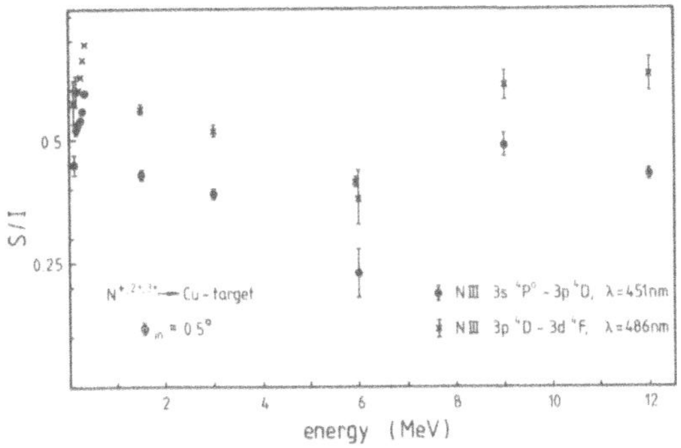

FIGURE 28. Dependence of S/I on projectile energy of light emitted in two transitions in N III.

Based on features of the "density gradient model" of Schröder and Kupfer, we can describe in a qualitative way the observed dependence of the orientation on energy (velocity). Trajectory and charge exchange at surface scattering result in a well-defined plane of scattering and effective impact parameter. Thus in the rest frame of the projectile the surface acts as a reservoir of electrons at a distance y_s from the ion core. These electrons have a distribution of the x components of velocities $-v_F \leqslant v_{e_x} \leqslant v_F$ that appears to be shifted in the ion rest frame according to a Galilei transformation by the parallel velocity $-v_x$.

Assuming in a semiclassical approach the generation of orbital angular momenta in capture or excitation processes results at finite velocities v_x in a net orientation in the spatial distribution of angular momenta with respect to $-\mathbf{x} \times \mathbf{y}$ (z axis). This orientation vanishes in the model for $v_x \approx 0$ since $\langle v_{e_x} \rangle \approx 0$, and for $v_x \gtrsim v_F$ we have $v_{e_x} < 0$ so that a large orientation is generated.

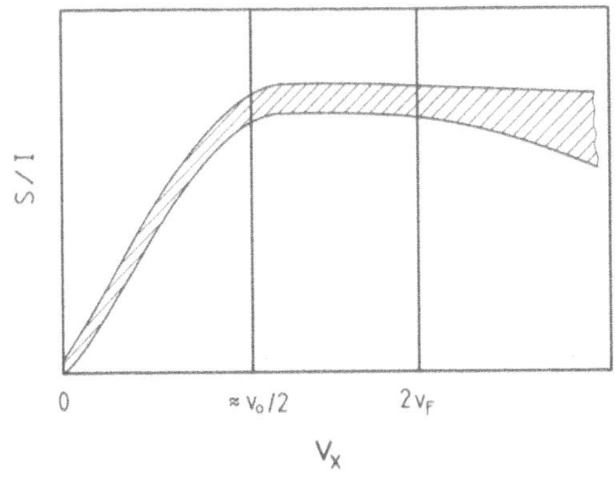

FIGURE 29. Sketch of the dependence of S/I on the projectile velocity.

Brako[35] and Andrä *et al.*[34] have shown that the concept of a shifted Fermi sphere as outlined in Section 3 can be well adopted to understand the production of orientation by grazing ion–surface scattering. To illustrate this feature we have to consider atomic states with finite orbital angular momenta and use in Figure 30 the same representation as in Figure 20 for the H I $2p$, $m = +1$ and $2p$, $m = -1$ states. For $k_a > k_F$ we get overlap between the resonance surface and the Fermi sphere for $|q| > k_a - k_F$.

The mechanism of polarization is caused by taking into account the transition matrix elements $W_{\mathbf{k}n}(y_s)$, which differ for the $2p$, $m = \pm 1$ states in momentum space. In Figure 30 the matrix elements are visualized by lines of constant values in the $k_x k_y$ plane, and it is obvious from the figure that $W_{\mathbf{k}n}$ is for the $2p$, $m = \pm 1$ asymmetric with respect to k_x. Thus for $q < 0$ (see the figure) the $2p$, $m = +1$ state is preferentially populated, whereas for $q > 0$ the same holds for the $2p$, $m = -1$ state. So this model explains the anisotropic nature of grazing ion–surface collisions and shows the relevance of the parallel velocity v_x for the production of orientation.

In the experiments a characteristic increase of orientation at $0 \leqslant v_x \leqslant v_0/2$ is observed. Because of the structure of the matrix elements (see Figure 30), however, we have for the whole domain of resonances $k_a - k_F < q < k_a + k_F$ the dominant contribution of one of the $2p$, $m = \pm 1$ substates so that at present this model does not provide an adequate description of the velocity (energy) dependence of the orientation for $v_x \leqslant v_0/2$. Possible mechanisms for the observed dependence are a matter of discussion, but no consistent interpretation has been given so far.

In the velocity domain $v_0/2 \leqslant v_x \leqslant 2v_F$ the model outlined in Section 3 and especially a more refined theory of charge exchange using the Newns–Anderson formalism by Brako and Newns[12,35] provides a quantitative description of the anisotropy after ion–surface scattering. Comparison

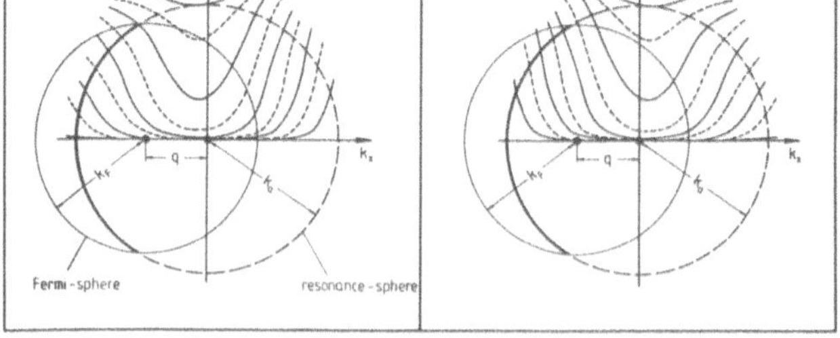

FIGURE 30. Model for charge exchange of a $2p$, $m = \pm 1$ state with a surface. Weighting the overlap between Fermi sphere and resonance surface with the matrix element $W_{\mathbf{k}n}(y_s)$ results in an orientation of $2p$.

of theory with experimental work of Tully *et al.*[36] at H I $n = 3$ yields good agreement, but the number of levels involved in the H_α transition causes some ambiguity in the evaluation of the data.

Therefore we manufactured an efficient polarization analyzer for Lyman-α radiation which can be used in UHV (Ref. 36a) and investigated the anisotropy of H I $2p$ after scattering of protons from a Ni(111) surface at $\Phi_{\text{in}} \approx 0.8°$ and a vacuum in the 10^{-11}-mbar regime. In Figure 31 we display the Stokes parameter of Lyman-α radiation emitted normal to the plane of scattering, which hardly depends on energy in the interval 7 keV $\leqslant E_0 \leqslant$ 29 keV (for experimental reasons we used D^+ projectiles to obtain the data below 15 keV/amu).

The solid lines in Figure 31 represent a theory of Brako and Newns[37] where a fixed distance of formation y_s is taken for all $2p$ sublevels, whereas the broken lines are obtained by using different y_s for the sublevels. The calculations of Brako are in reasonable agreement with the experimental data, but also this theoretical approach fails to reproduce the observed anisotropy for $v_x < v_0/2$.

Burgdörfer *et al.*[38,39] performed a two-channel calculation for the anisotropy of H I $2p$ after scattering at a Au surface. However, their computed Stokes parameters are in poorer agreement with our data.

All calculations ignore the observation that the probability for the population of H I $2p$ is about 10^{-3} in comparison to neutral fractions in the scattered beam of about 70%. So we have to assume a large population of the $1s$ term and possible channels to excite the $2p$ levels.

A very recent development in this field is based on the capture of spin-polarized electrons into excited terms. Rau *et al.*,[39a] who actually pioneered grazing ion-surface scattering, showed that in these collisions polarized electrons are captured into the groundterm of fast ($v \approx 2v_0$) deuterium atoms from magnetized targets. Despite the fact that the

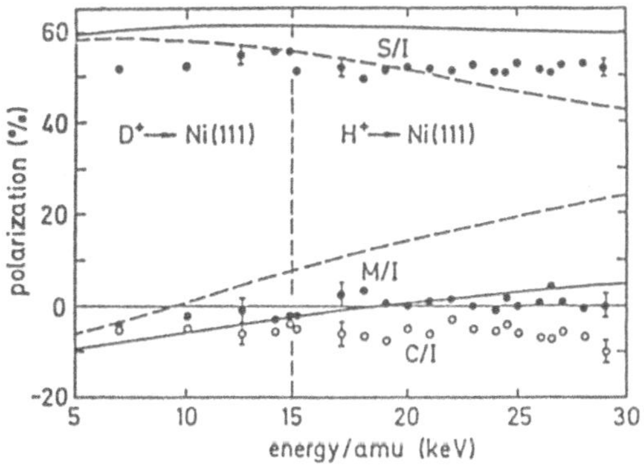

FIGURE 31. Dependence of Stokes parameter of Lyman-α radiation on projectile energy after the interaction of protons and deuterons with a Ni(111) surface. Solid and dashed lines represent theory by Brako (Ref. 37).

interaction process is only vaguely understood in this case, this technique has proven to be a powerful method in studying magnetism at surfaces.[39a]

A combination of the orbital angular momentum process and polarized electron pickup to enhance (reduce) the anisotropy of excited terms was proposed by Kupfer *et al.*[39c] In this case the density operator in equation (24) has the more general form

$$\rho(t_s) = \rho_L(t_s) \otimes \rho_s(t_s) \otimes \frac{1_I}{2I+1} \tag{28a}$$

Thus changing the magnetization of the target affects $\rho_s(t_s)$ and the resulting density matrix. A verification of this process is shown by experiments with a Fe(110) target in Figure 32, where S/I of the He I $2\,^3S$–$3\,^3P$ emission is recorded for different states of magnetization of the target after the interaction of 150 keV He^+ ions at grazing incidence.[39d] This method implies

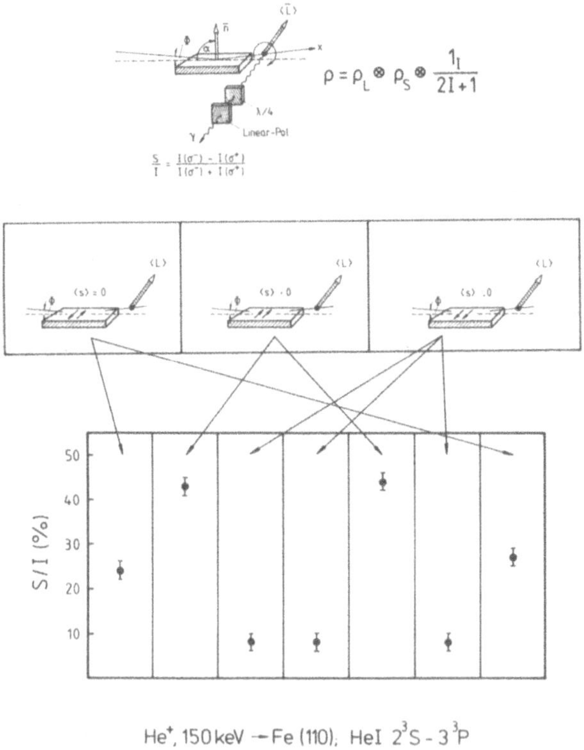

FIGURE 32. Dependence of the circular polarization fraction S/I of the light emitted in the He I $2\,^3S$–$3\,^3P$ transition after grazing scattering of 150 keV He^+-projectiles on the state of magnetization of the Fe(110) target. In the upper part a sketch of the experimental setup and concept is given.

some potential to investigate surface magnetism and eventually allows more detailed studies of the interaction process by "marking" target electrons.

So far, we have restricted our discussion to a structureless semi-infinite free electron gas which is appropriate for simple metals such as Al or *sp*-band electrons of noble metals. For transition metals this model is expected to be a crude approximation, and effects due to the crystallographic orientation may occur.

In grazing ion–surface scattering effects on the azimuthal orientation of monocrystalline targets are observed in the intensities of X-ray[39e] and light[39f] emission. Also in the polarization of fluorescence light pronounced effects are found.[39g] However, it is a matter of discussion if these effects are preferentially caused by trajectories or by asymmetries in the electronic structure of the surface with respect to the crystallographic direction. Recently, Easa and Modinos[46] calculated in detail an azimuthal variation of the neutral fraction of resonantly neutralized slow Na projectiles scattered from a W(110) surface.

5. STUDY OF STABLE TERMS VIA HYPERFINE INTERACTION

The anisotropy of magnetic sublevel populations of excited atomic terms is probed via the polarization of fluorescence light emitted in transitions from those terms. This technique fails for the detection of stable ground terms or any metastable terms. Since we expect a preferential population of these stable terms after surface scattering, it is of general interest to also investigate the anisotropy of these terms.

In the preceding section we briefly discussed the transfer of anisotropy of orbital angular momenta to the electronic and nuclear spin systems via fine and hyperfine interaction, respectively. Andrä *et al.*[31] first showed that nuclear polarized beams can be obtained by grazing ion–surface scattering, which clearly indicates that ground terms and metastable terms are predominantly populated and highly oriented.

The nuclear orientation of the scattered ions is detected by an additional beam–foil interaction that does not affect the nuclear spins, but rearranges the electronic shell and populates excited terms. In conclusion the polarization of fluorescence light is a direct measure of the nuclear orientation of the probed beam.[31] Since this chapter is focussed on interaction mechanisms, we refer for detailed discussions of experimental techniques and results to the literature[31,40] and concentrate on some aspects that provide interesting additional information on the interaction process.

For $^{14}N^+$ ions of some 100 keV the stable terms of the relevant charge states after surface scattering are given by $2p^n$ configurations. Therefore most of these terms in N I, II, III have nonzero orbital angular momenta and are polarized by the interaction with the surface. From the transfer of electronic orientation via hyperfine interaction results a nuclear vector polarization

$$P_I = \frac{\langle I_Z \rangle}{I} = \left(\frac{I+1}{3I}\right)^{1/2} \frac{{}^I\rho_0^1}{{}^I\rho_0^0} \tag{29}$$

with

$$\frac{{}^I\rho_0^1}{{}^I\rho_0^0} = \frac{C^1}{C^0} \frac{{}^L\rho_0^1}{{}^L\rho_0^0} \tag{30}$$

The coefficients of transfer C^k are given in Ref. 31. In the experiments nuclear orientation up to $P_I \approx 24\%$ are observed, and according to the contributing terms in the first three charge states comparable nuclear orientations are found. From equation (30) we deduce electronic orientations ${}^L\rho_0^1/{}^L\rho_0^0$ which are of comparable magnitude as for excited terms.

For the population of excited terms we expect (see Figure 7) dominant contributions from resonant transfer processes. Since ground terms of ions mostly have binding energies $|E_B| \geqslant \varepsilon_F + W$, Auger neutralization and violent collision processes dominate the charge exchange. Calculations of Taute[41] show also that Auger processes can populate electronic terms with large orientation after ion-surface scattering. The neutral fraction after interaction of N^+ ions with a Ni surface (see Figure 15), however, can be described by the model of resonant electron transfer incorporating the parallel velocity. A simple estimate with that model of the parallel velocity v_x, at which the neutral fraction amounts to 50% is in good agreement with the data in Figure 15; in this case possible contributions of Auger processes seem to be small.

The large nuclear orientation observed in the first three separated charge states clearly indicates that the electronic terms in the backscattered ions are highly oriented. This observation can only be understood by a pronounced interaction of the ion core of the projectile with the surface atoms at distances $y < y_s$.[10a,20] As already pointed out in the previous section, we expect some equilibrium of excitation and charge states at distances of about y_0 so that the identity of incoming projectiles is completely lost.

A further interesting aspect of grazing ion-surface interaction can also be studied via hyperfine interaction. By applying a longitudinal magnetic field H_x between surface and beam-foil interaction we induce a precession of the magnetic moments associated with electronic terms and identify from the observed g_F factor spectra the terms that contribute to the nuclear orientation observed, i.e., terms that are dominantly populated (and oriented).

From these experiments and estimates of depolarization due to unpolarized S terms we conclude that terms with large orbital angular momenta are preferentially populated. For the N I component we find the extreme case that from the ^4S, ^2P, ^2D terms of the $2p^3$ configuration only the ^2D term is significantly populated.[41a]

A possible explanation of this observation can be given by a unique feature of grazing ion-surface collisions. Since the electron capture in this type of collision is accompanied by a large anisotropy, consecutive capture of electrons on the exit part of the trajectory first leads to a preferential population of terms with larger angular momenta.

This process can be simply visualized in terms of a classical vector model or by considering the coupling of two angular momenta l_1, l_2 that are anisotropically distributed in the average on the ensembles. For the density operator components of the coupled angular momentum L we obtain[42]

$$ {}^{L}\rho_0^k = \sum_{k_1 k_2} (-1)^{k_1-k_2}[(2k_1+1)(2k_2+1)(2k+1)]^{1/2}(2L+1) \times \begin{pmatrix} k_1 & k_2 & k \\ 0 & 0 & 0 \end{pmatrix} \begin{Bmatrix} l_1 & l_1 & L \\ l_1 & l_2 & L \\ k_1 & k_2 & k \end{Bmatrix} {}^{l_1}\rho_0^{k_1}\, {}^{l_2}\rho_0^{k_2} \tag{31} $$

To demonstrate the selection with respect to population, we consider the angular momentum coupling of two p electrons ($l_1 = l_2 = 1$) and compute for the population of the resulting S, P, D terms

$$ \begin{aligned} {}^{S}\rho_0^0 &= 0.333({}^{P_1}\rho_0^0\, {}^{P_2}\rho_0^0 - {}^{P_1}\rho_0^1\, {}^{P_2}\rho_0^1 + {}^{P_1}\rho_0^2\, {}^{P_2}\rho_0^2) \\ {}^{P}\rho_0^0 &= 0.577\,{}^{P_1}\rho_0^0\, {}^{P_2}\rho_0^0 - 0.289({}^{P_1}\rho_0^1\, {}^{P_2}\rho_0^1 + {}^{P_1}\rho_0^2\, {}^{P_2}\rho_0^2) \\ {}^{D}\rho_0^0 &= 0.745\,{}^{P_1}\rho_0^0\, {}^{P_2}\rho_0^0 + 0.373\,{}^{P_1}\rho_0^1\, {}^{P_2}\rho_0^1 + 0.075\,{}^{P_1}\rho_0^2\, {}^{P_2}\rho_0^2 \end{aligned} \tag{32} $$

From these equations we obtain for unpolarized capture processes a "statistical population" (${}^{l}\rho_0^0 = 1/\sqrt{2l+1}$), whereas for polarized capture ($\rho_0^1, \rho_0^2 \neq 0$) a dominant population of the D term results. Taking the extreme limits of ρ_0^1, ρ_0^2 in equation (32) yields an exclusive population of the D term.

Applying this model to our observations in ^{14}N gives a consistent description of the data, irrespective to the fact that we have to take into account Coulomb repulsion and Pauli principle for the N I $2p^3$ configuration. Both effects cause level splittings in the eV domain which are much larger than any typical energies in the ion-solid interaction. So the factorization of the density operator as given in equation (31) is somewhat critical, and the assumption of a polarized core by Schröder[43] to better describe the polarization of excited Ar II multiplets is not applicable in this case. However, it seems to be reasonable to assume that the neutralization in nitrogen at grazing ion-surface scattering is strongly affected by consecutive anisotropic electron capture.

ACKNOWLEDGMENTS. Many valuable discussions with Professors J. Andrä, J. Burgdörfer, H. Schröder, E. Kupfer, J. Los, R. Brako, and D. Newns are gratefully acknowledged. We thank H. Hagedorn, H. W. Ortjohann,

A. Schirmacher, U. Schriek, and P. Strohmeier for their assistance during the measurements and A. Burroughs and W. Hassenmeier for processing the manuscript. This work is supported by the Sonderforschungsbereich 216 Bielefeld/Münster.

REFERENCES

1. H. J. ANDRÄ, *Phys. Rev. Lett.* **25**, 325 (1970).
2. H. G. BERRY, L. J. CURTIS, D. G. ELLIS, AND R. M. SCHECTMANN, *Phys. Rev. Lett.* **32**, 751 (1974).
3. D. G. ELLIS, *J. Opt. Soc. Am.* **63**, 1322 (1973); U. FANO AND J. H. MACEK, *Rev. Mod. Phys.* **45**, 553 (1973).
4. H. J. ANDRÄ, *Phys. Lett.* **54A**, 315 (1975).
5. U. SCHRIEK AND H. WINTER, unpublished.
6. D. S. GEMMELL, *Rev. Mod. Phys.* **46**, 129 (1974).
7. G. MOLIERE, *Z. Naturforsch.* **2a**, 133 (1947).
8. O. B. FIRSOV, *Sov. Phys. JETP* **33**, 534 (1958).
9. Y. H. OHTSUKI, K. KOYAMA, AND Y. YAMAMURA, *Phys. Rev.* **20**, 5044 (1979).
10. H. D. HAGSTRUM, *Phys. Rev.* **96**, 336 (1954); J. W. GADZUK, *Phys. Rev. B* **1**, 133 (1967).
10a. E. W. THOMAS, in *Applied Atomic Collision Physics*, edited by S. Datz, Vol. 4 (Academic Press, New York, 1983).
11. W. PAULI, in *Collected Scientific Papers*, edited by R. Kronig and V. F. Weisskopf (Interscience Publishers, New York, 1964), p. 549.
12. R. BRAKO AND D. M. NEWNS, *Vacuum* **32**, 39 (1982).
12a. K. L. SEBASTIAN, *Phys. Rev. B* **31**, 6976 (1985).
12b. R. BRAKO AND D. M. NEWNS, *Phys. Scr.* **32**, 451 (1985).
13. R. FRÖHLING AND H. J. ANDRÄ, *Z. Phys. A* **320**, 207 (1985).
14. W. GRASER AND C. VARELAS, *Phys. Scr.* **T6**, 153 (1983).
15. R. ZIMNY, H. WINTER, B. BECKER, A. SCHIRMACHER, AND H. J. ANDRÄ, *Nucl. Instrum. Methods B* **2**, 252 (1984).
16. W. PAULI, *Z. Phys.* **36**, 336 (1926); G. FLAMAND, *J. Math. Phys.* **7**, 1924 (1966).
17. R. ZIMNY, H. HAGEDORN, H. WINTER, AND H. J. ANDRÄ, *Nucl. Instrum. Methods B* **13**, 601 (1986).
17a. N. D. LANG AND W. KOHN, *Phys. Rev. B* **7**, 354 (1973); J. M. BARIBEAU, J. LOPEZ, AND J. C. LE BOSSE, *J. Phys. C* **18**, 3083 (1985); M. N. READ, *Phys. Rev. B* **32**, 2677 (1985).
18. T. P. GROZDANOV AND R. K. JANEV, *Phys. Lett. A* **65**, 396 (1978).
19. J. N. M. VAN WUNNIK, thesis, FOM-Institute, Amsterdam (1983); J. N. M. VAN WUNNIK, R. BRAKO, K. MAKASHI, AND D. M. NEWNS, *Surf. Sci.* **261**, 618 (1983); J. N. M. VAN WUNNIK AND J. LOS, *Phys. Scr.* **T6**, 27 (1983); H. GEERLINGS, thesis, FOM-Institute, Amsterdam (1986).
19a. H. NIENHAUS, R. ZIMNY, H. HAGEDORN, AND H. WINTER, *Surf. Sci.*, to be published.
19b. R. ZIMNY, thesis, University of Münster, 1988; H. WINTER, R. ZIMNY, B. BECKER, A. SCHIRMACHER, H. J. ANDRÄ, AND R. FROHLING, *Z. Phys. A* **311**, 167 (1983).
20. R. FRÖHLING, thesis, Free University, Berlin (1983).
21. J. NEUMANN, S. SCHUBERT, U. IMKE, P. VARGA, AND W. HEILAND, *Europhys. Lett.* **3**, 859 (1987).
22. J. D. JACKSON AND H. SCHIFF, *Phys. Rev.* **89**, 359 (1952).
23. S. Y. LEUNG, N. H. TOLK, W. HEILAND, J. C. TULLY, J. S. KRAUS, AND P. HILL, *Phys. Rev. A* **18**, 447 (1978).
24. J. MACEK, *Phys. Rev. A* **1**, 235 (1970).

25. P. STROHMEIER, unpublished.
26. M. W. LUCAS, W. STECKELMACHER, J. MACEK, AND J. E. POTTER, *J. Phys. B* **13**, 4833 (1980).
27. H. HAGEDORN, H. NIENHAUS, H. WINTER, G. MANK, R. VÖLPEL, AND E. SALZBORN, to be published; S. T. DE ZWART, T. FRIED, U. JELLEN, A. L. BOERS, AND A. G. DRENTJE, *J. Phys. B* **18**, L623 (1985).
28. K. BLUM, *Density Matrix Theory and Applications* (Plenum Press, New York, 1981).
29. H. J. ANDRÄ, R. FRÖHLING, AND H. J. PLÖHN, in *Inelastic Ion-Surface Collisions*, edited by N. H. Tolk, J. C. Tully, W. Heiland, and C. W. White (Academic, New York, 1977), p. 329.
30. H. J. ANDRÄ AND H. WINTER, *Hyperfine Int.* **5**, 403 (1978); H. WINTER AND H. J. ANDRÄ, *Phys. Rev. A* **21**, 581 (1980).
31. H. J. ANDRÄ, H. J. PLÖHN, A. GAUPP, AND R. FRÖHLING, *Z. Phys. A* **281**, 15 (1977); H. WINTER AND R. ZIMNY, *Hyperfine Int.* **22**, 237 (1985).
32. D. A. CHURCH, C. S. LEE, AND J. LENOIR, *Phys. Rev. Lett.* **53**, 1394 (1984).
32a. J. HERMANN, J. GEHRING, AND V. KEMPTER, *Surf. Sci.* **171**, 377 (1986).
33. N. H. TOLK, J. C. TULLY, J. S. KRAUS, W. HEILAND, AND S. H. NEFF, *Phys. Rev. Lett.* **41**, 643 (1978).
33a. J. LENOIR, J. LIU, AND D. A. CHURCH, *Nucl. Instrum. Methods B* **24/25**, 312 (1987).
34. H. J. ANDRÄ, R. ZIMNY, H. WINTER, AND H. HAGEDORN, *Nucl. Instrum. Methods B* **9**, 572 (1985).
35. R. BRAKO, *Phys. Rev. B* **30**, 5629 (1984).
36. J. C. TULLY, N. H. TOLK, J. S. KRAUSS, C. RAU, AND R. J. MORRIS, in *Inelastic Particle-Surface Collisions*, edited by E. Taglauer and W. Heiland, Springer Series in Chemical Physics 17 (Springer, Berlin, 1981), p. 196.
36a. H. WINTER AND H. W. ORTJOHANN, *Rev. Sci. Instrum.* **58**, 359 (1987).
37. R. BRAKO AND D. M. NEWNS, in *Proceedings of the ICPEAC*, edited by H. B. Gilbody *et al.* (Elsevier, Amsterdam, 1988), p. 801.
38. J. BURGDÖRFER AND E. KUPFER, *Phys. Rev. Lett.* **57**, 2649 (1986).
39. J. BURGDÖRFER, E. KUPFER, AND H. GABRIEL, *Phys. Rev. A* **35**, 4963 (1987).
39a. C. RAU AND R. SIZMANN, *Phys. Lett.* **43A**, 317 (1973).
39b. C. RAU, *J. Magn. Magn. Mater.* **30**, 14 (1982).
39c. E. KUPFER, H. GABRIEL, AND H. SCHRÖDER, *Z. Phys. A* **283**, 321 (1977).
39d. H. WINTER, H. HAGEDORN, R. ZIMNY, H. NIENHAUS, AND J. KIRSCHNER, *Phys. Rev. Lett.*, to be published.
39e. G. S. HARBINSON, B. W. FARMERY, H. J. PABST, AND M. W. THOMPSON, *Radiat. Eff.* **27**, 97 (1975).
39f. C. VARELAS, K. GOLTZ, AND R. SIZMANN, *Ned. Tijdschr. Vacuumtech.* **16**, 340 (1978); C. VARELAS AND R. SIZMANN, *Surf. Sci.* **71**, 51 (1978).
39g. N. H. TOLK, J. C. TULLY, J. S. KRAUS, W. HEILAND, AND S. H. NEFF, *Surf. Sci.* **90**, 447 (1979); H. H. ANDRÄ, H. WINTER, R. FRÖHLING, N. KIRCHNER, H. J. PLÖHN, W. WITTMANN, W. GRASER, AND C. VARELAS, *Nucl. Instrum. Methods* **170**, 527 (1980); H. OBERMEYER, K. SNOWDON, H. HEMME, AND W. HEILAND, *Z. Phys. B* **61**, 187 (1985).
40. H. WINTER AND H. J. ANDRÄ, *Hyperfine Int.* **24**, 277 (1985).
41. E. TAUTE, *Phys. Stat. Sol. B* **142**, 437 (1987).
41a. H. WINTER, M. LANGHEIM, A. SCHIRMACHER, R. ZIMNY, AND H. J. ANDRÄ, *Phys. Rev. Lett.* **52**, 1211 (1984).
42. M. DUCLOY AND M. DUMONT, *J. Phys. (Paris)* **31**, 419 (1970).
43. H. SCHRÖDER, *Nucl. Instrum. Methods B* **2**, 213 (1984).
44. E. G. OVERBOSCH, B. RASSER, A. D. TENNER, AND J. LOS, *Surf. Sci.* **92**, 310 (1980).
45. M. REMY, *J. Chem. Phys.* **53**, 2487 (1970).
46. S. I. EASA AND A. MODINOS, *Surf. Sci.* **183**, 531 (1987).

CHAPTER 9

THE TWO-PHOTON DECAY OF ATOMIC HYDROGEN AND TESTS OF BELL'S INEQUALITY

A. J. DUNCAN

1. INTRODUCTION

Ever since it was discovered that atoms have an internal structure, the theoretical and experimental study of atomic hydrogen has played a crucial role in the development of our knowledge and understanding of atomic physics and quantum mechanics. The states with principal quantum number $n = 2$ are and have been of special interest and importance, in particular with regard to the determination of the fine structure constant and the measurement of the Lamb shift. It is also possible that, in the future, precise measurement of the lifetime of the metastable $2\,^2S_{1/2}$ state could reveal the existence of time reversal and parity nonconserving effects due, for example, to the presence of a permanent electron dipole moment or neutral currents. It was, of course, the observation of the Lamb shift in 1947 by Lamb and Retherford(1) that, by demonstrating the nondegeneracy of the $2\,^2S_{1/2}$ and $2\,^2P_{1/2}$ states, confirmed that the $2\,^2S_{1/2}$ state would be metastable in experimentally realizable situations, and showed that it should be possible to observe the two-photon emission that is the main mode of decay of this state.

The existence of spontaneous two-photon decay processes had been predicted as early as 1931 by Göppert-Mayer(2) in a paper that pioneered the field of multiphoton transitions, and, in 1940, Breit and Teller,(3) with astrophysical applications in mind, applied this theory to the $2\,^2S_{1/2}$–$1\,^2S_{1/2}$ transition in atomic hydrogen. Further calculations of the characteristics of the two-photon decay process were carried out by Spitzer and Greenstein,(4)

A. J. DUNCAN • Atomic Physics Laboratory, University of Stirling, Stirling FK9 4LA, Scotland.

Shapiro and Breit,[5] Zon and Rapaport,[6] Klarsfeld,[7] and Johnson.[8] More recently, the two-photon decay process has continued to excite the attention of theoretical physicists.[9-14]

Despite the early theoretical work, the spontaneous two-photon decay process was not observed successfully in a laboratory until 1965 when Lipeles, Novick, and Tolk[15] carried out their experiment using singly ionized helium, further results being published by Artura, Tolk, and Novick[16] in 1969. Subsequently, the two-photon decay of hydrogenlike argon and sulfur ions was observed by Schmeider and Marrus[17] and Marrus and Schmeider[18] in a beam–foil-type experiment, and, in a similar experiment, Cocke, Curnutte, Macdonald, Bednar, and Marrus[19] measured the lifetime of the $2\,^2S_{1/2}$ state of hydrogenlike fluorine and oxygen. About the same time Prior[20] measured the lifetime of the $2\,^2S_{1/2}$ state of singly ionized helium using an ion-trapping technique, while Kocher, Clendenin, and Novick[21] and, later, Hinds, Clendenin, and Novick[22] did the same using a decay-in-flight method. In 1983, Gould and Marrus[23] carried out a further investigation of the two-photon decay process in hydrogenlike argon and derived a value for the Lamb shift by observing the quenching of the $2\,^2S_{1/2}$ state in an electric field. It is perhaps also worth noting here the experiment of Bannett and Freund[24] in which K-shell hole states in molybdenum were filled by $2S$ to $1S$ transitions with the accompanying emission of two X-ray photons, and the experiment of Braunlich, Hall, and Lambropoulos,[25] in which singly stimulated emission from the $2\,^2S_{1/2}$ state of atomic hydrogen was observed.

Because of the long lifetime of the $2\,^2S_{1/2}$ state and the problems involved in producing an intense enough source, it is very difficult experimentally to observe the two-photon decay in atomic hydrogen itself. That such an observation should be possible was suggested by Kleinpoppen while at Tübingen in the 1960s, but it was not until 1975 that the experiment was successfully completed at Stirling by O'Connell, Kollath, Duncan, and Kleinpoppen.[26] Almost simultaneously Kruger and Oed[27] at Tübingen published their results. These first experiments on atomic hydrogen concentrated on measurement of the lifetime of the metastable state as well as the angular correlation and spectral distribution of the two photons emitted in the decay process. However, considerable interest attaches to the measurement of the polarization correlation of the two photons both as a sensitive test of the theory of the two-photon decay process itself and as a test of Bell's inequality,[28] which allows a quantitative distinction to be made between, on the one hand, the predictions of quantum mechanics and, on the other, the predictions of local realistic or hidden variables theories. It is, therefore, fortunate that, among all one-electron atoms and ions, hydrogen itself is unique in that the wavelengths of the emitted photons fall mostly in a region of the spectrum where the polarization correlation can be measured in air using conventional optical components, and, at Stirling, measurements of the linear and circular polarization correlation have recently been completed and Bell's inequality put to the test.[29,30]

It is important to realize that experimental arrangements in which tests of Bell's inequality are possible are extremely rare and represent the only known situations where, making an apparently reasonable additional assumption, one is compelled to choose between the validity of quantum mechanics and local realism. Nearly all previous tests have been carried out by observing the polarization correlation either of the two γ rays that are emitted as a result of electron-positron annihilation or of the two photons emitted in succession in an atomic cascade when an excited atom decays to a state of lower energy via an intermediate state with a finite lifetime. Work in the field up to 1978 is reviewed in the articles by Pipkin[31] and Clauser and Shimony.[32] Since then a series of important experiments making use of a calcium atomic cascade has been reported by Aspect and his collaborators,[33-36] and Paramananda and Butt[37] have published further results involving an electron-positron annihilation source. However, all polarization correlation measurements so far used to test Bell's inequality can be criticized on the grounds that the analyzing polarizers themselves may introduce a bias or enhancement effect into the detection efficiencies for photons. To investigate such a possibility and to test further the quantum formalism, experiments have very recently been carried out at Stirling in which a third element, for example, a half-wave plate[38] or linear polarizer,[30] is inserted in the optical system. Before describing the Stirling experiments on the two-photon polarization correlation, however, the theory of the two-photon decay will be discussed and the appropriate form of Bell's inequality will be derived.

2. THEORY OF THE TWO-PHOTON DECAY OF HYDROGEN

Since, in this case, electric dipole and electric quadrupole transitions are forbidden, metastable atomic hydrogen [H(2S)] has a long lifetime of about 1/8 s and decays mainly by the simultaneous emission of the two photons resulting from a two-photon electric dipole transition to the ground state as shown in Figure 1. Other processes contribute to the decay. For example, since, because of the Lamb shift, the $2\,^2P_{1/2}$ state lies slightly below the $2\,^2S_{1/2}$ state in energy, a cascade process in which two photons are emitted sequentially is possible. However, owing to the smallness of the $2\,^2P_{1/2}$–$2\,^2S_{1/2}$ energy interval, in hydrogen the associated lifetime for such a cascade is about 5×10^9 s,[5] and, for most purposes, this mode of decay can be neglected. In addition, although, when the full relativistic

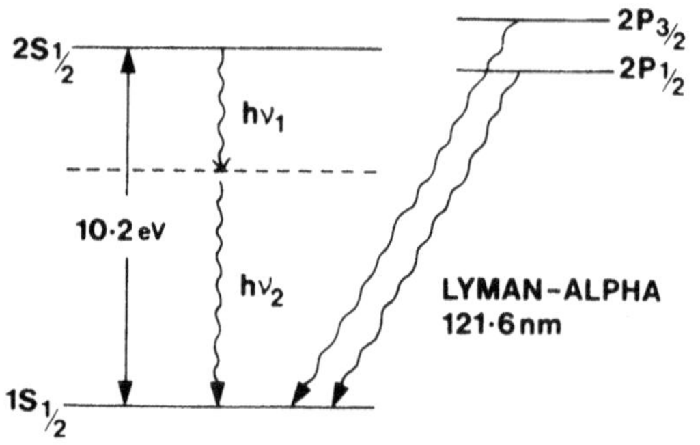

FIGURE 1. Energy level diagram, neglecting hyperfine structure, for atomic hydrogen, illustrating the two-photon decay process (not to scale).

Dirac wave functions are used, it turns out that single photon magnetic dipole transitions are then allowed, these transitions, in fact, only become important in high Z hydrogenic ions, and, in hydrogen itself, where the associated lifetime is about 4×10^5 s,[10] the process is of little significance. Finally, it should be pointed out that higher-order two-photon processes may occur. No lifetime estimates appear to exist for such processes, but Au[39] has shown that, taking the possibility of two-photon electric quadrupole transitions into account, the form of the angular correlation of the two photons would be altered from the two-photon dipole prediction only by terms quadratic in the fine structure constant.

The theory of the two-photon decay, first put forward by Göppert-Mayer using second-order perturbation theory, can be developed by expanding the scattering operator S to second order in terms of the interaction Hamiltonian H_i between the atomic electron and the radiation field, the relevant part of the interaction Hamiltonian being given by $H_i = -(e/m)\mathbf{p} \cdot \mathbf{A}$ where $-e$, m, and p are, respectively, the electron charge, mass, and canonical momentum, and $\mathbf{A}$ is the vector potential of the radiation field. This form for H_i gives rise to a matrix element describing the two-photon decay in the form

$$\frac{e^2}{m^2} \sum_j \frac{\langle \phi_f | \mathbf{p} \cdot \mathbf{A} | \phi_j \rangle \langle \phi_j | \mathbf{p} \cdot \mathbf{A} | \phi_i \rangle}{E_i - E_j} \tag{1}$$

where $|\phi_i\rangle$, $|\phi_f\rangle$ represent, respectively, the initial and final states of the atom and radiation field together, and E_i is the energy of the initial state. The summation is to be taken over all allowed virtual intermediate states $|\phi_j\rangle$ of energy E_j which, in the present case, contain one more photon than the initial state $|\phi_i\rangle$. In terms of the electronic states $|\psi_i\rangle$, $|\psi_f\rangle$, and $|\psi_j\rangle$ of energies W_i, W_f, and W_j, respectively, when photons of energies $h\nu_1$ and $h\nu_2$ are emitted, it is possible to write

$$|\phi_i\rangle = |\psi_i\rangle|0\rangle, \qquad |\phi_f\rangle = |\psi_f\rangle|h\nu_1\rangle|h\nu_2\rangle$$
$$|\phi_j\rangle = |\psi_j\rangle|h\nu_1\rangle \quad \text{or} \quad |\psi_j\rangle|h\nu_2\rangle \tag{2}$$

where $|0\rangle$ represents the initial state of the radiation field, $|h\nu_1\rangle$ ($|h\nu_2\rangle$) its state when a photon of energy $h\nu_1$ ($h\nu_2$) has been added. The intermediate states $|\psi_j\rangle$ are, of course, the P states of atomic hydrogen. If $\overline{\overline{A(\nu_1)}}\,d\nu_1$ denotes the probability per second for the spontaneous, simultaneous emission of two photons with one photon in the frequency range between ν_1 and $\nu_1 + d\nu_1$ irrespective of the directions of emission or polarization of the photons, then making the dipole approximation and changing from the dipole velocity to the dipole length form of the matrix element gives the result

$$\overline{\overline{A(\nu_1)}}\,d\nu_1 = \frac{1024\pi^6 e^4 \nu_1^3 \nu_2^3}{c^6} \left| \sum_j \frac{\langle\psi_f|\mathbf{r}\cdot\hat{e}_1|\psi_j\rangle\langle\psi_j|\mathbf{r}\cdot\hat{e}_2|\psi_i\rangle}{W_i - W_j - h\nu_2} + \sum_j \frac{\langle\psi_f|\mathbf{r}\cdot\hat{e}_2|\psi_j\rangle\langle\psi_j|\mathbf{r}\cdot\hat{e}_1|\psi_i\rangle}{W_i - W_j - h\nu_1} \right|^2_{\text{av}} d\nu_1 \tag{3}$$

where $\hat{e}_1$ and $\hat{e}_2$ represent unit vectors in the directions of polarization of the two photons and the average is taken over the directions of emission and polarization. The presence of two terms in equation (3) is a result of the fact that the emission of a complementary pair of photons of energies $h\nu_1$ and $h\nu_2$ can occur in two ways. Finally, before leaving this discussion of the formal theory it should be noted that in the description of the two-photon decay it is customary practice to talk in terms of the directions of emission, the directions of polarization, and the frequencies or energies of the photons. As the experiments to be described here forcefully demonstrate, the attribution of objective properties to photons in this way prior to an act of measurement, although semantically convenient, runs contrary to the fundamental concepts of quantum mechanics as expressed in the Copenhagen interpretation and, in some cases, can result in serious error.

The theoretical formalism outlined above makes certain predictions regarding the nature of the two-photon decay process. First of all, it is worth noting that, since the atom never really assumes the virtual intermediate states $|\psi_j\rangle$, the two photons are, in fact, emitted simultaneously. The energies of a complementary pair must, of course, conserve energy, so that the sum of their energies is always equal to 10.2 eV, but the energy of a single photon can be anything from 0 up to 10.2 eV, as shown in Figure 2. Since, too, the electric dipole operator is diagonal in the nuclear and

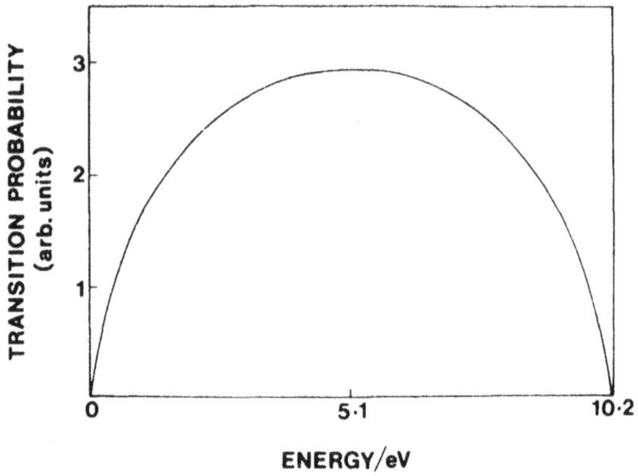

FIGURE 2. Spectral distribution predicted for the two photons emitted in the decay of H(2*S*). Note that the most probable occurrence is the emission of two photons each with energy 5.1 eV and wavelength 243 nm.

electronic spin, the presence of hyperfine and fine structure can be neglected in the determination of the properties of the two-photon decay. Also, if θ is the angle between the linear polarization unit vectors $\hat{e}_1$ and $\hat{e}_2$ describing the polarization of the photons of energy $h\nu_1$ and $h\nu_2$, respectively, emitted at an angle α relative to each other as shown in Figure 3, then it can be shown that the transition probability depends on the photon polarizations according to $|\hat{e}_1 \cdot \hat{e}_2|^2$, i.e., to $\cos^2 \theta$, while averaged over polarization, the transition probability has a dependence on the relative angle of emission given by $(1 + \cos^2 \alpha)$.

When the two photons are emitted in diametrically opposite directions, i.e., when $\alpha = \pi$, the polarization part of the two-photon state vector takes a particularly interesting form, which can easily be derived from the requirements of conservation of angular momentum and parity. For example, in order to conserve angular momentum when the two photons are emitted in opposite directions from a source whose angular momentum remains zero, it is necessary for the photons to have equal helicity. In terms of right-handed

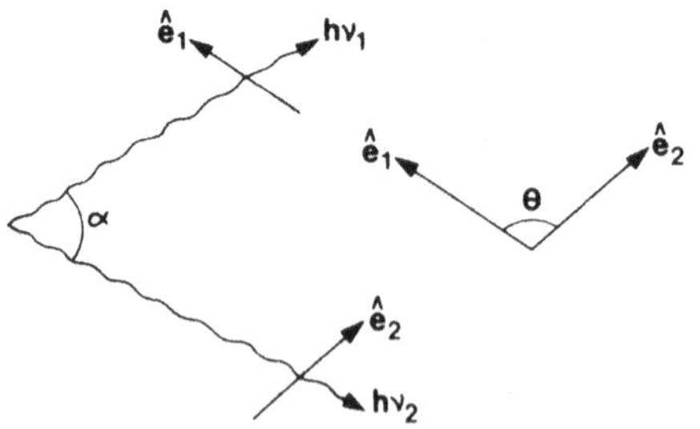

FIGURE 3. Diagram illustrating the angles involved in the description of the two-photon emission process.

($|R\rangle$) and left-handed ($|L\rangle$) helicity states it is thus expected that photon pairs will be found in states represented by $|R\rangle_1|R\rangle_2$, $|L\rangle_1|L\rangle_2$ or a superposition of these, where $|R\rangle_1$ denotes a photon of right-handed helicity propagating to the right, $|R\rangle_2$ a photon of right-handed helicity propagating to the left, and so on. In addition, since the initial and final states of the hydrogen atom both have even parity, the two-photon state vector must also possess even parity. Now, if P is the parity operator, $P|R\rangle_1 = |L\rangle_2$, $P|L\rangle_1 = |R\rangle_2$, etc. so that, to ensure even parity, the two-photon state vector must take the form

$$|\psi\rangle = 2^{-1/2}[|R\rangle_1|R\rangle_2 + |L\rangle_1|L\rangle_2] \tag{4}$$

Alternatively, assuming the photons propagate in the $+z$ and $-z$ directions, it is possible to describe the two-photon state vector in terms of the linear polarization basis states $|x\rangle$ and $|y\rangle$ using the relations

$$\begin{aligned} |R\rangle_1 &= 2^{-1/2}(|x\rangle_1 + i|y\rangle_1), \qquad & |R\rangle_2 &= 2^{-1/2}(|x\rangle_2 - i|y\rangle_2) \\ |L\rangle_1 &= 2^{-1/2}(|x\rangle_1 - i|y\rangle_1), \qquad & |L\rangle_2 &= 2^{-1/2}(|x\rangle_2 + i|y\rangle_2) \end{aligned} \tag{5}$$

Substituting these relations into equation (4) then gives

$$|\psi\rangle = 2^{-1/2}(|x\rangle_1|x\rangle_2 + |y\rangle_1|y\rangle_2) \tag{6}$$

as the appropriate state vector describing the polarization properties of the two photons. The fact that this state vector is invariant in form with respect to rotation about the z axis plays an important role in the formulation of the Einstein–Podolsky–Rosen (EPR)[40] argument, as applied to photons, concerning the incompleteness of quantum mechanics and in the discussion of the nonlocality and lack of realism inherent in the quantum formalism.

3. BELL'S INEQUALITY

In a typical ideal experimental arrangement, shown schematically in Figure 4, a linear polarizer and detector are placed on each side of a source that emits photons in diametrically opposite directions. The form of the state vector given in equation (6) then implies a strong correlation between the polarizations of the two photons detected on either side of the source, since detection of a photon polarized say, in the x direction on the left results in a collapse of the state vector to the form $|x\rangle_1|x\rangle_2$, and this implies that a photon polarized in the x direction will also be detected on the right. But, as has already been noted, the choice of x and y directions is quite

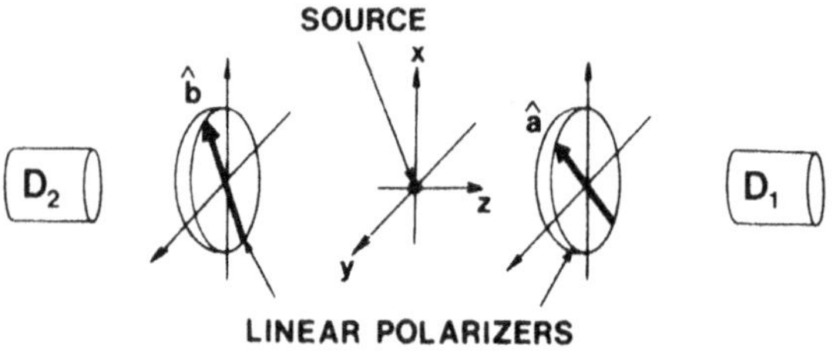

FIGURE 4. Schematic diagram of the typical ideal experimental arrangement using two linear polarizers and two detectors D_1 and D_2. The polarizers are set with their transmission axes in the directions $\hat{a}$ and $\hat{b}$.

arbitrary with respect to rotation about the axis of propagation, and thus the polarization of the photon, say, on the right can be considered to have been determined by the polarization measurement chosen to be made on the other member of the pair on the left, despite the fact that the detection events are spatially separated in the relativistic sense.

The above discussion, in fact, forms the essence of the Bohm[(41-43)] version of the Einstein-Podolsky-Rosen[(40)] argument against the completeness of quantum mechanics, for it implies that for every setting of the polarizer on one side of the source, there exists an "element of reality" corresponding to the polarization of the photon on the other side. That a photon can be linearly polarized in more than one direction at once is, of course, more than is allowed by quantum mechanics, and Einstein, Podolsky, and Rosen hoped that the situation might be resolved by the introduction of hidden variables, which would, at one and the same time, "complete" quantum mechanics and restore determinism to physics. However, the phenomenon is now seen more as a compelling example of the nonlocality inherent in nature as described by quantum mechanics, which, although it seems strange, is quite consistent with the Copenhagen interpretation in which the polarization correlation must be considered a property of the photon pair and measuring apparatus taken as an indivisible whole. It is, on this view, not possible to regard individual photons as particles with objectively real properties whose values are independent of any measurements that may be made on them. It appears, therefore, that quantum mechanics presents a picture of the world that is neither local nor realistic in nature. Many people have found such a picture unsatisfactory, and attempts have been made to find local realistic theories (local hidden variables theories) to replace or supplement quantum mechanics. However, in 1964 Bell[(28)] showed that, in a certain type of ideal experiment, local realistic theories were incapable of predicting the strong correlation given by quantum mechanics, and he set an upper bound on the strength of the correlation allowed by those theories. Subsequently, Clauser, Horne,

Shimony, and Holt[44] and, later, Clauser and Horne[45] extended the work of Bell so that it could be applied in actual experimental situations, where only a very small fraction of emitted photon pairs is detected, by introducing an additional assumption in either of the following two forms: (1) If a pair of photons emerges from two polarizers, the probability of their joint detection is independent of the orientation of the polarizers' axes. (2) For every atomic emission, the probability of a count with a polarizer in place is not larger than with the polarizer removed. The first form, due to Clauser, Horne, Shimony, and Holt[44] can, of course, be criticized on the grounds that it depends on the concept of emergence of individual photons from polarizers. The second form, due to Clauser and Horne[45] and not subject to this objection, was referred to by them as the no-enhancement assumption. Both forms are simply statements that the presence of the polarizers does not introduce any unsuspected bias or enhancement into the detection process for photon pairs.

Bell's inequality in the version suitable for use with the present series of experiments can be derived most easily following the method of Clauser and Horne.[45] Referring to Figure 4, consider a source emitting an ensemble of photon pairs, the state of each pair being described by a parameter λ (a hidden variable) of arbitrary complexity with a normalized probability density $\rho(\lambda)$. With reference to a pair described by λ, let the probability of detection on the right, when the corresponding polarizer is oriented with its transmission axis in the direction $\hat{a}$, be denoted by $p_1(\lambda, \hat{a})$ with a similar definition for $p_2(\lambda, \hat{b})$. The probability of joint detection of the two photons of a pair can be written $p_{12}(\lambda, \hat{a}, \hat{b})$. Then the ensemble probabilities $p_1(\hat{a})$, $p_2(\hat{b})$, and $p_{12}(\hat{a}, \hat{b})$ to detect a photon or a photon pair are given by

$$p_1(a) = \int_\Gamma p_1(\lambda, a)\rho(\lambda)\, d\lambda, \qquad p_2(b) = \int_\Gamma p_2(\lambda, b)\rho(\lambda)\, d\lambda$$
$$p_{12}(a, b) = \int_\Gamma p_{12}(\lambda, a, b)\rho(\lambda)\, d\lambda \tag{7}$$

where Γ is the space of the states λ and, for clarity, the unit vector signs on a and b have been omitted.

It should be noted that, since λ only determines the probability of detection, the inequality to be derived here applies to inherently stochastic local realistic theories. It is the question of locality, not strict determinism, that is of crucial importance to the argument, and the assumption of locality is made, in this case, by demanding that

$$p_{12}(\lambda, a, b) = p_1(\lambda, a)p_2(\lambda, b) \tag{8}$$

The no-enhancement assumption is introduced, for every λ and for all orientations a and b, in the form

$$\begin{aligned} 0 \leqslant p_1(\lambda, a) \leqslant p_1(\lambda, \infty) \leqslant 1 \\ 0 \leqslant p_2(\lambda, b) \leqslant p_2(\lambda, \infty) \leqslant 1 \end{aligned} \tag{9}$$

where the symbol ∞ is used to indicate the absence of a polarizer. Now, as shown by Clauser and Horne,[45] if x, x', y, y', X, and Y are real numbers such that $0 \leqslant x, x' \leqslant X$ and $0 \leqslant y, y' \leqslant Y$ then

$$-XY \leqslant xy - xy' + x'y + x'y' - Yx' - Xy \leqslant 0 \tag{10}$$

Thus, if a' and b' are alternative settings of the polarizers, the above result can be used to write the inequality

$$\begin{aligned} -p_1(\lambda, \infty)p_2(\lambda, \infty) \leqslant\ & p_1(\lambda, a)p_2(\lambda, b) - p_1(\lambda, a)p_2(\lambda, b') \\ & + p_1(\lambda, a')p_2(\lambda, b) + p_1(\lambda, a')p_2(\lambda, b') \\ & - p_2(\lambda, \infty)p_1(\lambda, a') - p_1(\lambda, \infty)p_2(\lambda, b) \\ \leqslant\ & 0 \end{aligned} \tag{11}$$

which, after multiplying by $\rho(\lambda)$ and integrating over λ, becomes

$$\begin{aligned} -p_{12}(\infty, \infty) \leqslant\ & p_{12}(a, b) - p_{12}(a, b') + p_{12}(a', b) + p_{12}(a', b') \\ & - p_{12}(a', \infty) - p_{12}(\infty, b) \leqslant 0 \end{aligned} \tag{12}$$

This inequality, which, conveniently, only contains probabilities p_{12} for joint detection, can be simplified if the source possesses rotational symmetry about the propagation axis since, then, $p_{12}(a', \infty)$ and $p_{12}(\infty, b)$ are independent of a' and b, respectively, and $p_{12}(a, b)$, $p_{12}(a, b')$, etc. depend only on the relative angle (a, b), (a, b'), etc. between the orientations of the transmission axes of the polarizers. By choosing the particular relative orientations $(a, b) = (a', b) = (a', b') = \theta$, $(a, b') = 3\theta$ for a, a', b, and b', inequality (12) can be expressed as

$$-p_{12}(\infty, \infty) \leqslant 3p_{12}(\theta) - p_{12}(3\theta) - p_{12}(a', \infty) - p_{12}(\infty, b) \leqslant 0 \tag{13}$$

Experimentally, of course, probabilities are deduced from observations carried out over an ensemble of emissions. If $R(a, b)$ denotes the number or rate of joint detections (coincidence counts) with the transmission axes of the polarizers set in the directions a and b, and R_0 the number or rate with both polarizers removed, then $p_{12}(a, b)/p_{12}(\infty, \infty) = R(a, b)/R_0$ with similar relations for the other probabilities. Inequality (13) then takes the form

$$-1 \leqslant [3R(\theta) - R(3\theta) - R(a', \infty) - R(\infty, b)]/R_0 \leqslant 0 \tag{14}$$

Writing inequality (14) for $\theta = 22.5°$ and $\theta = 67.5°$ and subtracting the two results gives Bell's inequality in the form

$$\eta = \left| \frac{R(22.5°) - R(67.5°)}{R_0} \right| \leqslant 0.25 \tag{15}$$

originally derived by Freedman.[46] The Freedman form of the inequality has proved particularly useful from the experimental point of view since it only necessitates making three measurements.

In contrast, it follows directly from consideration of the state vector given in equation (6), which predicts a $(1 + \cos 2\theta)/4$ variation in the ratio $R(\theta)/R_0$, that, in the ideal case, according to quantum mechanics, $\eta_{QM} = 0.354$, in clear violation of the Freedman form of Bell's inequality. However, in practice, various factors act to reduce η_{QM} and care must be taken experimentally to ensure that η_{QM} does not fall below 0.25. The main factors acting to reduce η_{QM} in this way are the transmission efficiencies of the polarizers and the finite solid collection angle subtended at the source by the lenses normally placed on either side of the source in front of the polarizers. If the transmission efficiencies of the polarizers (assumed identical) for light polarized parallel to and perpendicular to their transmission axes are ε_M and ε_m, respectively, and if the half-angle subtended at the source by each lens is δ, then it can be shown that according to quantum mechanics[44]

$$\frac{R(\theta)}{R_0} = \tfrac{1}{4}(\varepsilon_M + \varepsilon_m)^2 + \tfrac{1}{4}(\varepsilon_M - \varepsilon_m)^2 F(\delta) \cos 2\theta \tag{16}$$

where θ is the relative angle between the transmission axes and $F(\delta)$ is a geometrical factor, depending on δ, which takes into account the effect of the noncollinear emission of photon pairs. Thus, on the basis of quantum mechanics, in a real situation, it follows that

$$\eta_{QM} = \frac{F(\delta)}{2\sqrt{2}} (\varepsilon_M - \varepsilon_m)^2 \tag{17}$$

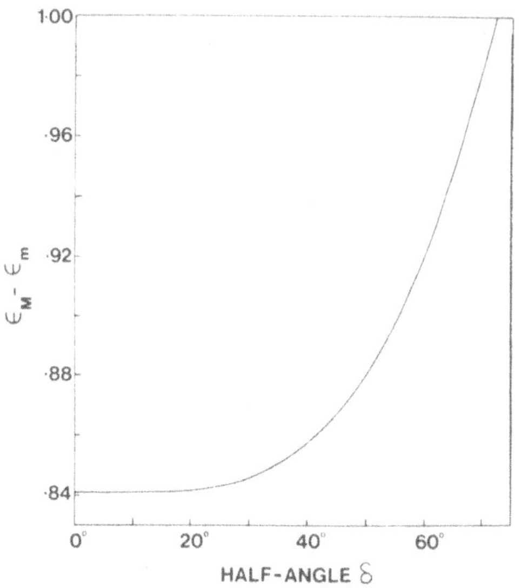

FIGURE 5. Lower limit on $(\varepsilon_M - \varepsilon_m)$ as a function of the half-angle δ subtended by the collection optics at the source. In order for a meaningful test of Bell's inequality to be made, the experiment must be carried out in the region above the curve.

and, clearly, this result places constraints on the values of ε_M, ε_m, and δ if a meaningful test of Bell's inequality is to be made. The requirement that $\eta_{QM} > 0.25$, in fact, sets a lower limit on the allowed value for $(\varepsilon_M - \varepsilon_m)$ which depends on δ in the way shown in Figure 5.

4. THE TWO-PHOTON APPARATUS

In the apparatus shown schematically in Figure 6 a 1-keV beam of metastable atomic deuterium [D(2S)], of density about 10^4 cm^{-3}, was produced by a charge exchange process in which deuterons extracted from a radiofrequency ion source passed through cesium vapor in a cell constructed according to a design by Bacal *et al.*[47,48] Deuterium was used rather than hydrogen since, for a given metastable density and hence two-photon signal, the noise generated by interaction of the beam with the background gas was less in the former case. Also, although it was possible to produce higher beam densities at higher energies, it was found, in practice, that the best statistical accuracy in a given time could be achieved at an energy of 1 keV.

After emerging from the charge exchange cell and being collimated down to about 10 mm in diameter, the D(2S) beam passed through electric field prequench plates which effectively allowed the metastable component of the beam to be switched on and off by making use of the effect of Stark mixing of the $2\,^2S_{1/2}$ and $2\,^2P_{1/2}$ states. In the same way, at the end of the apparatus the beam was totally quenched and the resulting Lyman-α signal

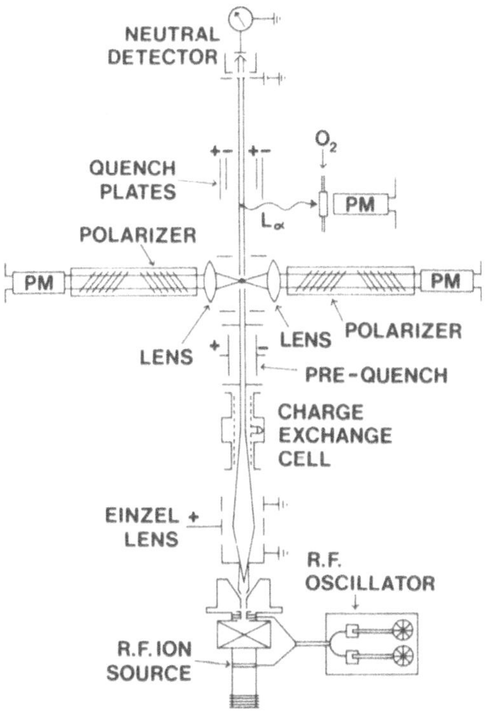

FIGURE 6. Schematic diagram of the Stirling two-photon apparatus (not to scale).

used to normalize the two-photon coincidence signal. The Lyman-α signal itself was detected by means of a solar blind uv photomultiplier used together with an oxygen filter cell with lithium fluoride windows through which dry oxygen flowed at a constant rate. A series of experiments was carried out to check that the two-photon coincidence signal was indeed proportional to the Lyman-α signal over the range of conditions encountered in practice.

At right angles to the metastable beam the two-photon radiation was collected and collimated by two 50-mm-diam lenses, each with a focal length of 43 mm at a wavelength of 243 nm. Each lens was placed close to 50 mm from the beam, and subtended a half angle $\delta \simeq 23°$ at the beam. The position of the lenses was chosen so that, for a wavelength of 243 nm, the center of the beam formed an image on the 46-mm-diam cathodes of the photomultipliers, 53 cm from the center of the beam. By considering the design of the optical system it was estimated that only a 4-mm-diam section of the beam acted as a source for two-photon radiation. For the linear polarization correlation measurements, two high-transmission ultraviolet polarizers, each made from 12 plates polished flat to 2λ at 243 nm and set nearly at Brewster's angle, were placed on either side of the source as shown in Figure 6. For other polarization correlation measurements additional optical elements such as quarter-wave plates, half-wave plates, and linear polarizers could be inserted in the optical system as required. The lenses and plates were made from high-quality fused silica with a short wavelength cutoff at 160 nm. However, because of absorption in oxygen, the short wavelength cutoff occurred at 185 nm, which in turn implied a long wavelength cutoff at the complementary wavelength of 355 nm. Hence, given that the quantum efficiency of the photomultipliers was fairly constant at about 20% over this range, all photons in the wavelength range from 185 to 355 nm were capable of contributing to the two-photon coincidence signal. Also, in order to compare the experimental results with theory it was necessary to know the values of the transmission efficiencies ε_M and ε_m for light polarized, respectively, parallel to and perpendicular to the transmission axes of the polarizers, and these quantities were measured carefully in a subsidiary experiment making use of the 254-nm line from a mercury lamp. In this experiment, using the apparatus as shown in Figure 6, the photomultiplier was removed from one arm and an almost parallel beam of polarized light was passed through the pile-of-plates polarizer (aligned with its transmission axis parallel to the plane of polarization of the light from the mercury source) to come to a focus at the position normally occupied by the center of the atomic source. The light emerging from this focus was then analyzed by the pile-of-plates polarizer and photomultiplier in the other detection arm. After application of a small correction to take into account the slight absorption in the fused silica between 185 and 200 nm, it was found that for two of the polarizers $\varepsilon_M = 0.908 \pm 0.013$ and $\varepsilon_m = 0.0299 \pm 0.0020$, while, for a third polarizer, constructed from plates provided by a different manufacturer, the values $\varepsilon_M = 0.938 \pm 0.010$ and $\varepsilon_m = 0.040 \pm 0.002$ were obtained.

The signal was detected on either side of the source by fast-rise-time photomultipliers, with bialkali cathodes and fused silica windows, which were specially selected for high gain, good single-photoelectron resolution, and good timing characteristics, and had a low dark count rate of about $10^2\,s^{-1}$. The photomultiplier pulses were fed to the coincidence circuit previously described by O'Connell *et al.*,[26] which consisted of the usual combination of constant fraction discriminators, a time-to-amplitude convertor, and a multichannel analyzer operated in the pulse-height analysis mode. The time correlation spectra obtained in this way with the metastable atoms present and quenched were stored in separate segments of the multichannel analyzer memory and subtracted at the end of run, which normally lasted for about 22 h. This procedure was adopted to ensure the elimination of any spurious true coincidence events due, for example, to cosmic rays and residual radioactivity in the apparatus. With the photomultiplier cathodes 1.06 m apart these events occurred at a rate of about $10^{-2}\,s^{-1}$,

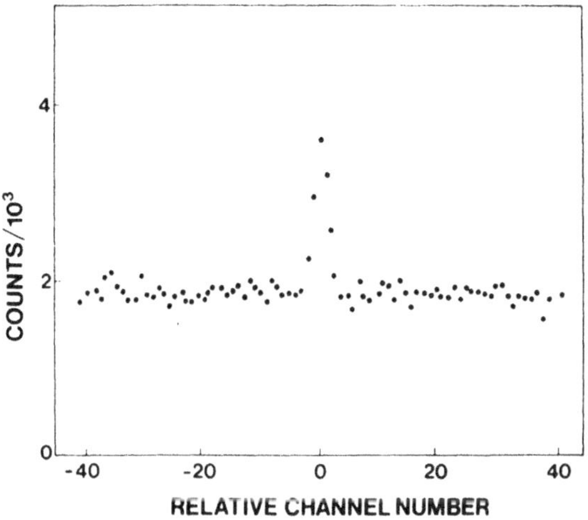

FIGURE 7. A typical time-correlation spectrum[29] after subtraction of the spectrum obtained with the metastable component of the beam quenched. Polarizer plates are removed. Time delay per channel is 0.8 ns, total collection time 21.5 h. Singles rate with metastables present (quenched) is about $1.15 \times 10^4\ s^{-1}$ ($0.85 \times 10^4\ s^{-1}$). True two-photon coincidence rate is $490\ h^{-1}$.

which decreased as the separation of the photomultipliers increased. A typical spectrum obtained in this way is shown in Figure 7, and, as expected for a simultaneous-emission process, the coincidence peak can be seen to be symmetrical. It should also be noted that the background signal resulting from the typical singles count rate of order $10^4\ s^{-1}$ was due mainly to radiation produced by interaction of the atomic beam with background gas in the vacuum system at a pressure of about 2×10^{-7} Torr, and only very little (about 0.01%) to uncorrelated photons from the two-photon decay process itself.

5. POLARIZATION CORRELATION MEASUREMENTS

5.1. Linear Polarization Correlation Experiment

With two polarizers placed as shown in Figure 6, measurements were made of the coincidence rate $R(\theta)$ for various angles θ between the transmission axes of the polarizers, and the rate R_0 with both sets of polarizer plates removed. The photomultipliers were linked to their corresponding polarizers and rotated with them so that any sensitivity to polarization that might exist in the photomultiplier windows or cathodes could be ignored. In addition, the rotational invariance of the detection system was checked by making sure that the background singles rate was constant as the angle of the polarizer transmission axis was varied in each arm independently. All coincidence rates were normalized to a typical Lyman-α count rate, and the resulting linear polarization correlation measurement is shown in Figure 8. The error bars represent one standard deviation and vary from point to point because of the different total counting times allocated to different angles. In particular, the points at 22.5° and 67.5° resulted from measurements that each involved a total counting time of 240 h. The solid curve, which represents the quantum mechanical prediction derived from equation (16) with $\varepsilon_M = 0.908$, $\varepsilon_m = 0.0299$, and $F(\delta) = 0.996$ corresponding to $\delta = 23°$, clearly agrees with the experimental results within the limits of error. Also, from the experimental points at 22.5° and 67.5°, it was found that $\eta = 0.268 \pm 0.011$, which violates the Freedman[46] form of Bell's inequality[28] by just less than two standard deviations, and agrees with the quantum mechanical prediction $\eta_{QM} = 0.272 \pm 0.008$ calculated from equation (17).

5.2. Circular Polarization Correlation Experiment

To measure the circular polarization correlation, quarter-wave plates were inserted in each detection arm of the apparatus between the collimating lens and linear polarizer. These plates were nominally achromatic with a retardation that varied by ±10% over the wavelength range from 180 to 300 nm provided the direction of the incident light did not deviate by more

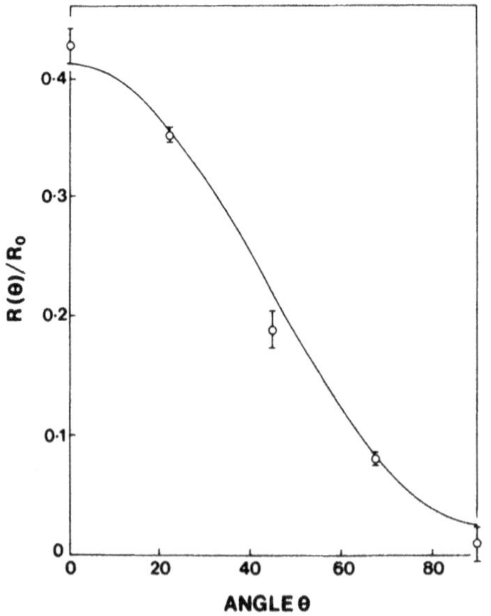

FIGURE 8. Coincidence signal $R(\theta)$ in the linear polarization correlation experiment as a function of the angle θ between the transmission axes of the two polarizers, relative to R_0, the coincidence signal with the polarizer plates removed. The solid curve represents the QM prediction using the median values for ε_M and ε_m.

than 1° from the normal. The presence of the plates reduced the detection solid angle and overall transmission efficiency of the system considerably so that a longer time was required to achieve a given degree of statistical accuracy.

In the experiment the transmission axes of the two linear polarizers were arranged to be fixed perpendicular to each other, and the fast axis of the quarter-wave plate in one detection arm was orientated at 45° to the transmission axis of the linear polarizer in that arm to form an analyzer for light of right-handed helicity. The fast axis of the quarter-wave plate in the other arm was then rotated from a position where that arm acted as an analyzer of light of right-handed helicity (fast axes parallel, $\phi = 0°$) to a position where it acted as an analyzer of light of left-handed helicity (fast axes perpendicular, $\phi = 90°$). If $R(\phi)$ denotes the coincidence rate when the angle between the fast axes of the quarter-wave plates was ϕ, and R_0 the rate when the linear polarizer plates were removed but the quarter-wave plates were kept in place, then, in the ideal case, on the basis of the collapse of the state vector given in equation (4) to the form $|R\rangle_1|R\rangle_2$, it was expected that the quantity $R(\phi)/R_0$ should vary as $(1 + \cos 2\phi)/4$. However, in reality, this result needed to be modified to take into account the imperfections of the linear polarizers and quarter-wave plates. Although the characteristics of the linear polarizers were known, since the light rays from the source, after collimation, could be at angle of ±2° to the detection axis, the retardation properties of the quarter-wave plates were uncertain, and it was difficult to produce a satisfactory quantum mechanical prediction for this case. The experimental results, shown in Figure 9, were, therefore, compared to a least-squares fit to the points of the form $A + B \cos 2\phi + C \cos 4\phi$ with $A = 0.2215$, $B = 0.1430$, and $C = -0.0005$. These results are clearly consistent with conservation of angular momentum for each photon pair along the detection axis but do not violate Bell's inequality since, in this case, $\eta = 0.195 \pm 0.016$. This failure to violate Bell's inequality is, however, most likely due to the aforementioned imperfections in the quarter-wave plates.

5.3. *Test for Enhancement Using a Half-Wave Plate*

Almost all the experiments making use of either a positronium annihilation or atomic cascade source have produced results that have violated Bell's inequality and have agreed with the predictions of quantum mechanics. However, the positronium annihilation experiments are open to the criticism[31,32] that the validity of quantum mechanics itself must be assumed in order to analyze the experimental results, and some discussion[36,49–51] has taken place regarding the validity of the results from atomic cascade experiments where the effects of absorption and reemission

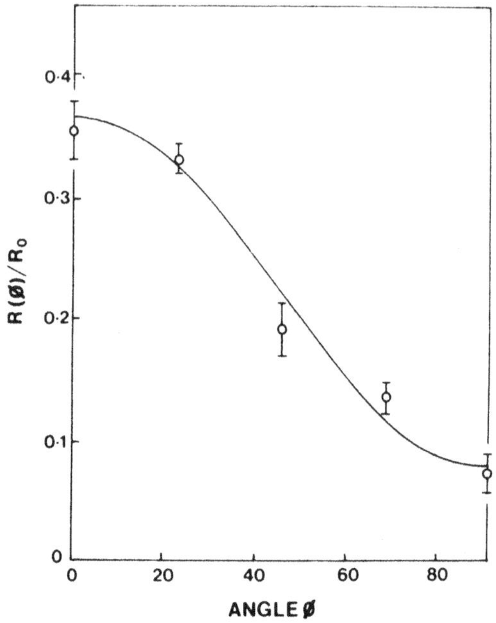

FIGURE 9. Coincidence signal $R(\phi)$ in the circular polarization correlation experiment as a function of the angle ϕ of the fast axis of the quarter-wave plate in one detection arm of the apparatus. The solid line represents a least-squares fit to the data of the form $A + B \cos 2\phi + C \cos 4\phi$ with $A = 0.2215$, $B = 0.1430$, and $C = -0.0005$.

could possibly be important. The metastable atomic deuterium experiment cannot be criticized on either of these grounds, and also the simultaneous nature of the decay ensured that the detection events for every pair were spacelike separated. Nevertheless, because of the low detection efficiency of photomultipliers in the ultraviolet and visible part of the spectrum and the finite solid angle subtended by the detection system at the source, the possibility remained that the results could be explained in local realistic terms if an additional assumption of the kind discussed in Section 3 regarding lack of bias or enhancement in the detection process was not made, and therefore from a purely logical viewpoint no existing experiment could be considered to have provided grounds for a complete rejection of local realistic theories. It followed that it was important to carry out experimental tests either in situations where a distinction could be drawn between the predictions of quantum mechanics and local realism without making any additional assumptions or in situations where the validity of the assumptions themselves could be investigated directly.

Of particular interest here is the work of Garuccio and Selleri,[52] who described how all existing results involving single-photon physics and two-polarizer-type experiments could be explained in local realistic terms if a detection vector **λ**, in addition to a polarization vector **l**, was attributed to each photon. The detection vector **λ** was assumed to be unaffected by passage through linear polarizers, but the detection probability was postulated to depend on the angle between **l** and **λ**. In effect, then, they denied the validity of the additional assumptions and assumed a particular mechanism leading to enhancement. The Clauser, Horne, Shimony, and Holt[(44)] form of the additional assumption and the above proposal of Garuccio and Selleri[(52)] involved the concept of photons emerging from polarizers, and both were tested in a fairly simple way by Haji-Hassan *et al.*[(38)] in an extension to the previously described linear polarization correlation measurements, in which a half-wave plate was inserted in one detection arm of the apparatus between the linear polarizer and photomultiplier as shown in Figure 10.

Assuming that only the relative orientation of the polarizers' axes was important and that the polarization state of each photon emerging from a polarizer was determined by the setting of the polarizer through which it had passed, the Clauser, Horne, Shimony, and Holt[(44)] form of the additional assumption could be expressed in the following form: If a pair of photons emerges from two polarizers, the probability of their joint detection is independent of the relative angle between the planes of polarization of the two photons just prior to detection. The approach of Garuccio and Selleri[(52)] was based precisely on the denial of this assumption since, if their detection vector **λ** was the same for both members of a photon pair, then the detection probability for the pair had to depend on the angle between their respective planes of polarization as they impinged on the cathodes of the photomultipliers. In the previous linear polarization correlation measurements, of course, this angle changed as the polarizers were rotated relative to each other. However, by adding a half-wave plate to the system, as shown in Figure 10, the plane of polarization of the photon incident on photomultiplier D_1 could be varied independently of the setting of polarizer P_1. Initially,

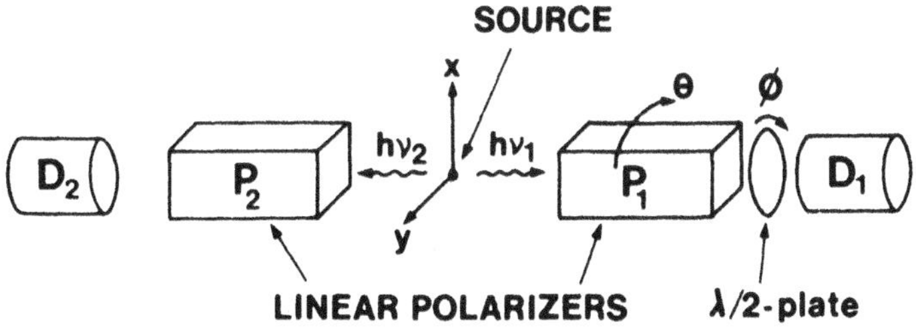

FIGURE 10. Arrangement for two-photon polarization correlation measurements with a half-wave plate placed between the linear polarizer and photomultiplier in one of the detection arms. The transmission axis of the right-hand linear polarizer is rotated through angle θ, the fast axis of the half-wave plate through angle $\phi = \theta/2$, relative to the x axis.

the transmission axes of both polarizers and the fast axis of the half-wave plate were arranged to be parallel. Then, when the polarizer P_1 was rotated through an angle θ, the fast axis of the half-wave plate was rotated through an angle $\phi = \theta/2$. In this way, it was arranged that, within the limits set by the imperfections of the nominally achromatic half-wave plate, which were similar to those of the quarter-wave plate described in Section 5.2, the orientation of the plane of polarization of the photon incident on the photomultiplier did not change as the rotation of the polarizer and half-wave plate took place. It followed that the relative angle between the planes of the polarization did not change significantly as the measurement proceeded and that any enhancement effect that might have existed in previous experiments could not now distort the results.

Results of the experiment are shown in Figure 11 where $R(\theta)$ represents the coincidence rate with angle θ between the transmission axes of the polarizers and R_0 the rate with the plates of both linear polarizers removed. The quantum mechanical prediction derived from equation (16), with the values of ε_M, ε_m, and $F(\delta)$ given in Section 5.1, is shown as a solid line and agrees with the experimental points within the limits of error.

In a further experiment, the transmission axes of the polarizers were fixed parallel to each other and the fast axis of the half-wave plate rotated through angle ϕ relative to the axes of the polarizers so that it could be argued that the relative angle between the planes of polarization of the photons impinging on the photomultipliers could be varied continuously.

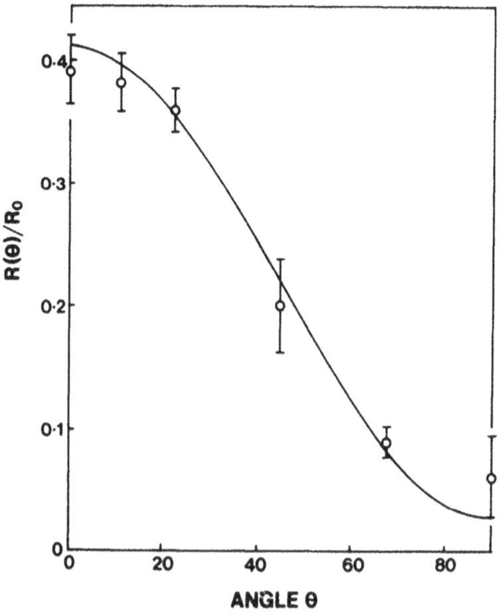

FIGURE 11. Coincidence signal $R(\theta)$ as a function of the angle θ between the transmission axes of the linear polarizers, relative to R_0, the coincidence signal with the linear polarizer plates removed but the half-wave plate still in place. The solid curve represents the quantum mechanical prediction, and is identical to that shown in Figure 8.

The results of this experiment are shown in Figure 12, from which it can be seen that, within the limits of error, there is no indication of any dependence of the signal on the orientation of the fast axis of the half-wave plate. Finally, it was verified that the singles rate did not vary as the half-wave plate was rotated.

These measurements using a half-wave plate indicate that, insofar as it is acceptable to talk in terms of individual photons emerging from the polarizers, the relative angle between the planes of polarization of the two photons just prior to detection plays no role in establishing the observed linear polarization correlation and, hence, that the additional assumption in the form proposed by Clauser, Horne, Shimony, and Holt[44] is valid. There is, also, no evidence here to support the idea of enhanced photon detection along the lines proposed by Garuccio and Selleri.[52]

If it is assumed that the presence of the half-wave plate did, indeed, have no effect on the results, then the measurements shown in Figure 11 can be used to test Freedman's[46] form of Bell's inequality,[28] equation (15). In this case $\eta = 0.271 \pm 0.021$, violating Bell's inequality, although only by one standard deviation. However, combining this result with that given for η in Section 5.1 gave $\eta = 0.269 \pm 0.009$ in violation of Bell's inequality by just more than two standard deviations.

5.4. *Experiment with Three Polarizers*

The experiments with the half-wave plate described in Section 5.3 did not rule out the possibility that enhancement might occur as a result of the setting of the linear polarizers themselves since, although the detection vector $\boldsymbol{\lambda}$ might not be affected by passage through a linear polarizer, it was possible to imagine that it was rotated in the same way as the polarization vector $\mathbf{l}$ by passage through a half-wave plate. Consequently, as suggested

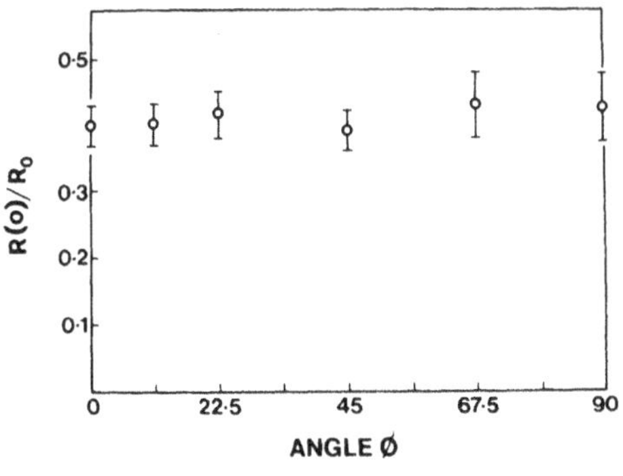

FIGURE 12. Coincidence signal $R(0)$ with the transmission axes of the linear polarizers parallel ($\theta = 0$), relative to R_0, the coincidence signal with the linear polarizer plates removed, as a function of the orientation ϕ of the fast axis of the half-wave plate relative to the transmission axes of the linear polarizers.

by Garuccio and Selleri,[52] it appeared that it would be interesting to carry out experiments with a linear polarizer in place of the half-wave plate, since, in particular, in this situation a finite and measurable difference was predicted to exist between quantum mechanics and their local realistic model. The introduction of a third linear polarizer also afforded an opportunity to test the quantum formalism in hitherto unexplored and novel circumstances.

The experiment was performed by Haji-Hassan *et al.*[53] with an additional linear polarizer inserted as shown in Figure 13. The orientation of the polarizer a was held fixed while polarizer b was rotated through an angle β in a clockwise sense and polarizer a′ through an angle α' in the opposite sense. The objective of the experiment was then to measure, for various angles β, the ratio $R(\beta, \alpha')/R(\beta, \infty)$, where $R(\beta, \alpha')$ was the coincidence rate with all polarizers in place, and $R(\beta, \infty)$ the rate with the plates of polarizer a′ removed. Results for $\beta = 0°$, 33°, and 67.5° are shown in Figure 14 along with the corresponding quantum mechanical prediction obtained by evaluating the trace of expressions of the form $(A'A \otimes B\rho_0 A^\dagger A'^\dagger \otimes B^\dagger)$ where A', A, and B are matrices representing the efficiencies and orientations of polarizers a′, a, and b, $A'^\dagger$, $A^\dagger$, and $B^\dagger$ are their Hermitian conjugates, ρ_0 the density matrix describing the two-photon state upon emission, and the symbol $\otimes$ denotes the direct product operation.[54] The quantum mechanical prediction for $\beta = 0°$ is extremely close to the form $\varepsilon_M \cos^2 \alpha' + \varepsilon_m \sin^2 \alpha'$, where ε_M and ε_m are the transmission efficiencies of polarizer a′, and the experimental results for this case can be considered to be a test of Malus' law for the transmission of polarized light from a very weak source through polarizer a′.

The model of Garuccio and Selleri[52] on the other hand, showed that, for any angle $\theta < 90°$, arranging that the angles of the transmission axes of the three polarizers satisfied the relations $\beta = 3\theta$, $\alpha' + \beta = \theta$, the ratio of the quantum mechanical prediction to their local realistic prediction must always be greater than some minimum value γ_L, as shown in Figure 15 for the parameters of the experiment of Haji-Hassan *et al.*[53] The important

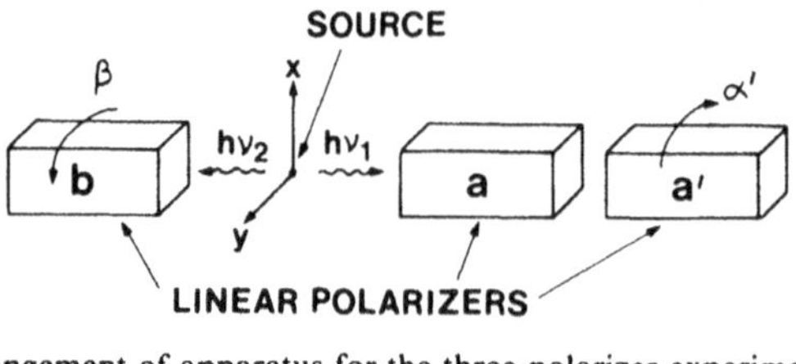

FIGURE 13. Arrangement of apparatus for the three-polarizer experiment. The orientation of polarizer a is fixed with its transmission axis parallel to the x axis, while the transmission axes of polarizers b and a′ are rotated, respectively, through angles β and α' relative to the x axis.

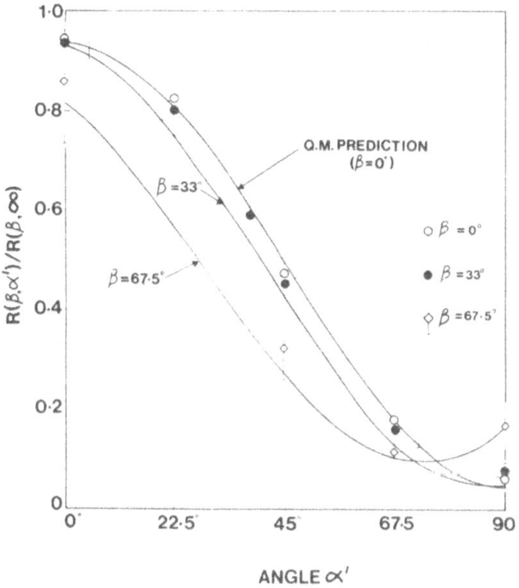

FIGURE 14. Variation of the ratio $R(\beta, \alpha')/(R(\beta, \infty)$ as a function of β and α' in the experiment of Haji-Hassan *et al.*[53] The solid lines represent the quantum mechanical predictions.

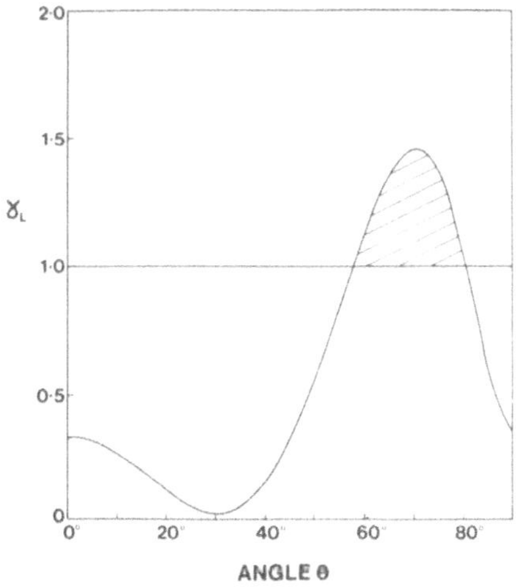

FIGURE 15. Lower limit γ_L of the ratio of the quantum mechanical prediction to the prediction of local realistic theories with enhancement of the type proposed by Garuccio and Selleri.[52]

range of angles occurred for $58° < \theta < 80°$, where $\gamma_L > 1$, since, in this region, a test can be made between quantum mechanics and the class of local realistic theories proposed by Garuccio and Selleri.[52] The maximum value of $\gamma_L = 1.447$ occurs for $\theta = 71°$ corresponding to the experimental angles $\beta = 33°$, $\alpha' = 38°$, and, at these angles, therefore, the approach of Garuccio and Selleri[52] set an upper limit on the predictions of local realistic theories for the ratio $R(\beta, \alpha')/R(\beta, \infty)$ of 0.413, whereas the actual experimental point has the value 0.585 ± 0.029, violating the Garuccio–Selleri model by over six standard deviations. More recently Selleri,[55] taking other factors into account, modified the prediction of the local realistic model to give $\gamma_L = 1.162$ with a corresponding upper limit on $R(\beta, \alpha')/R(\beta, \infty)$ of 0.514 which, however, is still violated by the experimental result by almost three standard deviations. The three-polarizer experiment, therefore, provides further strong evidence against the possibility of enhancement in the detection process and appears to rule out the class of local realistic theories proposed by Garuccio and Selleri.[52]

6. CONCLUSIONS

This chapter has described a series of experiments carried out at Stirling to measure the polarization correlation properties of the two photons emitted simultaneously in the spontaneous decay of metastable atomic deuterium. The measurements have, in general, confirmed the predictions of quantum mechanics for this true second-order process and, in particular, the linear polarization correlation results have violated Bell's inequality, thus rendering untenable any local realistic explanation of the observations, provided it is assumed there was no bias or enhancement in the detection process caused by the presence of the polarizers used in the experiment. The measurements of the circular polarization correlation, although consistent with conservation of angular momentum for each photon pair, failed to give results that violated Bell's inequality, probably because of the imperfections in the nominally achromatic quarter-wave plates used in the detection system. The experiments in which a half-wave plate or third linear polarizer was added to one detection arm of the apparatus produced results in agreement with the predictions of quantum mechanics, gave strong support to the assumption that the presence of the polarizers did not introduce any bias or enhancement into the detection process, and were able to eliminate a large class of local realistic theories in which enhancement was assumed as a central mechanism. Finally, it is probably worth noting how atomic hydrogen continues to prove such a fertile testing ground for fundamental ideas in quantum mechanics and atomic physics. No doubt there is more to come. For example, the effects of electric and magnetic fields that would remove the degeneracy of the virtual intermediate states through which the decay proceeds remains to be investigated. In such a study, because of the use of the coincidence technique, it would be possible to observe the polarization correlation properties of the two-photon decay independently, for example, of any quench radiation which might be produced. It might be argued, of course, that, since the two-photon absorption process in the presence of fields has been carefully investigated already, no further information would be forthcoming from a study of the above kind. However, it should be noted that it is not possible to reproduce the superposition state vector characteristic of the two-photon decay by combining polarized light from two lasers since this latter process would result in a mixed state that does not possess the subtle features of the superposition state vector.

REFERENCES

1. W. E. LAMB AND R. C. RETHERFORD, *Phys. Rev.* **72**, 241 (1947); **79**, 549 (1950).
2. M. GÖPPERT-MAYER, *Ann. Phys.* (*N.Y.*) **9**, 273 (1931).
3. G. BREIT AND E. TELLER, *Astrophys. J.* **91**, 215 (1940).
4. L. SPITZER AND J. L. GREENSTEIN, *Astrophys. J.* **114**, 407 (1951).
5. J. SHAPIRO AND G. BREIT, *Phys. Rev.* **113**, 179 (1959).
6. B. A. ZON AND L. P. RAPAPORT, *JETP Lett.* **7**, 52 (1968).
7. S. KLARSFELD, *Phys. Lett.* **30A**, 382 (1969).
8. W. R. JOHNSON, *Phys. Rev. Lett.* **29**, 1123 (1972).
9. S. P. GOLDMAN AND G. W. F. DRAKE, *Phys. Rev. A* **24**, 183 (1981).
10. F. A. PARPIA AND W. R. JOHNSON, *Phys. Rev. A* **26**, 1142 (1982).
11. J. H. TUNG, X. M. YE, G. J. SALAMO, AND F. T. CHAN, *Phys. Rev. A* **30**, 1175 (1984).
12. V. FLORESCU, *Phys. Rev. A* **30**, 2441 (1984).
13. A. COSTESCU, I. BRANDUS, AND N. MEZINCESCU, *J. Phys. B* **18**, L11 (1985).
14. G. W. F. DRAKE, *Phys. Rev. A* **34**, 2871 (1986).
15. M. LIPELES, R. NOVICK, AND N. TOLK, *Phys. Rev. Lett.* **15**, 690, 815 (1965).
16. C. J. ARTURA, N. TOLK, AND R. NOVICK, *Astrophys. J.* **157**, L181 (1960).
17. R. W. SCHMEIDER AND R. MARRUS, *Phys. Rev. Lett.* **25**, 1692 (1970).
18. R. MARRUS AND R. W. SCHMEIDER, *Phys. Rev. A* **5**, 1160 (1972).
19. C. L. COCKE, B. CURNETTE, J. R. MACDONALD, J. A. BEDNAR, AND R. MARRUS, *Phys. Rev. A* **9**, 2242 (1974).
20. M. H. PRIOR, *Phys. Rev. Lett.* **29**, 611 (1972).
21. C. A. KOCHER, J. E. CLENDENIN, AND R. NOVICK, *Phys. Rev. Lett.* **29**, 615 (1972).
22. E. A. HINDS, J. E. CLENDENIN, AND R. NOVICK, *Phys. Rev. A* **17**, 670 (1978).
23. H. GOULD AND R. MARRUS, *Phys. Rev. A* **28**, 2001 (1983).
24. Y. BANNETT AND I. FREUND, *Phys. Rev. Lett.* **49**, 539 (1982).
25. P. BRÄUNLICH, R. HALL, AND P. LAMBROPOULOS, *Phys. Rev. A* **5**, 1013 (1972).
26. D. O'CONNELL, K. J. KOLLATH, A. J. DUNCAN, AND H. KLEINPOPPEN, *J. Phys. B* **8**, L214 (1975).
27. H. KRUGER AND A. OED, *Phys. Lett.* **54A**, 251 (1975).
28. J. S. BELL, *Physics* (*N.Y.*) **1**, 195 (1964).
29. W. PERRIE, A. J. DUNCAN, H. J. BEYER, AND H. KLEINPOPPEN, *Phys. Rev. Lett.* **54**, 1790, 2647(E) (1985).

30. A. J. DUNCAN, in *Atomic Physics 10, Book of Invited Papers, Tenth International Conference on Atomic Physics (ICAP-X), Tokyo, Japan,* August 25–29, 1986, edited by H. Narumi and I. Shimamura, pp. 121–140 (Elsevier, New York, 1987).
31. F. M. PIPKIN, in *Advances in Atomic and Molecular Physics*, edted by D. R. Bates and B. Bederson (Academic, New York, 1978), Vol. 14, pp. 281–340.
32. J. F. CLAUSER AND A. SHIMONY, *Rep. Prog. Phys.* **41**, 1881 (1978).
33. A. ASPECT, P. GRANGIER, AND G. ROGER, *Phys. Rev. Lett.* **47**, 460 (1981).
34. A. ASPECT, P. GRANGIER, AND G. ROGER, *Phys. Rev. Lett.* **49**, 91 (1982).
35. A. ASPECT, J. DALIBARD, AND G. ROGER, *Phys. Rev. Lett.* **49**, 1804 (1982).
36. A. ASPECT AND P. GRANGIER, *Lett. Nuovo Cimento* **43**, 345 (1985).
37. V. PARAMANANDA AND D. K. BUTT, *J. Phys. G* **13**, 449 (1987).
38. T. HAJI-HASSAN, A. J. DUNCAN, W. PERRIE, H. J. BEYER, AND H. KLEINPOPPEN, *Phys. Lett.* **123A**, 110 (1987).
39. C. K. AU, *Phys. Rev. A* **14**, 531 (1976).
40. A. EINSTEIN, B. PODOLSKY, AND N. ROSEN, *Phys. Rev.* **47**, 777 (1935).
41. D. BOHM, *Quantum Theory* (Prentice-Hall, Englewood Cliffs, New Jersey, 1951).
42. D. BOHM, *Phys. Rev.* **85**, 166, 180 (1952).
43. D. BOHM AND Y. AHARONOV, *Phys. Rev.* **108**, 1070 (1957).
44. J. F. CLAUSER, M. A. HORNE, A. SHIMONY, AND R. A. HOLT, *Phys. Rev. Lett.* **23**, 880 (1969).
45. J. F. CLAUSER AND M. A. HORNE, *Phys. Rev. D* **10**, 526 (1974).
46. S. J. FREEDMAN, Ph.D. thesis, University of California, Berkeley (1972).
47. M. BACAL, A. TRUC, H. J. DOUCET, H. LAMAIN, AND M. CHRETIEN, *Nucl. Instrum. Methods* **114**, 407 (1974).
48. M. BACAL AND W. REICHELT, *Rev. Sci. Instrum.* **20**, 769 (1974).
49. T. W. MARSHALL, E. SANTOS, AND F. SELLERI, *Lett. Nuovo Cimento* **38**, 41 (1983).
50. F. SELLERI, *Nuovo Cimento* **39**, 252 (1984).
51. S. PASCAZIO, *Nuovo Cimento D* **5**, 23 (1985).
52. A. GARUCCIO AND F. SELLERI, *Phys. Lett.* **103A**, 99 (1984).
53. T. HAJI-HASSAN, A. J. DUNCAN, W. PERRIE, H. J. BEYER, H. KLEINPOPPEN, AND E. MERZBACHER, in *Book of Abstracts, Fifteenth International Conference on the Physics of Electronic and Atomic Collisions (XV ICPEAC)*, Brighton, United Kingdom, July 22–28 (1987), p. 74.
54. E. MERZBACHER, private communication; see, also, for example, M. J. Beran and G. B. Parrent Jr., *Theory of Partial Coherence* (Prentice Hall, Englewood Cliffs, New Jersey, 1964), pp. 138–142.
55. F. SELLERI, private communication.

INDEX

GPSR Compliance

The European Union's (EU) General Product Safety Regulation (GPSR) is a set of rules that requires consumer products to be safe and our obligations to ensure this.

If you have any concerns about our products, you can contact us on ProductSafety@springernature.com

In case Publisher is established outside the EU, the EU authorized representative is:

Springer Nature Customer Service Center GmbH
Europaplatz 3
69115 Heidelberg, Germany

Batch number: 10761588

Printed by Printforce, the Netherlands